Fundamental Biomedical Technologies

Series Editor
Mauro Ferrari, Ph.D.
The University of Texas
Houston, TX, USA

More information about this series at http://www.springer.com/series/7045

Brian G. Ballios • Michael J. Young
Editors

Regenerative Medicine and Stem Cell Therapy for the Eye

Springer

Editors
Brian G.Ballios
Department Ophthalmology
and Vision Sciences
University of Toronto
Toronto, ON, Canada

Michael J. Young
Schepens Eye Research Institute
Harvard Medical School
Boston, MA, USA

ISSN 1559-7083
Fundamental Biomedical Technologies
ISBN 978-3-319-98079-9 ISBN 978-3-319-98080-5 (eBook)
https://doi.org/10.1007/978-3-319-98080-5

Library of Congress Control Number: 2018959866

This Springer imprint is published by the registered company Springer Nature Switzerland AG
The registered company address is: Gewerbestrasse 11, 6330 Cham, Switzerland

Preface

Recent advances in the field of stem cell biology have led to a new field that has become known as regenerative ophthalmology. The aim of this field is to preserve, repair, or replace ocular cells that have been injured or lost.

We now better understand the complexity of how the eye develops, the cellular and regional heterogeneity that builds complex ocular tissues, and how these processes are disordered in disease. Ocular diseases affect the entire age spectrum, and loss of vision significantly affects the quality of life for those with blinding eye conditions. With our aging population, the burden of ocular disease is expected to rise over the next decades. As we enter the age of regenerative medicine, the eye has become the proving ground for first-in-man stem cell clinical trials. This is in part because of its accessibility for minimally invasive surgical approaches to deliver cellular therapeutics, the availability of multimodal behavioural, physiological, and imaging techniques, and the sheer number of patients that new treatments could potentially impact.

The purpose of the book is to provide those interested in the field of ocular regenerative medicine with a perspective on the various ways in which stem cells are being applied in the development of potential cures for blinding eye diseases. The aim is to provide a review-level understanding of the types, sources, and applications of stem cells in regenerating tissues in various parts of the eye, with a perspective on the promise of stem cell applications, as well as challenges confronting the field. It is intended for those with a background in the biological and/or medical sciences. The field will depend increasingly on multidisciplinary collaboration between scientists, physicians, and engineers as we develop the tools needed to realize the full potential of stem cells to heal the eye.

The scope of the text is broad and includes both preclinical and clinical applications, including stem cell-derived therapies based on endogenous tissue repair, stem cell transplantation/cell replacement therapy, the interface with gene therapy, and in vitro disease modelling. Applications in both anterior and posterior ocular disease are described, with a particular focus on diseases of the ocular surface/cornea/limbus, where stem cell transplantation has already found clinical application in patients, as well as diseases of the retina, where early clinical trials have begun for both RPE and photoreceptor replacement. Targets of these therapies include inherited retinal dystrophies as well as acquired diseases, such as age-related macular degeneration.

As editors, we have had the pleasure of working with these researchers to bring this volume together. We extend our gratitude and thanks to all of the authors for their time and commitment to this volume and for providing their perspectives on this rapidly expanding field.

Toronto, ON, Canada Brian G. Ballios
Boston, MA, USA Michael J. Young

Contents

Contributors

Brian G. Ballios Department of Ophthalmology and Vision Sciences, University of Toronto, Toronto, ON, Canada

Petr Baranov The Schepens Eye Research Institute of Massachusetts Eye and Ear, an affiliate of Harvard Medical School, Boston, MA, USA

Gilbert Bernier Stem Cell and Developmental Biology Laboratory, Maisonneuve-Rosemont Hospital, Montréal, QC, Canada

Department of Neuroscience, University of Montréal, Montréal, QC, Canada

Department of Ophthalmology, University of Montréal, Montréal, QC, Canada

Isabelle Brunette Maisonneuve-Rosemont Hospital Research Centre, Montreal, QC, Canada

Department of Ophthalmology, University of Montreal, Montreal, QC, Canada

Vianney Delplace Department of Chemical Engineering and Applied Chemistry, University of Toronto, Toronto, ON, Canada

Clarisse M. Fligor Department of Biology, Indiana University Purdue University Indianapolis, Indianapolis, IN, USA

Markus H. Frank Transplantation Research Center, Boston Children's Hospital, Boston, MA, USA

Department of Dermatology, Brigham and Women's Hospital, Boston, MA, USA

Natasha Y. Frank Department of Medicine, VA Boston Healthcare System, Harvard Medical School, Boston, MA, USA

Division of Genetics, Brigham and Women's Hospital, Boston, MA, USA

May Griffith Maisonneuve-Rosemont Hospital Research Centre, Montreal, QC, Canada

Department of Ophthalmology, University of Montreal, Montreal, QC, Canada

Kimberly T. Ho-A-Lim Department of Biology, Indiana University Purdue University Indianapolis, Indianapolis, IN, USA

Bruce R. Ksander Schepens Eye Research Institute, Massachusetts Eye and Ear Infirmary, Boston, MA, USA

Kirstin B. Langer Department of Biology, Indiana University Purdue University Indianapolis, Indianapolis, IN, USA

Christopher D. McTiernan Maisonneuve-Rosemont Hospital Research Centre, Montreal, QC, Canada

Department of Ophthalmology, University of Montreal, Montreal, QC, Canada

Jason S. Meyer Department of Biology, Indiana University Purdue University Indianapolis, Indianapolis, IN, USA

Department of Medical and Molecular Genetics, Indiana University, Indianapolis, IN, USA

Stark Neurosciences Research Institute, Indiana University, Indianapolis, IN, USA

Casey A. Miller Department of Biology, Indiana University Purdue University Indianapolis, Indianapolis, IN, USA

Sarah K. Ohlemacher Department of Biology, Indiana University Purdue University Indianapolis, Indianapolis, IN, USA

Julia Oswald The Schepens Eye Research Institute of Massachusetts Eye and Ear, an affiliate of Harvard Medical School, Boston, MA, USA

Jea Young Park Department of Ophthalmology, Yale School of Medicine, New Haven, CT, USA

Samantha L. Payne Department of Chemical Engineering and Applied Chemistry, University of Toronto, Toronto, ON, Canada

Molly S. Shoichet Department of Chemical Engineering and Applied Chemistry, University of Toronto, Toronto, ON, Canada

Allan R. Slomovic Department of Ophthalmology and Vision Sciences, University of Toronto, Toronto, ON, Canada

Toronto General Hospital, University Health Network, Toronto, ON, Canada

Department of Ophthalmology, Toronto Western Hospital, Toronto, ON, Canada

Akshayalakshmi Sridhar Department of Biology, Indiana University Purdue University Indianapolis, Indianapolis, IN, USA

Matthew Steinhart Medical Science Training Program, Indiana University, Bloomington, IN, USA

Michael J. Young Schepens Eye Research Institute, Harvard Medical School, Boston, MA, USA

Linyang Yu Department of Chemical Engineering and Applied Chemistry, University of Toronto, Toronto, ON, Canada

Chapter 1
Photoreceptor Cell Replacement Therapy from Stem Cells

Gilbert Bernier

Abstract Macular degenerations, retinitis pigmentosa, and retinal dystrophies affect millions of people worldwide. In most cases, loss of visual function results from the death of photoreceptors, the specialized cells involved in phototransduction. An innovative and efficient therapeutic solution for retinal degenerative diseases may be photoreceptor cell transplantation. Yet, the human eye contains about one hundred million photoreceptors, and cell replacement therapy would require at least a fraction of this, raising the issue of where to find an abundant source of healthy human photoreceptors to treat patients. Human pluripotent stem cells can be expanded quasi-indefinitely and differentiate into all cell types of the human body. Methods to direct the differentiation of human pluripotent stem cells into retinal cells and photoreceptors have been developed based on developmental biology principles. Here, we review the history and evolution of these methods, looking at two-dimensional and three-dimensional cell culture systems. We also analyze the current outcomes of photoreceptor cell transplantation therapy and explore the upcoming challenges for its clinical translation.

Keywords Photoreceptor · Rod · Cone · Embryonic development · Neural induction · Stem cell · Pluripotent stem cell · Human · Retinal organoid · Retinal sheet · Retinal degeneration · Transplantation · Cell therapy

---

G. Bernier (✉)
Stem Cell and Developmental Biology Laboratory, Maisonneuve-Rosemont Hospital, Montréal, QC, Canada

Department of Neuroscience, University of Montréal, Montréal, QC, Canada

Department of Ophthalmology, University of Montréal, Montréal, QC, Canada
e-mail: gbernier.hmr@ssss.gouv.qc.ca

B. G. Ballios, M. J. Young (eds.), *Regenerative Medicine and Stem Cell Therapy for the Eye*, Fundamental Biomedical Technologies,
https://doi.org/10.1007/978-3-319-98080-5_1

1.1 Development of Photoreceptors

The distinct competence of retinal progenitor cells to generate, in a sequential order, the diverse class of neurons and a single glial cell type during mammalian retinal development is thought to be modulated by an intrinsic transcription factor's molecular program and by extrinsic cues [1–4]. Loss- and gain-of-function studies in model organisms revealed that the transcription factors Pax6, Rax (also called Rx), Lhx2, Otx2, Sox2, Six6 (also called Optx2), and Six3 are involved in early eye patterning and retinal developmental processes [5–13]. Later on, specific sets of transcription factors define retinal cell type identity, including photoreceptors [14]. Photoreceptor progenitor and precursor cells express Otx2 and Crx, and conditional deletion of *Otx2* in the developing mouse retina impairs photoreceptor fate [15]. In turn, Crx is required for terminal differentiation and maintenance of photoreceptors and is mutated in human retinal degenerative diseases [16–21].

Photoreceptors exist in two types, rods and cones. Rods are the most abundant photoreceptors in the human retina and respond to dim-light. They are involved in night vision and important for peripheral vision. Cones respond to intense light and are required for color, daylight and high-resolution central vision [22, 23]. During development, photoreceptors follow an S-cone default pathway, which is determined by Crx and Thrß2; Crx induces the expression of *Opn1sw* (encoding for S-Opsin) by default while Thrß2 suppresses it and induces the expression of *Opn1mw* (encoding for M-Opsin) in the mouse [24, 25]. In turn, expression of *Nrl, RORbeta* and *Notch1* inhibits cone formation, while both Nrl and RORbeta promote rods genesis at the expense of cones [26–29]. Rods and cones express CRX, Recoverin, ABCA4, and RPGR (Fig. 1.1). RPGR is a protein that localizes to the connecting cilium, a specialized ciliated structure of the photoreceptor cytoskeleton involved in protein transport [30, 31]. Rods and cones express distinct but structurally related photosensitive opsins. Rods express Rhodopsin and cones express M/L--Opsin and S-Opsin (Fig. 1.1). Rods and cones outer segments are structurally distinct, and this is best viewed when using transmission electron microscopy [32]. Cones and rods also have a distinct nuclear organization, with rods displaying condensed chromatin at the center of their nuclei, in contrast with cones having a more relaxed chromatin (Fig. 1.2) [33].

1.2 Inherited Retinal Degenerative Diseases

Inherited retinal degenerative diseases are incurable eye disorders leading to reduced visual function and generally progressing toward complete blindness. Retinitis pigmentosa (RP) is one of the most common and genetically heterogeneous retinal degenerative diseases. Mutations in about 100 different genes can lead to RP, and the disease can be autosomal dominant (*RHO*, *PRPF3*, etc.), autosomal recessive (*ABCA4*, *USH2A*, etc.), X-linked (*RPGR*, *RP2*), or maternally

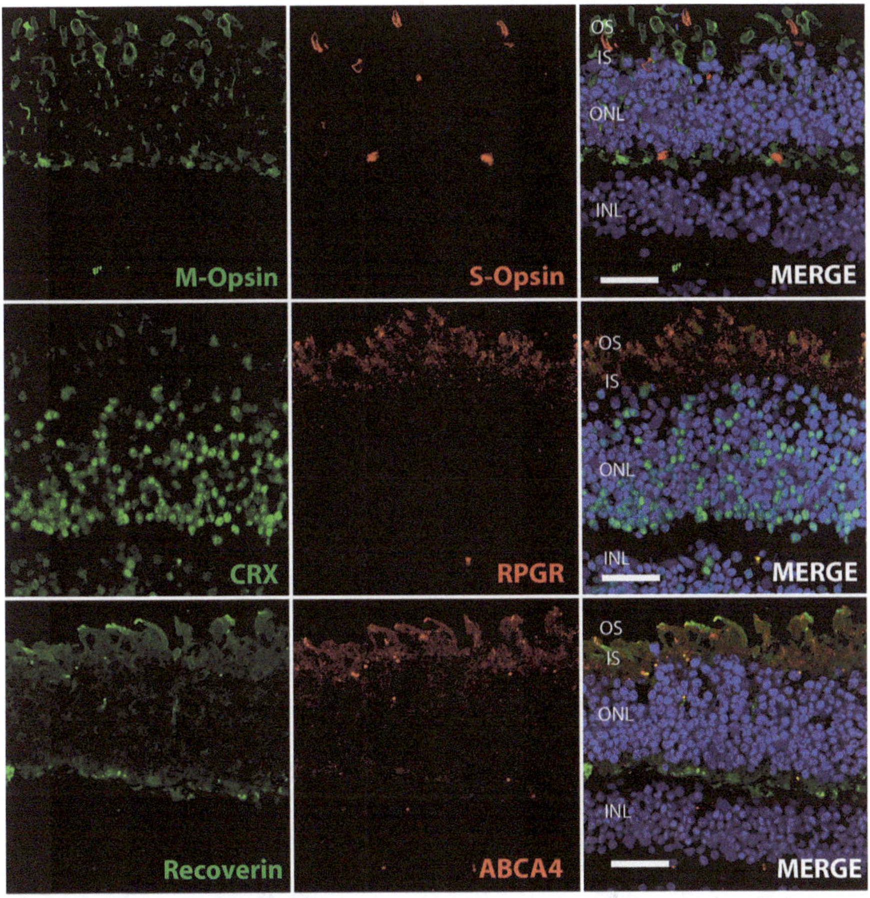

Fig. 1.1 The adult human retina labeled with photoreceptor-specific antibodies. Postmortem frozen sections of a human retina (80 years) labeled for S-Opsin (S-cone outer-segment marker), M-Opsin (M-cone outer-segment marker), CRX (pan-photoreceptor marker), RPGR (connecting cilium and outer-segment marker), Recoverin (pan-photoreceptors outer-segment and synaptic terminals marker), and ABCA4 (connecting cilium and outer-segment marker). OS (outer segment), IS (inner segment), ONL (outer nuclear layer), INL (inner nuclear layer). Scale bars: 40 μm (courtesy images from Andrea Barabino)

transmitted by mitochondria [34–36]. The disease also exists as non-syndromic and syndromic variants. Non-syndromic RP is the pure manifestation of the disease where only vision is affected (*RHO*, *ABCA4*) [23]. Syndromic RP involves additional neurologic and/or systemic manifestations such as hearing loss (e.g., Usher's syndrome) [36, 37]. First symptoms of RP are decreased night vision, followed by progressive loss of peripheral visual field. While the disease primarily affects rods, it usually progresses toward legal blindness only when cones located in the center of the retina are lost. Although the exact mechanism remains unclear, rods may provide trophic support to cones by secreting trophic factors such as rod-derived

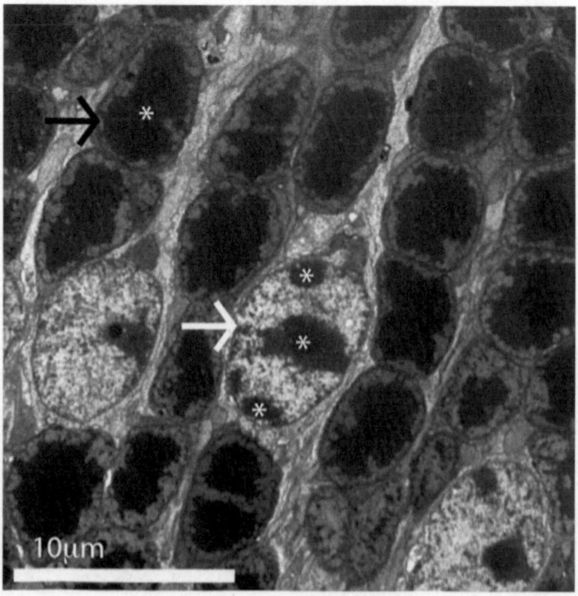

Fig. 1.2 Chromatin organization is distinct between rods and cones. The mouse retina's photoreceptor outer nuclear layer at post-natal day 30 was viewed using transmission electron microscopy. The condensed chromatin (asterisk) of rods (black arrow) is located in the center of the nucleus, while that of cones (white arrow) is dispersed throughout the nucleus (courtesy image from Gilbert Bernier)

cone viability factor (RdCVF) [38]. Notably, alteration of glucose metabolism seems to be involved since cones depend largely on the insulin/mTOR pathway and on aerobic glycolysis for survival [38, 39]. On the other hand, it is possible that rods are simply more susceptible to specific mutations than cones at early stages of the disease process since the mutated gene-products are frequently expressed in both rods and cones (Usherin, RPGR, RP2, etc.). Several other retinal degenerative diseases exist with pathologies involving rod and cone degeneration that leads to blindness. These include cone-rod dystrophy (*CRX*, *ABCA4*) and Leber congenital amaurosis (*CRX*, *CRB1*, *RPE65*, etc.), a form of infantile RP. Early loss of central vision is predominant in cone-specific diseases such as Stargardt's disease, cone dystrophies, and age-related macular degeneration (AMD). Though AMD is not an inherited genetic disease, there may be significant genetic susceptibility to the development of the disease.

1.3 Stem Cell Differentiation into Photoreceptors

In contrast to other mammals, the retinae of modern primates (apes, monkeys, and hominids) contains a unique circular structure called the macula, that is 4–5 mm in diameter and located near the center of the retina [40, 41]. The macula is highly enriched in cone photoreceptors and has a cone-only smaller region called the fovea, that is ~1.5 mm in diameter and involved in high-acuity vision [40, 41].

A high-acuity area somewhat comparable to the human fovea is also present in the retina of diurnal birds [42]. In comparison with model organisms generally used in experimental biology, human eyes are very large. For example, the average human retina contains 4.6 million cones and 92 million rods, with cone density reaching ~212,000 cones/mm^2 at the fovea [40]. In eye diseases where cones are severely affected, generation of cones is of particular interest for cell transplantation therapy [23]. Although most forms of RP primary affect rods, loss of cones and of central vision also occurs at later stages [23, 38, 43, 44]. Considering the human "eye-size problem," the efficiency of cell replacement therapy to treat retinal degenerative diseases will depend on our capacity to generate large numbers of rod and cone photoreceptors.

At the scale that will be needed to treat thousands of patients per year, photoreceptor cell transplantation therapy requires in principle a source of a quasi-unlimited number of human photoreceptors. These numbers can only be achieved by harnessing the immense expansion capability of human pluripotent stem cells. Human embryonic stem cells (hESCs) have been isolated from the inner cell mass of early stage human embryos (32–128 cells). These cells can generate all cell types of the human body (i.e., are pluripotent) and can be expanded and maintained almost indefinitely through proliferative symmetrical cell division [45, 46]. Ethical issues associated with the isolation of new hESC lines have led to the development of a revolutionary method for the induction of adult somatic cells into pluripotent stem cells in recent years [47, 48]. These human induced pluripotent stem cells (iPSCs) display hallmarks of genuine hESCs and can be, in principle, differentiated into any cell type in the human body. After these achievements, the next challenge was to direct the differentiation of human pluripotent stem cells specifically into retinal cells and/or photoreceptors using developmental biology principles.

Work performed in amphibians and chicks suggests that primordial cells adopt a neural fate in the absence of alternative cues [49]. The principles behind the default model of neural induction were first used to induce differentiation of mouse ESCs into primitive neurons [50] or into hypothalamic neurons [51]. The retina and cerebral cortex originate from the anterior portion of the neural plate, and hESCs spontaneously adopt an anterior positional identity when induced to differentiate into neurons [52, 53]. Learning from these developmental principles, several groups have attempted to induce the differentiation of hESCs into retinal neurons and photoreceptors. However, only a fraction of these cells actually differentiate into retinal neurons, possibly because active inhibition of bone morphogenic protein (BMP), Transforming growth factor β (TGFβ) superfamily, and Wingless (Wnt) signaling is required (Fig. 1.3).

This model was further supported by work performed with the highly related Six3 and Six6 "retinal" homeobox transcription factors, which revealed that they promote anterior neural and retinal fates in part through direct transcriptional inhibition of WNT and BMP-encoding genes [11, 54, 55]. Experimentally, this can be partially achieved by expressing Noggin and Chordin, two BMP antagonists, and Dickkopf-1 (Dkk), a Wnt antagonist. Furthermore, retinal fate can be promoted using Insulin Like Growth Factor-1 (IGF-1) [56, 57]. Application of this rationale

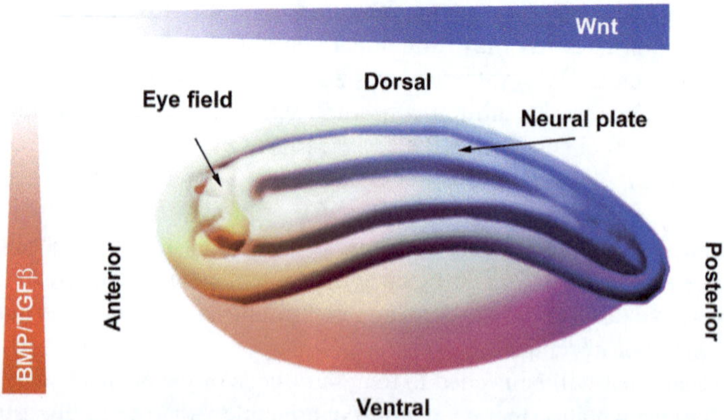

Fig. 1.3 Anterior neural and retinal cell fates require active inhibition of BMP, TGFβ, and WNT signaling. Illustration of the primitive vertebrate embryo body axis at the neural plate stage and showing the eye field region, which is located at the most anterior end of the neural plate (Courtesy artwork by Shufeng Zhou and Gilbert Bernier). Work using model organisms suggests that active inhibition of BMP, TGFβ, and WNT signaling is required for the formation of the anterior portion of the neural plate, which will give rise to the forebrain and the retina

has led to the differentiation of hES cells into retinal progenitor cells, where ~12% of cells express CRX [58]. CRX is expressed by photoreceptor progenitors, mature photoreceptors, and a subset of bipolar neurons [16, 17, 19]. Although ~4% of differentiated hESCs were reported to express Rhodopsin, a marker of rod photoreceptors, less than 0.01% express S-Opsin, a marker of cone photoreceptors. Notably, transplanted cells could adopt rod and cone photoreceptor phenotypes when grafted into the retina of normal and *Crx*-deficient mice [59]. The differentiation of hESCs into cone and rod photoreceptors at a frequency of 12–20% over a 150–200 days period was also achieved using Dkk and LEFTY, a Nodal antagonist, as well as retinoic acid and taurine, two factors that can promote terminal differentiation of photoreceptors [60]. In addition, directed differentiation of hESCs into retinal pigment epithelium was performed using nicotinamide [61]. These early successes in lineage-specific differentiation have paved the way for more sophisticated approaches.

1.4 Retinal Organoids and Retinal Sheets

When pluripotent stem cells are aggregated in three dimensions they form floating embryoid bodies that will rapidly differentiate to generate the three primordial germ layers (Fig. 1.4). Under specific cell culture conditions, embryoid bodies can generate tissue-specific organoids, which are small organ-like embryonic structures that can reach up to 4 mm in diameter [62–65].

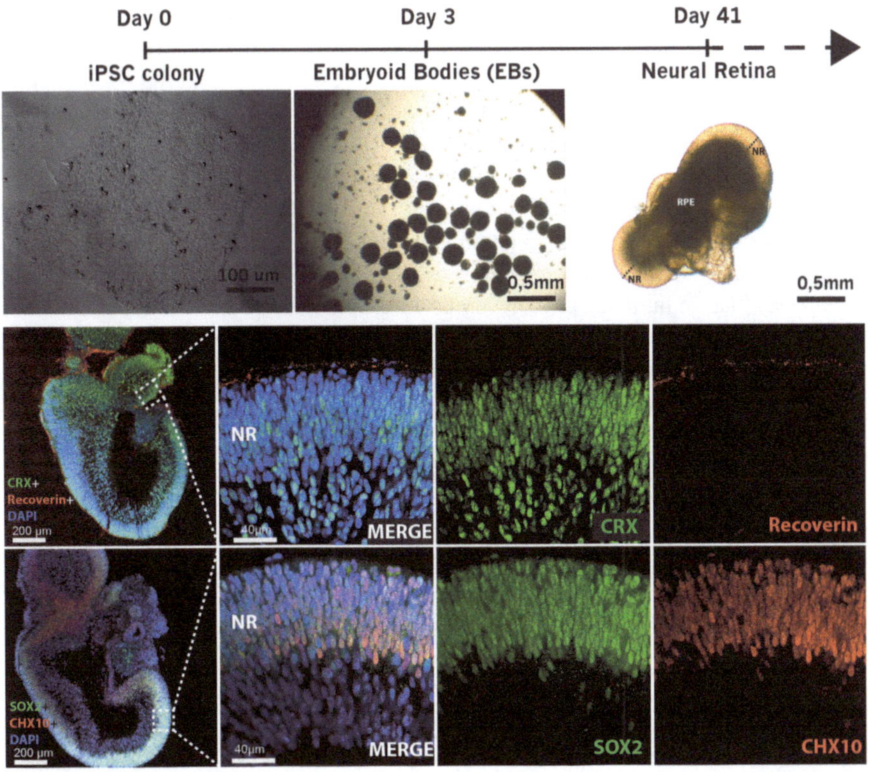

Fig. 1.4 Retinal organoids obtained from the differentiation of human iPSCs. iPSC colonies were dissociated and aggregated to form EBs in suspension cultures. After 41 days, pigmentation was observed suggesting the presence of retinal pigment epithelium (RPE) and neural retina-like (NR) structures. Organoids were sectioned and immunolabeled for photoreceptors markers CRX and Recoverin, and for retinal progenitors markers SOX2 and CHX10 (courtesy images from Andrea Barabino)

Using a cocktail of growth factors and matrigel—an extracellular matrix preparation containing laminin, collagen, and growth factors—for coating the embryoid bodies, pioneering work from Yoshiki Sasai's laboratory in Japan demonstrated the differentiation of mouse ESCs into retinal organoids. These amazing structures contained about all cell types present in the mammalian retina [66]. This was rapidly followed by the differentiation of hESCs into retinal organoids having also multiple retinal cell types, including rod and cone photoreceptors [67–71] (Fig. 1.4). Because they recapitulate organogenesis, retinal organoids follow the same temporal sequence of retinal cell type genesis and differentiation compared to the developing human retina [70]. One drawback of this recapitulation of human retinal ontogeny is that photoreceptor cell differentiation can take up to 140 days in these cultures.

As an alternative to retinal organoids, 2D/3D retinal sheets have been produced from hESCs. Multifunctional BMP, TGFβ, and WNT antagonists of the Cerberus/

DAN gene family are normally secreted by the primitive mesendoderm of the gastrula-stage embryo, promoting neural induction. Xenopus lacking the Cerberus homolog are headless, while Cerberus overexpression can lead to the formation of tadpoles with two heads [72–75]. In contrast to other gene-family members, Coco/Cerl2 is also expressed in the ectoderm and until the end of gastrulation [76]. In the mouse, COCO/CERL2 is expressed in the developing optic vesicle and forebrain, and its expression is maintained in the adult retina and photoreceptors [77]. When continually exposed to human recombinant COCO/CERL2, ~80% of hESCs differentiated into S-cone photoreceptors expressing CRX [77]. This supported the hypothesis that, in humans, photoreceptor development operates through an S-cone default pathway [14, 24–28]. Prolonging the cell culture time to 60 days resulted in the formation of a 150 μm thick polarized cone sheet with evidences of connecting cilium (RPGR) and primitive outer segment (S-Opsin and PNA) formation (Fig. 1.5). Self-organization of hESCs into a cone sheet was reminiscent of self-formation of hESCs into an optic cup-like structure [67].

In conclusion, findings made with retinal organoids and cone sheets opened the possibility of generating cones and rods in large numbers for cell transplantation therapy applications. They also constitute amenable platforms to study photoreceptor disease mechanisms and perform drug-screening assays using iPSCs generated from disease-affected patients [78].

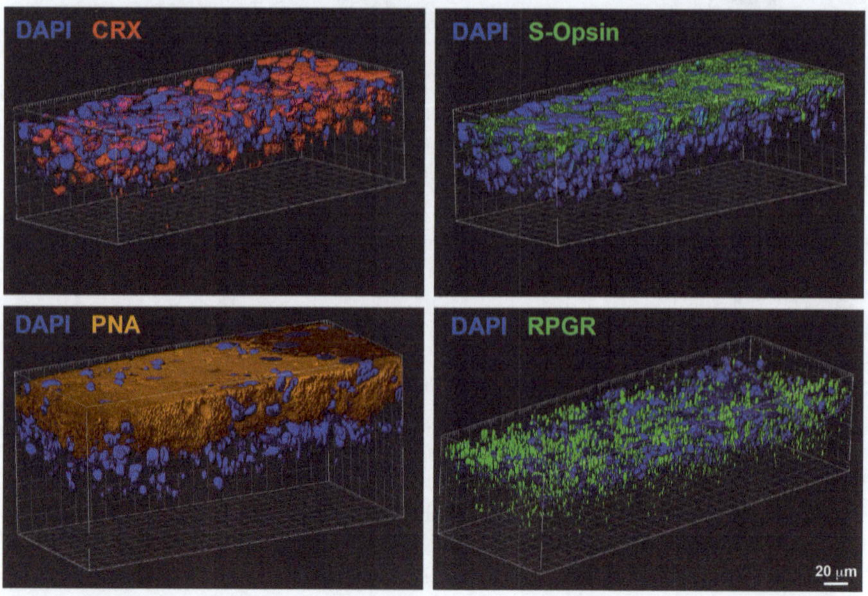

Fig. 1.5 hESCs can be differentiated into a cone sheet upon exposure to COCO. The cone sheet was obtained after exposing hESCs to COCO for 45 days. The sheet was labeled with photoreceptor-specific antibodies. Digital tissue reconstruction was made by the aggregation of multiple confocal images (courtesy images from Anthony Flamier)

1.5 Photoreceptor Cell Transplantation Therapy

Despite the major impact of vision loss on quality of life, few if any efficient treatments actually exist to treat or cure retinal degenerative diseases. Several drugs have been developed to correct the defective phototransduction visual cycle in disease-affected photoreceptors or to prevent photoreceptor cell death, with some of these showing promising results [79]. Gene therapy and gene editing techniques aiming at correcting or silencing the disease-causing mutation or replacing the defective gene also represent promising avenues. However, they generally rely on the injection of viruses and have shown limited success in patients to date, with the exception of patients affected by retinal degeneration linked to mutations in *RPE65* [80, 81]. Gene therapy may be also limited because each therapeutic focuses on a single gene or mutation at a time, when the RP phenotype can be caused by mutations in hundreds of different genes. As an alternative, prosthetic epi-retinal implants (e.g., Argus II) have been developed to transmit an electrical signal to retinal neurons mediated by the input of an external digital camera. Although interesting, the device is expensive and, as yet, can only provide a limited visual activity to patients [82–84].

One alternative to these approaches is photoreceptor cell transplantation therapy. However, conceptually, there have been many contradictory reports on the efficiency of photoreceptor transplantation therapy in animal models. Work using neonatal mouse rod precursors or mouse ESC-derived rod precursors grafted in the sub-retinal space of wild type (WT) or RP mouse models suggested efficient integration and terminal differentiation of rods into the host retina, with an improvement of some visual function in RP mice [85–87]. Comparable results were observed when using hESC-derived rods grafted in WT and *Crx−/−* mice [59]. On the other hand, recent work suggests that grafted rod precursors undergo rapid material transfer (RNA, protein, and exosomes) with the host photoreceptors, leading to the expression of the graft-carrying GFP reporter in host photoreceptors, potentially leading to non-cell-autonomous effects and false interpretation of previous cell engraftment efficiencies [88, 89]. That said, grafted photoreceptor precursors in the sub-retinal space could be useful to partially rescue endogenous disease-affected photoreceptors by the transfer of missing proteins or RNAs. Also, these observations do not necessarily prove that grafted photoreceptors are nonfunctional. Work from another group suggests that dissociated photoreceptors derived from hESCs and grafted in the sub-retinal space can integrate and differentiate efficiently into the mouse retina photoreceptor cell layer but only when the animal is partially immune-deficient [90]. Visual function of the *Crx−/−/IL2Rγ−/−* mice (which are completely blind) could be also improved, and extensive analyses of grafted cells confirmed that material transfer or cell fusion does not occur in this context [90]. Notably, human iPSC-derived retinal tissue grafted into the retina of late-stage *rp1* mice (which have no more photoreceptors) could form synapses with endogenous bipolar neurons and improve visual function, also excluding the thesis of material transfer to endogenous photoreceptors as a mechanism of visual function improvement in this context [91, 92]. Results from these two experiments suggested that even in congenitally blind mice, the central connections from the retina are still intact and functional. Photoreceptor-containing retinal organoids produced from human iPSCs have been also transplanted into the retina of cynomolgus monkeys (*Macaca fascicularis*) after photoreceptor depletion secondary to an injection of cobalt chloride in the sub-retinal space [93]. This method resulted in complete loss of peripheral rod and cone photoreceptors, while generally sparing inner retinal neurons and the retinal pigment epithelium, representing a preclinical model of RP. In these experiments, retinal organoids did not form clear synapses with the host bipolar neurons and accumulated as rosette-shaped structures within the host sub-retinal space [93]. These problems may originate from the heterogeneous nature of retinal organoids and from their tendency to generate embryoid body-like structures. On the other hand, rod-containing mouse retinal sheets also produced using the organoid method could form synaptic connections with the host bipolar neurons when transplanted into an RP mouse model [91]. Future experiments using a human cone sheet as a retinal patch to replace the macula in non-human primates will be critical to evaluate the clinical potential of this method.

One of the historical fears of cell transplantation therapy applications using cells generated from the differentiation of hESCs is the possibility of tumor formation. This is because hESCs can generate teratomas in immune compromise animals and this test is normally used to validate their pluripotency in vivo [47]. However, this problem is apparently over-stated for cell transplantation therapy using post-mitotic neural precursors or differentiated neurons. Clinical trials performed in the past years using dissociated retinal pigment epithelium cells or retinal pigment epithelium sheets derived from hESCs and iPSCs did not report a single case of benign or malignant cell mass formation [94–96]. Tumor formation or abnormal cell proliferation was also not observed in the retina of mice grafted with mouse ESC-derived photoreceptor organoids [91, 93, 96]. In all cases, one possible safeguard mechanism would be to stably integrate a drug-inducible "killer gene" within the genome of the therapeutic cell line, allowing the destruction of the grafted cells with a drug injection in the event of uncontrolled proliferation.

The major problem of transplantation therapy in general is the immune compatibility of the donor cells with the host [97]. Despite the dogma that the retina is immune-privileged, recent work revealed that immune reaction of the host against grafted photoreceptors is the number one obstacle to graft survival and integration [90, 98]. Although autologous transplantation of photoreceptors derived from the patient own iPSCs will prevent immune reaction, the generation of GMP-grade iPSC lines for every patient is time consuming and would be extremely expensive for clinical trial approval [99]. Importantly, the cell lines will also require genetic correction of the disease-causing mutation for all patients with inherited diseases [100]. One possibility to overcome immune reaction is to generate a bank of iPSC lines covering most HLA haplotypes of an entire population and this project is now ongoing in Japan. Alternatively, new gene editing technologies may allow the engineering of a "universal donor" iPSC line compatible with any patient. In addition to acquired immunity, microglial activation and pro-inflammatory microenvironments are characteristic of retinal degenerations [101, 102]. While neuro-inflammation and microglia activation can be attenuated with drugs such as minocyclin, the surgical trauma of transplantation therapy may exacerbate innate immunity and lead to the destruction of the graft. Notably, a small protein of 18 kDa (MANF), structurally related to cerebral dopamine neurotrophic factor, was shown to promote photoreceptor survival after transplantation in the sub-retinal space of mice [98]. MANF is a natural immune-modulator molecule that activates positive innate immunity cells for tissue repair [98]. Many new options now exist to potentially overcome immune reactions associated with photoreceptor cell transplantation therapy.

1.6 Conclusion

The important technological advances made in the last decade in pharmacology, gene therapy, gene editing, bioengineering, and stem cell biology have demonstrated that inherited human retinal degenerative diseases may be treatable or cured. It is over the next decade that we shall see the application of these technologies to cure blindness.

Acknowledgements I would like to thank the Foundation Fighting Blindness Canada for their financial support. Special thanks to Anthony Flamier, Andrea Barabino, and Shufeng Zhou for the images and to Roy Hanna for critical reading of the text.

References

1. Belecky-Adams, T., Cook, B., & Adler, R. (1996). Correlations between terminal mitosis and differentiated fate of retinal precursor cells in vivo and in vitro: Analysis with the "window-labeling" technique. *Developmental Biology, 178*, 304–315.
2. Cepko, C. L., Austin, C. P., Yang, X., Alexiades, M., & Ezzeddine, D. (1996). Cell fate determination in the vertebrate retina. *Proceedings of the National Academy of Sciences of the U.S.A., 93*, 589–595.
3. Reh, T. A., & Kljavin, I. J. (1989). Age of differentiation determines rat retinal germinal cell phenotype: Induction of differentiation by dissociation. *Journal of Neuroscience, 9*, 4179–4189.
4. Watanabe, T., & Raff, M. C. (1990). Rod photoreceptor development in vitro: Intrinsic properties of proliferating neuroepithelial cells change as development proceeds in the rat retina. *Neuron, 4*, 461–467.
5. Carl, M., Loosli, F., & Wittbrodt, J. (2002). Six3 inactivation reveals its essential role for the formation and patterning of the vertebrate eye. *Development, 129*, 4057–4063.
6. Mathers, P. H., Grinberg, A., Mahon, K. A., & Jamrich, M. (1997). The Rx homeobox gene is essential for vertebrate eye development. *Nature, 387*, 603–607.
7. Taranova, O. V., Magness, S. T., Fagan, B. M., Wu, Y., Surzenko, N., Hutton, S. R., et al. (2006). SOX2 is a dosedependent regulator of retinal neural progenitor competence. *Genes & Development, 20*, 1187–1202.

8. Chow, R. L., Altmann, C. R., Lang, R. A., & Hemmati-Brivanlou, A. (1999). Pax6 induces ectopic eyes in a vertebrate. *Development, 126*, 4213–4222.
9. Marquardt, T., Ashery-Padan, R., Andrejewski, N., Scardigli, R., Guillemot, F., & Gruss, P. (2001). Pax6 is required for the multipotent state of retinal progenitor cells. *Cell, 105*, 43–55.
10. Porter, F. D., Drago, J., Xu, Y., Cheema, S. S., Wassif, C., Huang, S. P., et al. (1997). Lhx2, a LIM homeobox gene, is required for eye, forebrain, and definitive erythrocyte development. *Development, 124*, 2935–2944.
11. Lagutin, O. V., Zhu, C. C., Kobayashi, D., Topczewski, J., Shimamura, K., Puelles, L., et al. (2003). Six3 repression of Wnt signaling in the anterior neuroectoderm is essential for vertebrate forebrain development. *Genes & Development, 17*, 368–379.
12. Loosli, F., Winkler, S., & Wittbrodt, J. (1999). Six3 overexpression initiates the formation of ectopic retina. *Genes & Development, 13*, 649–654.
13. Bernier, G., Panitz, F., Zhou, X., Hollemann, T., Gruss, P., & Pieler, T. (2000). Expanded retina territory by midbrain transformation upon overexpression of Six6 (Optx2) in Xenopus embryos. *Mechanisms of Development, 93*, 59–69.
14. Swaroop, A., Kim, D., & Forrest, D. (2010). Transcriptional regulation of photoreceptor development and homeostasis in the mammalian retina. *Nature Reviews. Neuroscience, 11*, 563–576.
15. Nishida, A., Furukawa, A., Koike, C., Tano, Y., Aizawa, S., Matsuo, I., et al. (2003). Otx2 homeobox gene controls retinal photoreceptor cell fate and pineal gland development. *Nature Neuroscience, 6*, 1255–1263.
16. Chen, S., Wang, Q. L., Nie, Z., Sun, H., Lennon, G., Copeland, N. G., et al. (1997). Crx, a novel Otx-like pairedhomeodomain protein, binds to and transactivates photoreceptor cell-specific genes. *Neuron, 19*, 1017–1030.
17. Freund, C. L., Gregory-Evans, C. Y., Furukawa, T., Papaioannou, M., Looser, J., Ploder, L., et al. (1997). Cone-rod dystrophy due to mutations in a novel photoreceptor-specific homeobox gene (CRX) essential for maintenance of the photoreceptor. *Cell, 91*, 543–553.
18. Freund, C. L., Wang, Q. L., Chen, S., Muskat, B. L., Wiles, C. D., Sheffield, V. C., et al. (1998). De novo mutations in CRX homeobox gene associated with Leber congenital amaurosis. *Nature Genetics, 18*, 311–312.
19. Furukawa, T., Morrow, E. M., & Cepko, C. L. (1997). Crx, a novel otx-like homeobox gene, shows photoreceptor-specific expression and regulates photoreceptor differentiation. *Cell, 91*, 531–541.
20. Furukawa, T., Morrow, E. M., Li, T., Davis, F. C., & Cepko, C. L. (1999). Retinopathy and attenuated circadian entrainment in Crx-deficient mice. *Nature Genetics, 23*, 466–470.
21. Swaroop, A., Wang, Q. L., Wu, W., Cook, J., Coats, C., Xu, S., et al. (1999). Leber congenital amaurosis caused by a homozygous mutation (R90W) in the homeodomain of the retinal transcription factor CRX: Direct evidence for the involvement of CRX in the development of photoreceptor function. *Human Molecular Genetics, 8*, 299–305.
22. Aboshiha, J., Dubis, A. M., Carroll, J., Hardcastle, A. J., & Michaelides, M. (2016). The cone dysfunction syndromes. *British Journal of Ophthalmology, 100*(1):115–121.
23. Michaelides, M., Hardcastle, A. J., Hunt, D. M., & Moore, A. T. (2006). Progressive cone and cone-rod dystrophies: Phenotypes and underlying molecular genetic basis. *Survey of Ophthalmology, 51*, 232–258.
24. Yanagi, Y., Takezawa, S., & Kato, S. (2002). Distinct functions of photoreceptor cell-specific nuclear receptor, thyroid hormone receptor beta2 and CRX in one photoreceptor development. *Investigative Ophthalmology & Visual Science, 43*, 3489–3494.
25. Ng, L., Hurley, J. B., Dierks, B., Srinivas, M., Salto, C., Vennstrom, B., et al. (2001). A thyroid hormone receptor that is required for the development of green cone photoreceptors. *Nature Genetics, 27*, 94–98.
26. Jadhav, A. P., Mason, H. A., & Cepko, C. L. (2006). Notch 1 inhibits photoreceptor production in the developing mammalian retina. *Development, 133*, 913–923.
27. Mears, A. J., Kondo, M., Swain, K. S., Takada, Y., Bush, R. A., Saunders, T. L., et al. (2001). Nrl is required for rod photoreceptor development. *Nature Genetics, 29*, 447–452.

28. Yaron, O., Farhy, C., Marquardt, T., Applebury, M., & Ashery-Padan, R. (2006). Notch1 functions to suppress cone photoreceptor fate specification in the developing mouse retina. *Development, 133*, 1367–1378.
29. Jia, L., Oh, E. C., Ng, L., Srinivas, M., Brooks, M., Swaroop, A., et al. (2009). Retinoid-related orphan nuclear receptor ROR beta is an early-acting factor in rod photoreceptor development. *Proceedings of the National Academy of Sciences of the U.S.A., 106*, 17534–17539.
30. Meindl, A., Dry, K., Herrmann, K., Manson, F., Ciccodicola, A., Edgar, A., et al. (1996). A gene (RPGR) with homology to the RCC1 guanine nucleotide exchange factor is mutated in X-linked retinitis pigmentosa (RP3). *Nature Genetics, 13*, 35–42.
31. Rachel, R. A., Li, T., & Swaroop, A. (2012). Photoreceptor sensory cilia and ciliopathies: Focus on CEP290, RPGR and their interacting proteins. *Cilia, 1*, 22.
32. Narayanan, K., & Wadhwa, S. (1998). Photoreceptor morphogenesis in the human retina: A scanning electron microscopic study. *Anatomical Record, 252*, 133–139.
33. Solovei, I., Kreysing, M., Lanctot, C., Kosem, S., Peichl, L., Cremer, T., et al. (2009). Nuclear architecture of rod photoreceptor cells adapts to vision in mammalian evolution. *Cell, 137*, 356–368.
34. Adams, N. A., Awadein, A., & Toma, H. S. (2007). The retinal ciliopathies. *Ophthalmic Genetics, 28*, 113–125.
35. Clarke, G., Goldberg, A. F., Vidgen, D., Collins, L., Ploder, L., Schwarz, L., et al. (2000). Rom-1 is required for rod photoreceptor viability and the regulation of disk morphogenesis. *Nature Genetics, 25*, 67–73.
36. Malm, E., Ponjavic, V., Moller, C., Kimberling, W. J., Stone, E. S., & Andreasson, S. (2011). Alteration of rod and cone function in children with Usher syndrome. *European Journal of Ophthalmology, 21*, 30–38.
37. Papal, S., Cortese, M., Legendre, K., Sorusch, N., Dragavon, J., Sahly, I., et al. (2013). The giant spectrin beta V couples the molecular motors to phototransduction and Usher syndrome type I proteins along their trafficking route. *Human Molecular Genetics, 22*, 3773–3788.
38. Ait-Ali, N., Fridlich, R., Millet-Puel, G., Clerin, E., Delalande, F., Jaillard, C., et al. (2015). Rod-derived cone viability factor promotes cone survival by stimulating aerobic glycolysis. *Cell, 161*, 817–832.
39. Punzo, C., Kornacker, K., & Cepko, C. L. (2009). Stimulation of the insulin/mTOR pathway delays cone death in a mouse model of retinitis pigmentosa. *Nature Neuroscience, 12*, 44–52.
40. Curcio, C. A., Sloan, K. R., Kalina, R. E., & Hendrickson, A. E. (1990). Human photoreceptor topography. *Journal of Comparative Neurology, 292*, 497–523.
41. Franco, E. C., Finlay, B. L., Silveira, L. C., Yamada, E. S., & Crowley, J. C. (2000). Conservation of absolute foveal area in New World monkeys. A constraint on eye size and conformation. *Brain, Behavior and Evolution, 56*, 276–286.
42. da Silva, S., & Cepko, C. L. (2017). Fgf8 expression and degradation of retinoic acid are required for patterning a high-acuity area in the retina. *Developmental Cell, 42*, 68–81 e66.
43. Mohand-Said, S., Hicks, D., Dreyfus, H., & Sahel, J. A. (2000). Selective transplantation of rods delays cone loss in a retinitis pigmentosa model. *Archives d'Ophtalmologie, 118*, 807–811.
44. Sancho-Pelluz, J., Arango-Gonzalez, B., Kustermann, S., Romero, F. J., van Veen, T., Zrenner, E., et al. (2008). Photoreceptor cell death mechanisms in inherited retinal degeneration. *Molecular Neurobiology, 38*, 253–269.
45. Reubinoff, B. E., Pera, M. F., Fong, C. Y., Trounson, A., & Bongso, A. (2000). Embryonic stem cell lines from human blastocysts: Somatic differentiation in vitro. *Nature Biotechnology, 18*, 399–404.
46. Thomson, J. A., Itskovitz-Eldor, J., Shapiro, S. S., Waknitz, M. A., Swiergiel, J. J., Marshall, V. S., et al. (1998). Embryonic stem cell lines derived from human blastocysts. *Science, 282*, 1145–1147.
47. Takahashi, K., Tanabe, K., Ohnuki, M., Narita, M., Ichisaka, T., Tomoda, K., et al. (2007). Induction of pluripotent stem cells from adult human fibroblasts by defined factors. *Cell, 131*, 861–872.

48. Takahashi, K., & Yamanaka, S. (2006). Induction of pluripotent stem cells from mouse embryonic and adult fibroblast cultures by defined factors. *Cell, 126*, 663–676.
49. Munoz Sanjuan, I., & Brivanlou, A. H. (2002). Neural induction, the default model and embryonic stem cells. *Nature Reviews Neuroscience, 3*, 271–280.
50. Tropepe, V., Hitoshi, S., Sirard, C., Mak, T. W., Rossant, J., & van der Kooy, D. (2001). Direct neural fate specification from embryonic stem cells: a primitive mammalian neural stem cell stage acquired through a default mechanism. *Neuron, 30*, 65–78.
51. Wataya, T., Ando, S., Muguruma, K., Ikeda, H., Watanabe, K., Eiraku, M., et al. (2008). Minimization of exogenous signals in ES cell culture induces rostral hypothalamic differentiation. *Proceedings of the National Academy of Sciences of the U.S.A., 105*, 11796–11801.
52. Couly, G., & Le Douarin, N. M. (1988). The fate map of the cephalic neural primordium at the presomitic to the 3-somite stage in the avian embryo. *Development, 103*, 101–113.
53. Banin, E., Obolensky, A., Idelson, M., Hemo, I., Reinhardtz, E., Pikarsky, E., et al. (2006). Retinal incorporation and differentiation of neural precursors derived from human embryonic stem cells. *Stem Cells, 24*, 246–257.
54. Liu, W., Lagutin, O., Swindell, E., Jamrich, M., & Oliver, G. (2010). Neuro retina specification in mouse embryos requires Six3-mediated suppression of Wnt8b in the anterior neural plate. *Journal of Clinical Investigation, 120*, 3568–3577.
55. Gestri, G., Carl, M., Appolloni, I., Wilson, S. W., Barsacchi, G., & Andreazzoli, M. (2005). Six3 functions in anterior neural plate specification by promoting cell proliferation and inhibiting Bmp4 expression. *Development, 132*, 2401–2413.
56. Pera, E. M., Wessely, O., Li, S. Y., & De Robertis, E. M. (2001). Neural and head induction by insulin-like growth factor signals. *Development Cell, 1*, 655–665.
57. Rorick, A. M., Mei, W., Liette, N. L., Phiel, C., El-Hodiri, H. M., & Yang, J. (2007). PP2A:B56epsilon is required for eye induction and eye field separation. *Developmental Biology, 302*(2), 477–493.
58. Lamba, D. A., Karl, M. O., Ware, C. B., & Reh, T. A. (2006). Efficient generation of retinal progenitor cells from human embryonic stem cells. *Proceedings of the National Academy of Sciences of the U.S.A., 103*, 12769–12774.
59. Lamba, D. A., Gust, J., & Reh, T. A. (2009). Transplantation of human embryonic stem cell-derived photoreceptors restores some visual function in Crx-deficient mice. *Cell Stem Cell, 4*, 73–79.
60. Osakada, F., Ikeda, H., Mandai, M., Wataya, T., Watanabe, K., Yoshimura, N., et al. (2008). Toward the generation of rod and cone photoreceptors from mouse, monkey and human embryonic stem cells. *Nature Biotechnology, 26*, 215–224.
61. Idelson, M., Alper, R., Obolensky, A., Ben-Shushan, E., Hemo, I., Yachimovich-Cohen, N., et al. (2009). Directed differentiation of human embryonic stem cells into functional retinal pigment epithelium cells. *Cell Stem Cell, 5*, 396–408.
62. Sasai, Y., Eiraku, M., & Suga, H. (2012). In vitro organogenesis in three dimensions: Self-organising stem cells. *Development, 139*, 4111–4121.
63. Freedman, B. S., Brooks, C. R., Lam, A. Q., Fu, H., Morizane, R., Agrawal, V., et al. (2015). Modelling kidney disease with CRISPR-mutant kidney organoids derived from human pluripotent epiblast spheroids. *Nature Communications, 6*, 8715.
64. Lancaster, M. A., Renner, M., Martin, C. A., Wenzel, D., Bicknell, L. S., Hurles, M. E., et al. (2013). Cerebral organoids model human brain development and microcephaly. *Nature, 501*, 373–379.
65. Morizane, R., Lam, A. Q., Freedman, B. S., Kishi, S., Valerius, M. T., & Bonventre, J. V. (2015). Nephron organoids derived from human pluripotent stem cells model kidney development and injury. *Nature Biotechnology, 33*, 1193–1200.
66. Eiraku, M., Takata, N., Ishibashi, H., Kawada, M., Sakakura, E., Okuda, S., et al. (2011). Self-organizing optic-cup morphogenesis in three-dimensional culture. *Nature, 472*, 51–56.
67. Nakano, T., Ando, S., Takata, N., Kawada, M., Muguruma, K., Sekiguchi, K., et al. (2012). Self-formation of optic cups and storable stratified neural retina from human ESCs. *Cell Stem Cell, 10*, 771–785.

68. Tucker, B. A., Mullins, R. F., Streb, L. M., Anfinson, K., Eyestone, M. E., Kaalberg, E., et al. (2013). Patient-specific iPSC derived photoreceptor precursor cells as a means to investigate retinitis pigmentosa. *eLife, 2*, e00824.
69. Mellough, C. B., Collin, J., Khazim, M., White, K., Sernagor, E., Steel, D. H., et al. (2015). IGF-1 signaling plays an important role in the formation of three-dimensional laminated neural retina and other ocular structures from human embryonic stem cells. *Stem Cells, 33*, 2416–2430.
70. Gonzalez-Cordero, A., Kruczek, K., Naeem, A., Fernando, M., Kloc, M., Ribeiro, J., et al. (2017). Recapitulation of human retinal development from human pluripotent stem cells generates transplantable populations of cone photoreceptors. *Stem Cell Reports, 9*, 820–837.
71. Gonzalez-Cordero, A., West, E. L., Pearson, R. A., Duran, Y., Carvalho, L. S., Chu, C. J., et al. (2013). Photoreceptor precursors derived from three-dimensional embryonic stem cell cultures integrate and mature within adult degenerate retina. *Nature Biotechnology, 31*, 741–747.
72. Bouwmeester, T., Kim, S., Sasai, Y., Lu, B., & De Robertis, E. M. (1996). Cerberus is a head-inducing secreted factor expressed in the anterior endoderm of Spemann's organizer. *Nature, 382*, 595–601.
73. Glinka, A., Wu, W., Onichtchouk, D., Blumenstock, C., & Niehrs, C. (1997). Head induction by simultaneous repression of Bmp and Wnt signalling in Xenopus. *Nature, 389*, 517–519.
74. Piccolo, S., Agius, E., Leyns, L., Bhattacharyya, S., Grunz, H., Bouwmeester, T., et al. (1999). The head inducer Cerberus is a multifunctional antagonist of Nodal, BMP and Wnt signals. *Nature, 397*, 707–710.
75. Silva, A. C., Filipe, M., Kuerner, K. M., Steinbeisser, H., & Belo, J. A. (2003). Endogenous Cerberus activity is required for anterior head specification in Xenopus. *Development, 130*, 4943–4953.
76. Bell, E., Munoz-Sanjuan, I., Altmann, C. R., Vonica, A., & Brivanlou, A. H. (2003). Cell fate specification and competence by Coco, a maternal BMP, TGFbeta and Wnt inhibitor. *Development, 130*, 1381–1389.
77. Zhou, S., Flamier, A., Abdouh, M., Tetreault, N., Barabino, A., Wadhwa, S., et al. (2015). Differentiation of human embryonic stem cells into cone photoreceptors through simultaneous inhibition of BMP, TGFbeta and Wnt signaling. *Development, 142*, 3294–3306.
78. Jin, Z. B., Okamoto, S., Osakada, F., Homma, K., Assawachananont, J., Hirami, Y., et al. (2011). Modeling retinal degeneration using patient-specific induced pluripotent stem cells. *PLoS One, 6*, e17084.
79. Koenekoop, R. K., Sui, R., Sallum, J., van den Born, L. I., Ajlan, R., Khan, A., et al. (2014). Oral 9-cis retinoid for childhood blindness due to Leber congenital amaurosis caused by RPE65 or LRAT mutations: An open-label phase 1b trial. *Lancet, 384*, 1513–1520.
80. Russell, S., Bennett, J., Wellman, J. A., Chung, D. C., Yu, Z. F., Tillman, A., et al. (2017). Efficacy and safety of voretigene neparvovec (AAV2-hRPE65v2) in patients with RPE65-mediated inherited retinal dystrophy: A randomised, controlled, open-label, phase 3 trial. *Lancet, 390*, 849–860.
81. Takkar, B., Bansal, P., & Venkatesh, P. (2018). Leber's congenital amaurosis and gene therapy. *Indian Journal of Pediatrics, 85*(3), 237–242.
82. Cheng, D. L., Greenberg, P. B., & Borton, D. A. (2017). Advances in retinal prosthetic research: A systematic review of engineering and clinical characteristics of current prosthetic initiatives. *Current Eye Research, 42*, 334–347.
83. Duncan, J. L., Richards, T. P., Arditi, A., da Cruz, L., Dagnelie, G., Dorn, J. D., et al. (2017). Improvements in vision related quality of life in blind patients implanted with the Argus II epiretinal prosthesis. *Clinical and Experimental Optometry, 100*, 144–150.
84. Parmeggiani, F., De Nadai, K., Piovan, A., Binotto, A., Zamengo, S., & Chizzolini, M. (2017). Optical coherence tomography imaging in the management of the Argus II retinal prosthesis system. *European Journal of Ophthalmology, 27*, e16–e21.
85. MacLaren, R. E., Pearson, R. A., MacNeil, A., Douglas, R. H., Salt, T. E., Akimoto, M., et al. (2006). Retinal repair by transplantation of photoreceptor precursors. *Nature, 444*, 203–207.

86. Pearson, R. A., Barber, A. C., Rizzi, M., Hippert, C., Xue, T., West, E. L., et al. (2012). Restoration of vision after transplantation of photoreceptors. *Nature, 485*, 99–103.
87. Barnea-Cramer, A. O., Wang, W., Lu, S. J., Singh, M. S., Luo, C., Huo, H., et al. (2016). Function of human pluripotent stem cell-derived photoreceptor progenitors in blind mice. *Scientific Reports, 6*, 29784.
88. Ortin-Martinez, A., Tsai, E. L., Nickerson, P. E., Bergeret, M., Lu, Y., Smiley, S., et al. (2017). A reinterpretation of cell transplantation: GFP transfer from donor to host photoreceptors. *Stem Cells, 35*, 932–939.
89. Pearson, R. A., Gonzalez-Cordero, A., West, E. L., Ribeiro, J. R., Aghaizu, N., Goh, D., et al. (2016). Donor and host photoreceptors engage in material transfer following transplantation of post-mitotic photoreceptor precursors. *Nature Communications, 7*, 13029.
90. Zhu, J., Cifuentes, H., Reynolds, J., & Lamba, D. A. (2017). Immunosuppression via loss of IL2rgamma enhances long-term functional integration of hESC-derived photoreceptors in the mouse retina. *Cell Stem Cell, 20*, 374–384 e375.
91. Assawachananont, J., Mandai, M., Okamoto, S., Yamada, C., Eiraku, M., Yonemura, S., et al. (2014). Transplantation of embryonic and induced pluripotent stem cell-derived 3D retinal sheets into retinal degenerative mice. *Stem Cell Reports, 2*, 662–674.
92. Mandai, M., Fujii, M., Hashiguchi, T., Sunagawa, G. A., Ito, S. I., Sun, J., et al. (2017). iPSC-derived retina transplants improve vision in rd1 end-stage retinal-degeneration mice. *Stem Cell Reports, 8*, 1112–1113.
93. Shirai, H., Mandai, M., Matsushita, K., Kuwahara, A., Yonemura, S., Nakano, T., et al. (2016). Transplantation of human embryonic stem cell-derived retinal tissue in two primate models of retinal degeneration. *Proceedings of the National Academy of Sciences of the U.S.A., 113*, E81–E90.
94. Schwartz, S. D., Regillo, C. D., Lam, B. L., Eliott, D., Rosenfeld, P. J., Gregori, N. Z., et al. (2015). Human embryonic stem cell-derived retinal pigment epithelium in patients with age-related macular degeneration and Stargardt's macular dystrophy: Follow-up of two open-label phase 1/2 studies. *Lancet, 385*, 509–516.
95. Song, W. K., Park, K. M., Kim, H. J., Lee, J. H., Choi, J., Chong, S. Y., et al. (2015). Treatment of macular degeneration using embryonic stem cell-derived retinal pigment epithelium: Preliminary results in Asian patients. *Stem Cell Reports, 4*, 860–872.
96. Mandai, M., Watanabe, A., Kurimoto, Y., Hirami, Y., Morinaga, C., Daimon, T., et al. (2017). Autologous induced stem-cell-derived retinal Cells for macular degeneration. *New England Journal of Medicine, 376*, 1038–1046.
97. Xian, B., & Huang, B. (2015). The immune response of stem cells in subretinal transplantation. *Stem Cell Research & Therapy, 6*, 161.
98. Neves, J., Zhu, J., Sousa-Victor, P., Konjikusic, M., Riley, R., Chew, S., et al. (2016). Immune modulation by MANF promotes tissue repair and regenerative success in the retina. *Science, 353*, aaf3646.
99. Wiley, L. A., Burnight, E. R., DeLuca, A. P., Anfinson, K. R., Cranston, C. M., Kaalberg, E. E., et al. (2016). cGMP production of patient-specific iPSCs and photoreceptor precursor cells to treat retinal degenerative blindness. *Scientific Reports, 6*, 30742.
100. Borooah, S., Phillips, M. J., Bilican, B., Wright, A. F., Wilmut, I., Chandran, S., et al. (2013). Using human induced pluripotent stem cells to treat retinal disease. *Progress in Retinal and Eye Research, 37*, 163–181.
101. Zhao, L., Zabel, M. K., Wang, X., Ma, W., Shah, P., Fariss, R. N., et al. (2015). Microglial phagocytosis of living photoreceptors contributes to inherited retinal degeneration. *EMBO Molecular Medicine, 7*, 1179–1197.
102. Peng, B., Xiao, J., Wang, K., So, K. F., Tipoe, G. L., & Lin, B. (2014). Suppression of microglial activation is neuroprotective in a mouse model of human retinitis pigmentosa. *Journal of Neuroscience, 34*, 8139–8150.

Chapter 2
Human Pluripotent Stem Cells as In Vitro Models for Retinal Development and Disease

Akshayalakshmi Sridhar, Kirstin B. Langer, Clarisse M. Fligor, Matthew Steinhart, Casey A. Miller, Kimberly T. Ho-A-Lim, Sarah K. Ohlemacher, and Jason S. Meyer

Abstract Human pluripotent stem cells (hPSCs) provide unprecedented access to the earliest stages of retinogenesis that remain inaccessible to investigation, thereby serving as powerful tools for studies of retinal development. Additionally, the ability to derive hPSCs from patient sources allows for the modeling of retinal degenerative diseases in vitro, with the potential to facilitate cell replacement strategies in advanced stages of disease. For these purposes, many studies over the past several years have directed the differentiation of hPSCs to generate retinal cells using stochastic methods of differentiation, yielding all major cell types of the retina. In particular, these studies have favored the derivation of RPE, photoreceptors, and more recently retinal ganglion cells for disease modeling, drug screening as well as cell replacement purposes. More recently, advances in retinal differentiation methods have led to the generation of three-dimensional retinal organoids that recapitulate key developmental and morphological features of the retina, including the stratified organization of retinal cells into a tissue-like structure. This review provides an overview of retinal differentiation from hPSCs and their potential use for studies of retinogenesis as well as diseases that affect the retina.

A. Sridhar · K. B. Langer · C. M. Fligor · C. A. Miller · K. T. Ho-A-Lim · S. K. Ohlemacher
Department of Biology, Indiana University Purdue University Indianapolis, Indianapolis, IN, USA
e-mail: akshsrid@uw.edu; klanger@iupui.edu; cfligor@iupui.edu; cam26@iu.edu; kth37@case.edu; sarohlem@iupui.edu

M. Steinhart
Medical Science Training Program, Indiana University, Bloomington, IN, USA
e-mail: steinmat@iu.edu

J. S. Meyer (✉)
Department of Biology, Indiana University Purdue University Indianapolis, Indianapolis, IN, USA
e-mail: meyerjas@iupui.edu

Department of Medical and Molecular Genetics, Indiana University, Indianapolis, IN, USA

Stark Neurosciences Research Institute, Indiana University, Indianapolis, IN, USA

B. G. Ballios, M. J. Young (eds.), *Regenerative Medicine and Stem Cell Therapy for the Eye*, Fundamental Biomedical Technologies,
https://doi.org/10.1007/978-3-319-98080-5_2

Keywords Human pluripotent stem cells · Retina · Organoids · Development · Disease

2.1 Introduction

The human retina is a multilayered tissue composed of an intricate network of several types of retinal neurons that function in an integrated manner to convert the incoming light stimulus into an electrical impulse, which will be propagated to the brain to be converted into an image. Consequently, any disease or injury affecting retinal neurons disrupts this visual circuit, resulting in blindness. Hence, a thorough understanding of the development and functions of the human retina will facilitate the development of successful therapies for retinal degenerative diseases. However, studies of the human retina are especially challenging as retinogenesis occurs early in gestation and remains largely inaccessible to investigation [1]. In this regard, human pluripotent stem cells (hPSCs), including human embryonic stem cells [2] and human induced pluripotent stem cells [3–5], provide a unique in vitro model capable of recapitulating the growth and diversification of developing retinal neurons.

hPSCs are self-renewing cells analogous to the inner cell mass/blastocyst stage of human development, which possess the ability to generate all cell types of the body. Therefore, hPSCs can be used to study even the earliest events of retinogenesis and generate limitless numbers of retinal neurons for translational applications [6–15]. While advancements in hPSC-retinal differentiation protocols over the last decade have led to the successful generation of all types of retinal neurons [9, 16–27], these cells have traditionally been differentiated in a manner that lacked the ability to assemble into a multilayered retinal-like structure. This lack of cellular organization not only affects the ability to faithfully recapitulate the events of retinogenesis as an in vitro model, but may also impact the quality and functionality of retinal cells generated for future translational applications, including disease modeling and cell replacement.

More recently, a fundamental shift in retinal differentiation protocols has developed which allows for the organization of hPSC-derived retinal neurons into an organized, multi-layered retinal-like structure [21, 28–30]. These resultant populations, known as retinal organoids, are composed of retinal neurons arranged in a stratified manner that recapitulates the spatial and temporal patterning of native retinal tissue [21, 23, 28–39]. Thus, such hPSC-derived retinal organoids will likely serve as more effective in vitro models with which to recapitulate earliest events of retinogenesis. Furthermore, these retinal organoids may enhance the application of hPSCs for disease modeling and cell replacement. To serve in these capacities, however, refinements in the differentiation of retinal organoids will be needed, with improvements in these protocols likely to be inspired by our growing understanding of the regulatory factors at play in the developing retina in vivo.

2.2 Development and Organization of the Vertebrate Retina

The retina is a complex multilayered tissue that originates from the developing diencephalon and consists of six neuronal cell types that work in a coordinated fashion to perceive and interpret incoming visual information [40, 41]. Based on the orientation of retinal cells, the retina can be broadly classified into three layers: (1) the outer nuclear layer consisting of the photoreceptor cells, including the rods and cones, (2) the inner nuclear layer consisting of the interneurons, namely bipolar cells, amacrine cells, and horizontal cells, and (3) the ganglion cell layer consisting of the retinal ganglion cells whose axons extend to form the optic nerve [42, 43]. In addition to these neuronal cell types, Muller glia are the primary glial cells of the retina, with cell bodies in the inner nuclear layer and processes traversing the length of the retina, providing necessary architectural and functional support. Additionally, photoreceptors are supported and nourished by the retinal pigmented epithelium (RPE), a sheet-like layer of epithelium located below the photoreceptor layer. Retinal neurons are intricately connected through a network of synapses, with connections between the photoreceptors, bipolar cells, and horizontal cells, referred to as the outer plexiform layer. Similarly, the inner plexiform layer represents the dense fibrils between the ganglion cells, bipolar cells, and amacrine cells.

This structure forms a highly regulated pathway for visual transduction, which is critical to the functioning of the retina [44]. Briefly, incoming light is focused onto the retina via the cornea and lens, where it first interacts with the photoreceptors in the outermost layers of the retinal tissue. These photoreceptors convert the visual light into an electrical stimulus via the phototransduction pathway, which is then transmitted to the retinal ganglion cells via the interneurons of the retina. Finally, the ganglion cells extend their long axons via the optic nerve and synapse with their postsynaptic targets, including the superior colliculus and the lateral geniculate nucleus. Relays to cortical areas responsible for signal integration enable vision. Overall, the function of the retina depends on all its components working in a sequential manner to integrate and transmit the visual information to the brain. Consequently, any disruption in this visual circuit due to injury or disease results in loss of vision or blindness. As such, the use of hPSCs provides a powerful tool with which to study the development of the retina, as well as disruptions to retinal function resulting in vision loss [45]. However, modeling the functions of the retina and its pathophysiology requires the differentiation and organization of these cells in a manner which closely recapitulates the native retina, necessitating a thorough understanding of mechanisms associated with retinal development in vivo.

Retinal development is determined by the combinatorial actions of growth factors as well as transcription factors, which not only specify retinal cell types but also determine their spatio-temporal location. Retinogenesis begins early in gestation and the first morphological evidence of the retina is seen during neurulation [46]. As the developing neural plate forms the neural tube, optic grooves emerge on either side of the diencephalon. These grooves, now known as optic vesicles, evaginate toward the surface ectoderm, resulting in reciprocal signaling between these structures. This reciprocal exchange of signals leads to the induction of the retina from the distal optic vesicle and the formation of the lens placode from the ectoderm. Consequently, the proximal optic vesicle is induced by the surrounding mesenchyme to form the retinal pigmented epithelium (RPE). Following specification of the optic vesicle, these cells acquire a multipotent progenitor identity and will subsequently multiply and differentiate into all cell types of the retina (Fig. 2.1). Retinal cell genesis is specified in an evolutionally-conserved order, which is dictated by the competence of retinal progenitors, and a combination of exogenous signaling gradients as well as endogenous transcriptional regulation [47–49]. Based on this model, studies of retinal development in model systems have demonstrated that ganglion cells, horizontal cells, and cone photoreceptors are the earliest-born retinal cell types. Amacrine cells are specified slightly later in development, followed by rod photoreceptors, while bipolar and muller glia cells are the last cell types to develop in the retina. Retinal development and maturation continues throughout gestation and visual synapses continue to mature after birth. Overall, the specification of the retina from its early diencephalic origins follows a tightly conserved order of events.

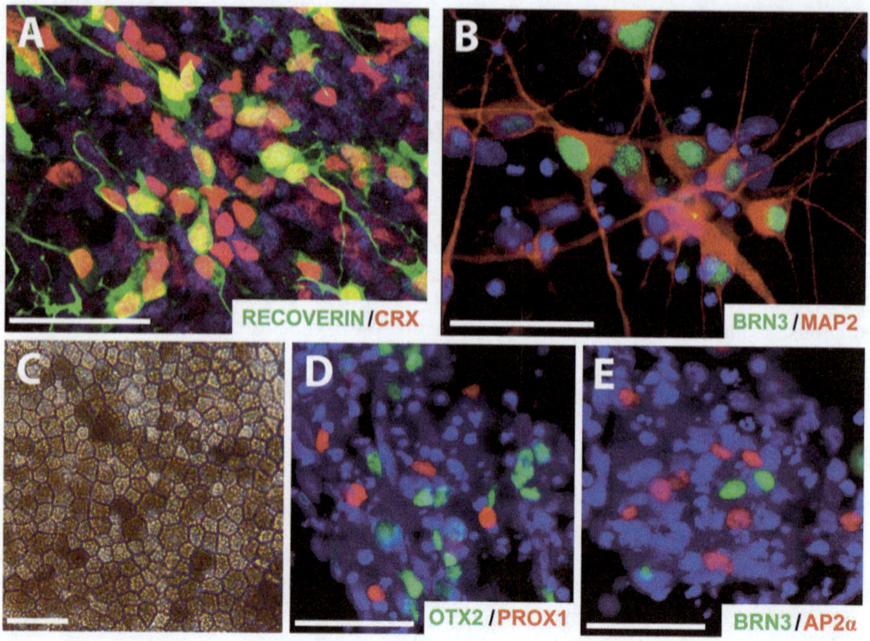

Fig. 2.1 hPSCs generate retinal cells using stochastic methods of differentiation. hPSCs were directed to a retinal fate using a stepwise protocol, where retinal neurons were specified in a conserved, temporal sequence. Within 70 days of differentiation, photoreceptors were readily identified by the expression of photoreceptor-specific markers CRX/RECOVERIN (**a**), while ganglion cells expressed BRN3 and extended MAP2 positive neurites (**b**). hPSCs-derived RPE demonstrate characteristic pigmentation and hexagonal morphology, seen via bright field microscopy (**c**). Additionally, hPSC-derived interneurons such as horizontal cells and amacrine cells could be identified via PROX1 and AP2α expression respectively. Scale bars equal 40 μm

Likewise, to properly and reliably direct the differentiation of hPSCs to a retinal fate, knowledge and application of these developmental events to cellular differentiation protocols is essential.

2.3 In Vitro Studies of Retinal Development Using hPSCs

The development of the human retina is initiated at some of the earliest stages of gestation, making the study of these critical cell fate determination events difficult. Given their pluripotent nature, hPSCs may provide a unique and novel tool for the study of these early developmental events by serving as comprehensive models of the major stages of human retinogenesis, even at stages that would be otherwise inaccessible to investigation in the embryo. With the resultant retinal cells, the potential then exists for their translational application, including cell replacement approaches as well as the ability to model and study retinal degenerative diseases in a dish when derived from specific patient sources. In order to serve in this capacity, however, these cells must be directed to differentiate toward a retinal lineage in a step-wise process that faithfully recapitulates the major stages of retinogenesis in vivo [50]. As such, numerous efforts have been made over the last decade focused on the derivation of retinal cells from hPSCs, often adopting critical principles of developmental biology to guide the differentiation process (Table 2.1). Initial work in this field focused upon the differentiation of retinal cells by inhibiting BMP and WNT signaling in the presence of IGF-1 [9, 19, 68]. Similarly, other groups have been successful in achieving retinal differentiation through the inhibition of WNT and Nodal signaling [17, 18, 24, 25, 69]. Subsequently, efforts relying upon the default adoption of a rostral neural fate in the absence of specific morphogenic factors led to the development of discrete retinal progenitor cell populations, which would later give rise to retinal neurons in a temporally appropriate manner [21–23, 26, 27, 63].

While these protocols provided the ability to generate all the major cell types of the retina, most of the early focus has favored the generation of RPE and photoreceptor cells as many retinal diseases primarily affect these outer retinal cells, resulting in their degeneration and subsequent loss of vision. Additionally, RPE and photoreceptors possess unique phenotypic markers and functional properties that enables their ease of identification in vitro, which is often lacking for other cell types of the retina. Among the earliest retinal differentiation studies, RPE was first observed to be spontaneously differentiated from hPSCs in relatively high numbers [51]. These cells were initially identifiable in cultures of differentiating stem cells due to the accumulation of melanin pigment within these cells that could be readily visualized. Further confirmation of RPE differentiation was provided by their characteristic hexagonal shape and upon isolation of these cells, they commonly expressed a full complement of RPE-associated features [21, 52, 54–56, 59, 60, 70–77]. Similarly, photoreceptors were among the first retinal neurons to be

Table 2.1 Summary of stochastic methods of retinal differentiation from hPSCs

Authors	Cell types observed	Signaling factors	Functional characteristics
Klimanskaya et al. [51]	RPE	bFGF LIF KOSR	
Lamba et al. [19]	Amacrine cells Bipolar cells Horizontal cells Photoreceptors Retinal ganglion cells	Dkk1 IGF1 Noggin	Calcium response to glutamate and NMDA
Banin et al. [16]	Photoreceptors	bFGF EGF Noggin	
Osakada et al. [24]	Amacrine cells Bipolar cells Horizontal cells Immature muller cells Photoreceptors RPE Retinal ganglion cells	Activin-A aFGF bFGF DAPT Dkk1 FBS Lefty Retinoic acid Shh Taurine	
Carr et al. [7, 52]	RPE	bFGF Hydrocortisone KOSR Taurine T3	Photoreceptor outer segment phagocytosis
Meyer et al. [22]	Photoreceptors RPE	Default neural specification	
Idelson et al. [53]	RPE	KOSR Activin-A TGFβ1 SB431542 bFGF	
Buchholz et al. [54]	RPE	bFGF KOSR FBS	Photoreceptor outer segment phagocytosis
Mellough et al. [20]	Photoreceptors RPE	Activin A bFGF Dkk1 IGF1 Lefty Noggin Retinoic acid Shh T3 Taurine	

(continued)

Table 2.1 (continued)

Authors	Cell types observed	Signaling factors	Functional characteristics
Singh et al. [55, 56]	RPE	bFGF EGF FBS FGF2	Photoreceptor outer segment phagocytosis Polarized secretion of growth factors Calcium responses to ATP ATP, TER measurements
Tucker et al. [57]	Photoreceptors RPE	aFGF bFGF DAPT Dkk1 IGF1 Lefty Noggin Retinoic acid Shh T3 Taurine	
Riazifar et al. [58]	Retinal ganglion cells	bFGF DAPT FBS KOSR	Inward/outward currents Action potential response to CNQX
Ferrer et al. [59]	RPE	Activin A Hydrocortisone KOSR Nicotinamide Noggin SB431542 Taurine Triiodothyronine	Intracellular calcium responses Respond to changes in potassium and ATP concentrations
Maruotti et al. [60]	RPE	Chetomin Nicotinamide	
Zhou et al. [61]	Photoreceptors	COCO Dkk1 FGF2 IGF1 Noggin T3	
Sluch et al. [62]	Retinal ganglion cells	FGF-A FBS Forskolin Taurine	Firing of action potentials Response to AMPA/NBQX Mitochondrial movement through RGC axons
Ohlemacher et al. [63]	Retinal ganglion cells	Default neural specification	Firing of action potentials Hyperpolarized resting membrane potential Inward/outward ionic currents

(continued)

Table 2.1 (continued)

Authors	Cell types observed	Signaling factors	Functional characteristics
Barnea-Cramer et al. [64]	Photoreceptors	BDNF CNTF DAPT Insulin Noggin Retinoic acid	
Gill et al. [65]	Retinal ganglion cells	bFGF Dkk1 IGF1 KOSR Noggin	Firing of action potentials Inward/outward currents Axonal transport of mitochondria
Teotia et al. (2017)	Retinal ganglion cells	BDNF cAMP CNTF Cyclopamine DAPT FGF8 Follistatin Forskolin NT4 Shh Y27632	Inward/outward currents Fire action potentials
Sluch et al. [66]	Retinal ganglion cells	DAPT Dorsomorphin Forskolin Glutamax IDE2 LDN-193189 Nicotinamide Noggin SB431542	Efficient immunopurification of RGCs In vitro axonal injury model
Langer et al. [67]	Retinal ganglion cells	Default neural specification	Identification of RGC subtypes Use of single cell RNAseq to elucidate RGC subtypes and novel subtype markers

identified due to the large number of photoreceptor-specific markers that have been previously identified in retinal development studies [9, 19, 20, 22, 24, 25, 78].

More recently, some efforts have focused upon the differentiation of retinal ganglion cells from hPSCs. These cells have been somewhat more difficult to definitively identify in differentiating cultures as they lack any truly specific markers to separate them from some other neuronal populations. However, the ability to identify these cells has been facilitated in recent years by following their differentiation through a retinal progenitor intermediary or via the use of fluorescent reporters [58, 62, 63, 65, 79–82]. Additionally, some studies have demonstrated the ability to derive all the major neuronal cell types of the retina, including interneurons, although these cells have not been extensively characterized to date [21, 22, 25–27, 29, 61, 63, 68, 69, 83].

While the above methodologies have been highly successful for the derivation of all of the major types of retinal neurons, this differentiation often occurred as a somewhat heterogenous population of retinal cells. This differentiation allows for the ability to study many features within individual cells, but does not account for the critical interactions between neurons of the retina which are necessary for their proper maturation and function. Furthermore, many disorders of the retina result from the loss of connectivity between cells, making the study of these disorders more difficult in heterogeneously arranged cultures. To overcome these shortcomings, efforts have been directed toward the differentiation of retinal cells from hPSCs in a manner which closely mimics the development and three-dimensional organization of the retina. Initially, studies described the ability of hPSCs to differentiate toward a retinal lineage in a step-wise fashion, yielding three-dimensional structures closely resembling the optic vesicle and optic cup stages of retinogenesis [21–23, 29]. Subsequently, further efforts expanded upon these early results to generate three-dimensional structures termed retinal organoids that were found to effectively recapitulate the spatial and temporal organization of the various neuronal cell types of the retina, resulting in a stratified, multilayered structure [28, 30, 34–38].

2.4 Applications of Retinal Organoids for Modeling Human Development

With the goal of effectively recapitulating the complex organization and interplay between the different types of neurons of the retina, studies within the past few years have described the ability of hPSCs to differentiate toward a retinal lineage in a step-wise fashion [17, 22, 68]. The resultant populations of cells have yielded structures that closely resemble the developing optic cup, with enriched populations of retinal progenitor cells discretely arranged into a cup-like structure (Fig. 2.2). Subsequent efforts have expanded upon these early results to generate retinal organoids that effectively recapitulate the spatial and temporal organization of the various neuronal cell types of the retina (Table 2.2). As a result, these retinal organoids provide a powerful and novel tool for studies of the earliest stages of human retinal development.

As compared to early methods of retinal differentiation from pluripotent cells, retinal organoids offer several advantages as an in vitro model of retinogenesis. Importantly, these organoids can self-assemble into discrete three-dimensional structures with major classes of retinal neurons arranged into distinct layers similar to their organization within the retina [21, 28–31, 33, 35–39]. The differentiation of these retinal organoids progresses through all the major stages of retinogenesis, including stages analogous to the eye field, optic vesicle, and optic cup, thereby allowing for the ability to visualize some of the earliest events of human retinal development. Similar to embryonic retinogenesis [50], differentiation of resultant cells within retinal organoids has been demonstrated to follow a conserved sequence of events, with early-born cell types such as RGCs among the first retinal neurons

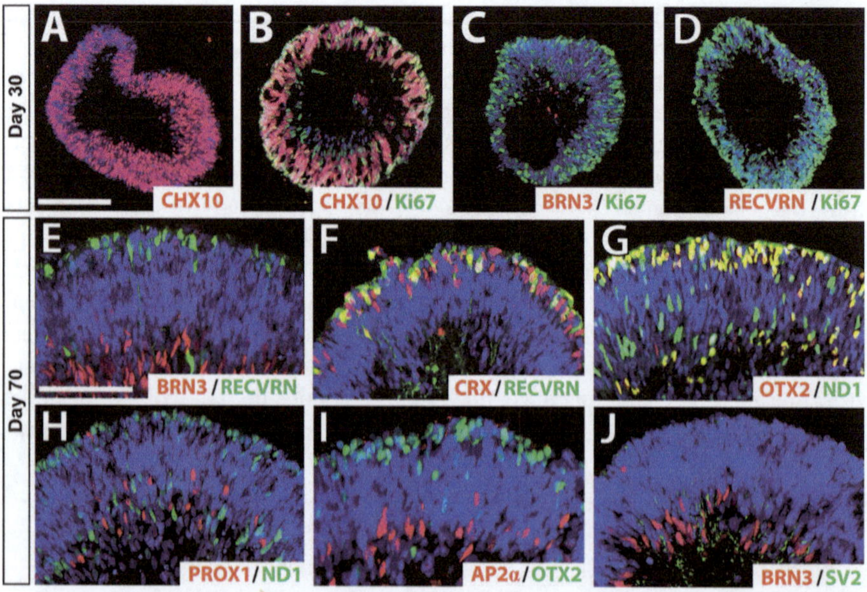

Fig. 2.2 hPSCs can be directed to generate retinal organoids using three-dimensional differentiation approaches. hPSCs were directed to generate retinal organoids in a stepwise manner analogous to major stages of retinogenesis. Optic vesicle-like retinal organoids expressed retinal progenitor markers, CHX10, PAX6, and cell proliferation marker Ki67 after 1 month of differentiation (**a–d**) while photoreceptor marker RECOVERIN and ganglion cells marker BRN3 were seldom seen at this early stage of development. After 2 months of differentiation, retinal organoids acquired a cup-like appearance and retinal cells were arranged in a stratified manner. Photoreceptor markers (CRX, RECOVERIN, OTX2, and ND1) occupied apical layers of the organoids (**e–i**), horizontal and amacrine cells (PROX1, AP2α) in the middle and ganglion cells (BRN3, SV2) occupied basal layers within the organoids (**e–j**). Scale bars equal 150 μm for **a–d** and 100 μm for **e–j**

to be specified, while later-born cell types such as rod photoreceptors among the last [21, 23, 28–31, 33, 34, 37, 85].

Retinal cells occupy strategic positions within the adult retina, with ganglion cells residing in the innermost layers of the retina, whereas photoreceptor cells closely associate with RPE and form the outermost layers. The spatial arrangement of retinal neurons and their synaptic connections linking them together are critical to their proper function and as such, retinal cells derived from hPSCs should similarly recapitulate this level of organization. While traditional methods of differentiation have allowed for the successful generation of all the major cell types of the retina, these approaches have lacked the ability of retinal cells to assemble into a layered structure. These shortcomings of traditional approaches have been overcome by the development of retinal organoids, which allow for the maintenance of cell–cell contacts between retinal neurons [21, 23, 28–30, 32, 33, 35, 37, 38, 85]. These organoids formed a pseudostratified epithelium-like structure which allows the retinal cells to

Table 2.2 Summary of retinal organoid differentiation from hPSCs

Authors	Cell type-specific markers expressed				Signaling factors	Structural characteristics	Functional characteristics
	Photoreceptors	Ganglion cells	Amacrine/ horizontal cells	Bipolar cells			
Meyer et al. [21]	CRX Recoverin	BRN3	Calretinin	VSX2 PKCα	Default neural specification	Establishment of optic vesicle-like organoids	Photoreceptors responsive to 8-Br-cGMP
Philips et al. [29]	CRX Recoverin Rhodopsin S-opsin	BRN3 HuC/D TUJ1	Calretinin	PKCα	Default neural specification	Expression of synaptic and gap junction proteins	
Nakano et al. [28]	CRX NRL Recoverin Rxr-gamma	BRN3 PAX6 TUJ1	Calretinin Ptf1a	PKCα	CHIR99021 DAPT IWR1 Retinoic acid SAG Y-27632	Self-organization of hPSCs into optic cup like structures Multilayered retina with roads and cones	
Zhong et al. [30]	CNG α-and β $G_{T1\alpha}$ Guanylate cyclase 1 OTX2 PDE6-α Recoverin Rhodopsin S-opsin	BRN3 HuC/D	AP2α HuC/D PROX1	VSX2	FBS Taurine Retinoic acid	Outer segment disc formation Outer limiting membrane Inner segments Mitochondria and basal bodies in photoreceptors	Photoreceptors exhibited rudimentary responses to light

(continued)

Table 2.2 (continued)

Authors	Cell type-specific markers expressed				Signaling factors	Structural characteristics	Functional characteristics
	Photoreceptors	Ganglion cells	Amacrine/ horizontal cells	Bipolar cells			
Reichman et al. (2017)	CD73 Cone arrestin CRX L/M-Opsin OTX2 Recoverin Rhodopsin	BRN3A	AP2α Calretinin LIM1	Glutamine Synthase PKCα SOX9	DAPT FGF2	Laminar organization of retinal cells	
Mellough et al. [34]	Basoon Calbindin28 CRX L/M-Opsin Recoverin Rhodopsin S-Opsin	CALB1 HuC/D Islet1 TUJ1	CALB1 HuC/D TUJ1	CALB1	IGF-1	Cilia and melanosomes in RPE, nascent outer segment discs, basal body complex in photoreceptors.	Photoreceptors responsive to 8-Br-cGMP
Ohlemacher et al. [23]	CRX OTX2 Recoverin	BRN3 MAP2			Default neural specification	Morphological isolation of retinal organoids apart from forebrain populations	
Singh et al. [36, 76]	CRX Recoverin	BRN3B	Calretinin LGR5 NF200		bFGF Dkk1 IGF-1 FGF9 Noggin	Laminar organization of retinal cells Migration of axons within retinal tissue	Electrophysiological responses in inner retinal neurons

Kaewkhaw et al. [31]	BLIMP1 CRX NRL S-Opsin Recoverin	BRN3B CALB1	CALB1 CALB2	CALB1 PKCα	Matrigel FBS Retinoic acid SAG	Laminar organization of retinal cells	
Völkner et al. [37]	CRX OTX2 Recoverin Rhodopsin SALL3	BRN3 HuC/D Islet1 TUJ1	BHLHB5 CALB1 CALB2 EBF3 HuC/D PROX1 TFAP2A TUJ1	PRKCA SOX9 VSX2	DAPT	Laminar organization of retinal cells	
Lowe et al. [33]	L/M-Opsin OTX2 Recoverin Rhodopsin	CALB1 ISLET1/2 TUJ1	CALB1 TUJ1	CALB1 VSX2	Y-27632	Cilia, basal bodies, mitochondria and membrane discs in photoreceptors	
Parfitt et al. [84]	Arrestin L/M-Opsin Recoverin Rhodopsin S-opsin	BRN3 HuC/D	HuC/D		CHIR99021 IWR1 Retinoic acid SAG Y-27632	Cilia and mitochondria in inner segments	

(continued)

Table 2.2 (continued)

Authors	Cell type-specific markers expressed				Signaling factors	Structural characteristics	Functional characteristics
	Photoreceptors	Ganglion cells	Amacrine/ horizontal cells	Bipolar cells			
Wiley et al. [39]	CRX L/M-Opsin NRL Recoverin Rhodopsin S-Opsin	HuC/D	HuC/D	PKCα	CHIR99021 DAPT ECM IWR1 SAG	Laminar organization of retinal cells	
Wahlin et al. [38]	CRX L/M-Opsin OTX2 Recoverin Rhodopsin Opsin rod RET-P1 S-Opsin	BRN3 NF200			DAPT IWR1 Retinoic acid SAG	Microvilli on RPE Photoreceptor inner segment with cilia, basal body, mitochondria Rudimentary outer segment with discs Photoreceptor outer limiting membrane Synaptic ribbons	Excitatory and inhibitory neurotrasmitter responses HCN-type current in cones Voltage-dependent release of synaptic vesicles
Gonzalez-Cordero et al. (2017)	ABCA4 Arrestin CRX L/M Opsin NRL PRPH2 Onecut-1 OTX2 Recoverin Rhodopsin RETGC S opsin				FBS Glutamax Glutamine Retinoic Acid Taurine	Inner segments and connecting cilia Nascent outer segments and presynaptic structures Transplantable cone photoreceptors	

mature in both a temporal and spatial fashion, with ganglion cells specified in basal laminae of the organoids while photoreceptors occupy apical regions.

The three-dimensional nature of organoids also likely aids in the functional maturation of retinal neurons, which has been largely limited in retinal cells derived using traditional differentiation methods. While these retinal cells differentiated by traditional approaches commonly express a variety of features associated with all of the major cell types of the retina, these cells lacked the structural and functional differentiation typically associated with more mature retinal neurons. The use of three-dimensional retinal organoids allows for the acquisition of more advanced features of differentiation within these cells including enhanced outer segments and the ability to respond to light stimuli, presumably due to their ability to interact and self-organize with neighboring cells. Further refinements of these organoid cultures have also involved the addition of external signaling molecules in long-term cultures to further guide their differentiation [28, 30, 32]. This has been particularly true for photoreceptors, which have been the most extensively studied cell type derived within retinal organoids. The experimental manipulation of critical signaling pathways within retinal organoids has led to refinements in photoreceptor differentiation, including accelerated differentiation as well as increased expression of phototransduction proteins. Photoreceptors derived in this fashion exhibited characteristic bulb-like structures at their tips, demonstrated membranous disc-like structures in regions resembling outer segments, and occasionally displayed electrophysiologic responses to light stimuli [30, 38].

2.5 Application of hPSC-Derived Cells for Retinal Disease Modeling

Beyond the applications of hPSCs for modeling retinal development, these cells also serve as powerful and unique platforms for the study of human retinal degenerative diseases. Due to the degeneration of specific populations of retinal neurons, these diseases are characterized by loss of vision and eventual blindness. Retinal degenerative diseases can be most readily classified into diseases that affect cells of the outer retina or those affecting the inner retina, most notably age-related macular degeneration and glaucoma, respectively [86, 87]. Traditionally, the ability to study the progression of these disease states has been limited to animal models. While these animal models have led to significant advances in our understanding of retinal disease progression [88–94], important differences exist between the retinas of animal models and humans, including the prevalence of rods and cones as well as the presence of a macula in humans. Furthermore, studies in humans have been largely limited to postmortem retinal tissue or to retinal imaging approaches that lack the resolution to examine individual cells. While these studies have been informative about the end-result of disease pathology, the approach necessarily limits the ability to better understand disease progression within individual cells.

In order to overcome these shortcomings for studies of retinal degenerative diseases, recent research has focused on the use of hPSCs to model and understand disease progression (Table 2.3). When generated from patients with a known genetic basis for retinal degeneration, hPSCs provide an infinite supply of cells for the derivation of the affected cell type, and can thereby serve as powerful tools to study the disease phenotype [85, 110, 111]. Over the last several years, studies have utilized hPSCs for studies of degenerative diseases of the retina, with a particular focus on those diseases that affect RPE and photoreceptors [11, 15, 21, 56, 60, 70–72, 95, 96, 98, 101, 105, 109, 112–119]. These cells are often affected in retinal degenerative diseases such as age-related macular degeneration, and the derivation of these cells has been extensively characterized through hPSC retinal differentiation protocols. Such approaches have helped to demonstrate the improper function and/or reduced survival of RPE and photoreceptors in patient-derived cells, thereby providing insight into potential mechanisms underlying the loss of these retinal cell types [11, 21, 53, 56, 101, 105, 115, 116]. Furthermore, patient-derived hPSCs have also been utilized to identify novel genetic variants underlying retinal degeneration, highlighting the potential to target this area for the development of therapies [96, 98].

While diseases affecting cells of the outer retina have been extensively studied with hPSCs, studies related to diseases affecting inner retinal neurons have been largely limited. Of the diseases affecting inner retinal neurons, the most common is glaucoma with a current incidence of greater than 60 million people worldwide [120, 121]. Glaucoma results in the degeneration of retinal ganglion cells (RGCs), leading to a decreased connectivity between the eye and the brain and subsequent

Table 2.3 Selected demonstrations of retinal disease modeling with hPSCs

Authors	Disease modeled	Cell types studied	Disease phenotype	Therapeutic approach	
				Gene correction	Drug screening
Jin et al. [95]	Retinitis Pigmentosa	Photoreceptors	ER stress Oxidative stress		✓
Meyer et al. [21]	Gyrate Atrophy	RPE	Enzymatic defect	✓	✓
Tucker et al. [96]	Retinitis Pigmentosa	Photoreceptors	Alu mutation in MAK gene		
Singh et al. [55, 56]	Best disease	RPE	Phagocytosis defects Oxidative stress Altered calcium homeostasis		
Minegishi et al. [97]	Glaucoma	Neural cells	Increased protein deposits and insolubility		✓
Lustremant et al. [98]	LCA	RPE	Gene polymorphism Oxidative stress		
Tucker et al. [99]	Glaucoma	Retinal ganglion cells	Autophagy defects		
Yang et al. [100]	AMD	RPE	Oxidative stress		

(continued)

Table 2.3 (continued)

Authors	Disease modeled	Cell types studied	Disease phenotype	Therapeutic approach	
				Gene correction	Drug screening
Cereso et al. [101]	Choroideremia	RPE	Biochemical defect Under prenylation of protein	✓	
Yoshida et al. [102]	Retinitis pigmentosa	Photoreceptors	Increased ER stress Increased apoptosis Autophagy defects	✓	✓
Burnight et al. [6]	LCA	Photoreceptors	Fewer and shorter cilia	✓	
Li et al. [103]	Retinitis pigmentosa	RPE	Disorganized RPE Loss of apical microvilli Reduced pigmentation	✓	
Lukovic et al. [11]	Retinitis pigmentosa	RPE	Phagocytosis defects		
Schwarz et al. [104]	Retinitis pigmentosa	RPE	Trafficking defects in cilia Loss of protein Disrupted golgi cohesion	✓	
Singh et al. [36, 76]	Best disease	RPE	Increased oxidation Phagocytosis defects		✓
Moshfegh et al. [105]	Best disease	RPE	Chloride channel defect		
Chen at al. [106]	Dominant optic atrophy	RGCs	Increased apoptosis		✓
Ohlemacher et al. [63]	Glaucoma	Retinal ganglion cells	Increased apoptosis		✓
Parfitt et al. [84]	LCA	RPE Photoreceptors	Cilia defects Abnormal protein splicing	✓	
Saini et al. [107]	AMD	RPE	Upregulated complement and inflammatory markers		✓
Ramsden et al. [108]	Retinitis pigmentosa	RPE	Phagocytosis defects	✓	✓
Teotia et al. [82]	Glaucoma	Retinal ganglion cells	Shortened neurites, immature activity, reduced expression of guidance cues		
Hallam et al. [109]	AMD	RPE	Increased inflammation and cell stress, accumulation of liquid droplets, impaired autophagy		

loss of vision. The ability to derive RGCs from hPSCs has been a more recent area of investigation, which now allows for the application of these cells for studies of retinal degenerative diseases affecting the inner retina [58, 62, 63, 81, 82, 97, 99, 122]. Recently, efforts have focused on the use of hPSCs from patients with genetic determinants of degenerative diseases that directly affect the RGCs, such as gene mutations underlying some forms of normal tension glaucoma and dominant optic atrophy. Interestingly, upon the differentiation of these cells, RGCs from patient sources exhibited increased apoptosis, thereby allowing for subsequent studies of disease mechanisms leading to degeneration of RGCs [63, 106].

While traditional retinal differentiation protocols have been highly successful in modeling certain features of some retinal degenerative diseases, the resultant retinal cells differentiate in a manner that lacks any three-dimensional organization that mimics how cells are arranged into retinal tissue. Retinal organoids may serve as an improved model for studies of retinal disease modeling, allowing for the interaction between different cell types and therefore providing the ability to assess the effects of degeneration on the entire tissue. While such an ability has yet to be demonstrated for retinal organoids, the use of cerebral organoids for disease modeling has provided an important proof of principle and have been particularly successful for some of the effects of cerebral diseases, such as microcephaly and lissencephaly [123–125]. In the near future, it is likely that retinal organoids will be applied for the study of retinal degenerative diseases. As recent studies have demonstrated the successful organization and maturation of photoreceptors within retinal organoids [21, 29, 30, 34, 38, 76], disease-modeling approaches will most likely be applied for outer retinal diseases. Recent studies utilizing hPSC-derived retinal organoids have primarily utilized a genetic basis versus an idiopathic basis for retinal degenerative diseases [39, 76, 84]. Additionally, further improvements to retinal organoids will likely be necessary to be able to apply them to a wide variety of retinal degenerative diseases. For example, hPSC-derived retinal organoids do not demonstrate a macula-like region or a functioning RPE layer, they are currently suited to model diseases that affect peripheral photoreceptors [39, 76, 84]. Further improvements in the differentiation methods to also include the characterization and maturation of inner retinal neurons will enable the study of diseases to affect ganglion cells with retinal organoids.

2.6 Drug Screening with hPSC-Derived Retinal Cells

When derived from individual patient populations, particularly those with a known genetic basis underlying retinal disease, hPSCs possess the ability to recreate certain features of the disease phenotype and model the degeneration associated with retinal diseases. With the resulting data accumulated from such studies, these cells can then be utilized for the development of therapeutic approaches for retinal degenerative diseases [111, 119, 126–129]. Following the directed differentiation of patient-derived hPSCs to a retinal fate, drug screening efforts can be targeted to an affected retinal cell type, providing a platform for assessing the ability of candidate compounds to rescue the disease phenotype.

The use of patient-derived hPSCs for drug screening has been particularly successful for degenerative diseases that affect the outer retina, whose cells have been routinely derived and extensively characterized from hPSCs [6, 11, 21, 76, 84, 102–104, 107, 108, 116, 117, 130]. Photoreceptors and RPE are the most common cell types affected in many retinal degenerative diseases such as age-related macular degeneration (AMD), where the loss of photoreceptors combined with dysfunctions in RPE leads to loss of vision. As patient-derived RPE has been shown to recapitulate some of the hallmark features of AMD, including elevated expression of inflammatory factors and defective oxidative stress responses, recent studies have utilized hPSC-derived RPE as a platform for the screening of candidate drugs to assess the ability to improve their survival [56, 59, 76, 100, 131]. The results of these studies have enabled the identification of select compounds as potential neuroprotective agents that can alleviate RPE degeneration [107]. Similarly, hPSC-derived retinal cells have also been utilized for drug screening purposes as a means to alleviate photoreceptor loss due to retinitis pigmentosa, with results indicating that hPSC-derived photoreceptors were able to recapitulate the disease phenotype and upregulate markers of oxidative stress, lipid oxidation, and apoptosis [11, 57, 102, 103, 108]. Treatment of the degenerating rod photoreceptors with antioxidant vitamins effectively increased photoreceptor survival.

While hPSC differentiation strategies initially emphasized the cells of the outer retina, recent refinements in differentiation protocols have enabled the stepwise differentiation and identification of inner retinal neurons, particularly RGCs [58, 62, 63, 79, 81, 82, 106]. RGCs serve as the critical connection between the eye and the brain to transmit visual information, and their degeneration is part of a spectrum of diseases known as optic neuropathies, resulting in vision loss and eventual blindness. RGCs differentiated from hPSCs, particularly when derived from patient-specific sources, allow for the ability to screen new drug compounds and develop personalized treatment profiles for optic neuropathies [63, 106]. As a proof of principle, recent studies have successfully demonstrated the ability to faithfully recapitulate some of the degenerative processes associated with optic neuropathies in hPSC-derived RGCs, with subsequent drug screening approaches enabling the identification of neurotrophic factors capable of rescuing RGC degeneration [63].

While a number of studies have successfully demonstrated the ability to screen compounds for their neuroprotective effects on hPSC-derived retinal cells, these approaches have focused on isolated cells lacking any three-dimensional organization reminiscent of retinal tissue. With the advent of retinal organoids, hPSCs can be directed to differentiate in a manner that recapitulates the architecture, spatial connectivity and functioning of the retina, and may therefore be better suited for drug screening purposes. Given the more detailed demonstration to date of photoreceptor differentiation and organization in the outer layers of retinal organoids, these cells are likely better suited for drug screening applications for photoreceptor diseases. In contrast to outer retinal diseases, retinal organoids can also be used to test and develop therapies for inner retinal neurons such as RGCs, which are primarily affected in optic neuropathies.

2.7 hPSC-Derived Retinal Cells as a Vehicle for Cell Replacement

While early stages of retinal degenerative diseases may be effectively studied with hPSCs, and subsequently drug screening approaches may aid in the neuroprotection of these degenerating cells, the irreversible loss of retinal neurons in later stages renders such measures ineffective, resulting in severe vision loss and blindness. In such cases, attempts to replace degenerated cells through transplants of healthy retinal cells constitute the only remaining effective option to restore some visual function [129, 132]. The transplantation of cells into the retina represents a more feasible option for cell replacement when compared to other cells of the nervous system, as the relative ease of accessibility of the retina and its reduced immunological response will likely facilitate cell replacement [133, 134]. To aid in this goal, hPSCs can serve as a renewable source of stem cells for the differentiation of retinal cells for a variety of translational approaches to retinal repair. Transplants of hPSC-derived retinal cells can assist in neuroprotection, particularly at earlier stages of the disease process, and can lead to potential delay in disease progression. At later stages of the degenerative process, hPSC-derived retinal cells can serve as a source for repopulation of the retina following the loss of host neurons.

Several studies have examined the use of hPSC-derived photoreceptors for cell replacement in diseases that affect the outer retina, with the goal to replace the degenerating neurons with their functional equivalents [9, 39, 64, 135, 136]. Initial studies focused on transplantation of undifferentiated retinal stem cells into animal models, which could integrate into many layers of the retina and exhibit neuronal morphologies [137–139]. However, these cells were often limited in number and their ability to be expanded, and rarely exhibited any ability to give rise to photoreceptor cells. As an alternative, more recent efforts have focused upon the ability of hPSC-derived photoreceptor cells for cell replacement. Upon transplantation, several groups have demonstrated the ability of these cells to integrate into the host retina and form connections with other retinal neurons, in some cases leading to improved visual function and restoration of light sensitivity [135, 140, 141]. Further investigations into the transplantation of hPSC-derived photoreceptors have demonstrated the use of immunodeficient mouse models to improve survival of hPSC-derived photoreceptors [136].

As the RPE provides essential support for photoreceptors, similar approaches for cell replacement have also been developed for RPE loss in retinal degenerative diseases, often associated with the secondary loss of photoreceptors. hPSC-derived RPE has been utilized in the development of cell replacement strategies for diseases such as age-related macular degeneration [15, 21, 76, 107, 109, 114, 131]. In this capacity, the transplant of RPE cells has been accomplished by either subretinal injection as a cell suspension or as RPE sheet transplantation [142–151]. The latter approach may offer numerous advantages, as the cells retain their polarization and are arranged in a discrete monolayer, allowing better integration within the host retina. The success of the above-named transplantation strategies has paved the way for hPSC-derived RPE in clinical trials for AMD and Stargardt's disease, where transplanted cells were shown to improve visual acuity in patients, illustrating the ability of hPSCs to rescue visual defects in retinal degenerative diseases [14, 152].

Many of the cell replacement strategies developed to date have focused on the transplantation of RPE and/or photoreceptors due to their ease of differentiation and more limited need of these cells to extend neurites to form synaptic connections, which will likely make replacement efforts easier. However, the development of replacement strategies for inner retinal neurons such as RGCs is more complicated, largely due to their more elaborate nature and need to extend long axonal projections to form synaptic connections in the optic tectum [153–155]. As such, pharmacologic strategies to combat RGC degeneration have focused on early stages of the disease process where neuroprotection is feasible [63, 106]. The goal is both to improve RGC survival, as well as potentially regrow axons to reestablish central synaptic connections. Similar efforts have not been widely adopted yet for hPSC-derived RGCs, although early studies have demonstrated the ability of hPSC-derived RGCs to survive following intravitreal transplantation [80]. Further studies into the use of hPSC-derived RGCs are certainly warranted, as several recent reports have demonstrated the differentiation and enrichment of RGCs from hPSCs in vitro [63, 66, 67, 80, 82].

Efforts for cell replacement to date have often focused on the transplantation of a single type of retinal neuron. At late stages of retinal degeneration, other retinal neurons are often damaged and lost, leading to the need to replace multiple types of cells. Retinal organoids represent an exciting option for cell replacement at these late stages of retinal degeneration, as these organoids possess the relevant retinal cells pre-assembled into a stratified structure, and can serve as "mini-retinas" for replacement of retinal tissue [21, 23, 28–30, 32, 39, 63]. Early attempts at these strategies have recently been demonstrated in mouse models of retinitis pigmentosa, where retinal organoids were transplanted and retained transplants of retinal organoids in mice led to the retention of their three-dimensional architecture and formed presumptive synaptic connections with host bipolar cells [142]. Similar experiments have also been conducted in nonhuman primates, with the transplantation of hPSC-derived retinal organoids resulting in increased visual acuity [156].

2.8 Conclusions and Future Directions

Overall, research over the past several years has established hPSCs as a powerful tool for studying some of the earliest stages of human development that would otherwise remain inaccessible to investigation [19, 22, 23, 25, 27, 37]. This has encouraged the establishment of efficient differentiation protocols to generate all major cell types of the retina, including photoreceptors, RPE, and retinal ganglion cells [9, 20, 21, 30, 53, 55, 58, 61, 63, 70, 78, 81, 82, 113, 157, 158]. These hPSC-derived retinal cells have assisted in modeling retinal degenerative diseases, especially when generated from patients with inherited retinal dystrophies. For this purpose, patient-derived hPSCs have helped in understanding disease progression and mechanisms, and have subsequently enabled the identification of candidate neuroprotective factors to combat the degeneration of retinal neurons [11, 21, 39, 56, 57, 63, 76, 84, 98, 100, 103, 106, 115, 116, 128, 148]. However, these measures have limited utility at late-stage disease, where the loss of multiple retinal cell types is irreversible, resulting in severe loss of vision. As a source of cell replacement therapies, hPSC-derived retinal cells have been shown to integrate within the host retina, form synaptic connections as well as demonstrate functional rescue. Such strategies have been extensively studied in the context of RPE and photoreceptor degeneration [6, 7, 10, 39, 136, 141, 142, 149, 156, 159–161], and is finding application in current clinical trials in AMD and Stargardt's disease using hPSC-derived RPE [14, 152].

While tremendous progress has been made in the differentiation of retinal neurons from hPSCs [17, 19, 20, 22, 26, 53, 63, 69, 71, 81, 162], these cells often fail to fully differentiate into functionally relevant phenotypes which would better mimic the structure and functionality of the retina. Therefore, recent advances have led to the development of a three-dimensional approach to retinal differentiation, where hPSCs are directed to yield discrete populations closely analogous to the developing optic cup and eventually giving rise to a pseudostratified structure resembling the retina [21, 28–31, 33, 35–39]. With these advances, retinal organoids follow predicted stages of retinal development, and have led to enhanced differentiation and maturation of photoreceptors, facilitating the application of these approaches for studies of retinal development and pathogenesis in both normal and diseased states.

Patient-derived organoids may be best suited for assessing the effects of disease-related neurodegeneration on specific retinal cell types, as well as their interactions with each other. Currently, retinal organoids are likely better suited for studies of photoreceptor diseases, as photoreceptor development and maturation has been extensively characterized in retinal organoids, leading to rod-dominant retinal domains similar to peripheral regions of the retina [29, 30, 35, 38, 39]. Therefore, rod-cone dystrophies like retinitis pigmentosa, which begins as a peripheral retinal degeneration, can be most effectively modeled with retinal organoids, with the goal of developing neuroprotective strategies. Moreover, future efforts to characterize inner retinal neurons within retinal organoids will help to model and develop therapies for RGC degeneration in optic neuropathies. In addition to studies of retinal development and disease, the most exciting feature of retinal organoids may be their ability to serve as a replacement for retinal tissue in severely degenerated retinas. The interconnected structure composed of multiple retinal neurons may facilitate integration and replacement of multiple cell types within the degenerated retina.

References

1. Barnstable, C. J. (1987). A molecular view of vertebrate retinal development. *Molecular Neurobiology, 1*, 9–46.
2. Thomson, J. A., Itskovitz-Eldor, J., Shapiro, S. S., Waknitz, M. A., Swiergiel, J. J., Marshall, V. S., et al. (1998). Embryonic stem cell lines derived from human blastocysts. *Science, 282*, 1145–1147.
3. Takahashi, K., Tanabe, K., Ohnuki, M., Narita, M., Ichisaka, T., Tomoda, K., et al. (2007). Induction of pluripotent stem cells from adult human fibroblasts by defined factors. *Cell, 131*, 861–872. https://doi.org/10.1016/j.cell.2007.11.019
4. Takahashi, K., & Yamanaka, S. (2006). Induction of pluripotent stem cells from mouse embryonic and adult fibroblast cultures by defined factors. *Cell, 126*, 663–676. https://doi.org/10.1016/j.cell.2006.07.024
5. Yu, J., Vodyanik, M.A., Smuga-Otto, K., Antosiewicz-Bourget, J., Frane, J.L., Tian, S., et al. (2007). Induced pluripotent stem cell lines derived from human somatic cells. *Science, 318*, 1917–1920. https://doi.org/10.1126/science.1151526
6. Burnight, E. R., Wiley, L.A., Drack, A.V., Braun, T.A., Anfinson, K.R., Kaalberg, E.E., et al. (2014). CEP290 gene transfer rescues Leber congenital amaurosis cellular phenotype. *Gene Therapy, 21*, 662–672. https://doi.org/10.1186/1756-6606-7-45
7. Carr, A. J., Vugler, A.A., Hikita, S.T., Lawrence, J.M., Gias, C., Chen, L.L., et al. (2009). Protective effects of human iPS-derived retinal pigment epithelium cell transplantation in the retinal dystrophic rat. *PLoS One, 4*, e8152. https://doi.org/10.1371/journal.pone.0008152
8. Gonzalez-Cordero, A., Kruczek, K., Naeem, A., Fernando, M., Kloc, M., Ribeiro, J., et al. (2017). Recapitulation of Human Retinal Development from Human Pluripotent Stem Cells Generates Transplantable Populations of Cone Photoreceptors. *Stem Cell Reports*, 9(3), 820–837. https://doi.org/10.1016/j.stemcr.2017.07.022
9. Lamba, D. A., McUsic, A., Hirata, R. K., Wang, P. R., Russell, D., & Reh, T. A. (2010). Generation, purification and transplantation of photoreceptors derived from human induced pluripotent stem cells. *PLoS One, 5*, e8763. https://doi.org/10.1371/journal.pone.0008763
10. Li, Y., Tsai, Y.T., Hsu, C.W., Erol, D., Yang, J., Wu, W.H., et al. (2012). Long-term safety and efficacy of human-induced pluripotent stem cell (iPS) grafts in a preclinical model of retinitis pigmentosa. *Molecular Medicine, 18*, 1312–1319. https://doi.org/10.2119/molmed.2012.00242
11. Lukovic, D., Artero Castro, A., Delgado, A.B., Bernal Mde, L., Luna Pelaez, N., Diez Lloret, A., et al. (2015). Human iPSC derived disease model of MERTK-associated retinitis pigmentosa. *Scientific Reports, 5*, 12910. https://doi.org/10.1016/j.trsl.2015.08.007
12. Pearson, R. A. (2014). Advances in repairing the degenerate retina by rod photoreceptor transplantation. *Biotechnology Advances, 32*, 485–491. https://doi.org/10.1016/j.biotechadv.2014.01.001
13. Rowland, T. J., Buchholz, D. E., & Clegg, D. O. (2012). Pluripotent human stem cells for the treatment of retinal disease. *Journal of Cellular Physiology, 227*, 457–466. https://doi.org/10.1002/jcp.22814

14. Schwartz, S. D., Hubschman, J.P., Heilwell, G., Franco-Cardenas, V., Pan, C.K., Ostrick, R.M., et al. (2012). Embryonic stem cell trials for macular degeneration: A preliminary report. *Lancet, 379*, 713–720. https://doi.org/10.1016/s0140-6736(12)60028-2
15. Tsai, Y., Lu, B., Bakondi, B., Girman, S., Sahablan, A., Sareen, D., et al. (2015). Human iPSC-derived neural progenitors preserve vision in an amd-like model. *Stem Cells (Dayton, OH), 33*, 2537–2549. https://doi.org/10.1002/stem.2032
16. Banin, E., Obolensky, A., Idelson, M., Hemo, I., Reinhardtz, E., Pikarsky, E., et al. (2006). Retinal incorporation and differentiation of neural precursors derived from human embryonic stem cells. Stem cells. *Stem Cells (Dayton, OH), 24*, 246–257. https://doi.org/10.1634/stemcells.2005-0009
17. Hirami, Y., Osakada, F., Takahashi, K., Okita, K., Yamanaka, S., Ikeda, H., et al. (2009). Generation of retinal cells from mouse and human induced pluripotent stem cells. *Neuroscience Letters, 458*, 126–131. https://doi.org/10.1016/j.neulet.2009.04.035
18. Ikeda, H., Osakada, F., Watanabe, K., Mizuseki, K., Haraguchi, T., Miyoshi, H., et al. (2005). Generation of Rx+/Pax6+ neural retinal precursors from embryonic stem cells. *Proceedings of the National Academy of Sciences of the United States of America, 102*, 11331–11336. https://doi.org/10.1073/pnas.0910012107
19. Lamba, D. A., Karl, M. O., Ware, C. B., & Reh, T. A. (2006). Efficient generation of retinal progenitor cells from human embryonic stem cells. *Proceedings of the National Academy of Sciences of the United States of America, 103*, 12769–12774. https://doi.org/10.1073/pnas.0601990103
20. Mellough, C. B., Sernagor, E., Moreno-Gimeno, I., Steel, D. H., & Lako, M. (2012). Efficient stage specific differentiation of human pluripotent stem cells towards retinal photoreceptor cells. *Stem Cells (Dayton, OH)*. https://doi.org/10.1002/stem.1037
21. Meyer, J. S., Howden, S. E., Wallace, K. A., Verhoeven, A. D., Wright, L. S., Capowski, E. E., et al. (2011). Optic vesicle-like structures derived from human pluripotent stem cells facilitate a customized approach to retinal disease treatment. *Stem Cells (Dayton, OH), 29*, 1206–1218. https://doi.org/10.1002/stem.674
22. Meyer, J. S., Shearer, R. L., Capowski, E. E., Wright, L. S., Wallace, K. A., McMillan, E. L., et al. (2009). Modeling early retinal development with human embryonic and induced pluripotent stem cells. *Proceedings of the National Academy of Sciences of the United States of America, 106*, 16698–16703. https://doi.org/10.1073/pnas.0905245106
23. Ohlemacher, S. K., Iglesias, C. L., Sridhar, A., Gamm, D. M., & Meyer, J. S. (2015). Generation of highly enriched populations of optic vesicle-like retinal cells from human pluripotent stem cells. *Current Protocols in Stem Cell Biology, 32*, 1H 8 1–1H 8 20. https://doi.org/10.1002/9780470151808.sc01h08s32
24. Osakada, F., Ikeda, H., Mandai, M., Wataya, T., Watanabe, K., Yoshimura, N., et al. (2008). Toward the generation of rod and cone photoreceptors from mouse, monkey and human embryonic stem cells. *Nature Biotechnology, 26*, 215–224. https://doi.org/10.1038/nbt1384
25. Osakada, F., Ikeda, H., Sasai, Y., & Takahashi, M. (2009). Stepwise differentiation of pluripotent stem cells into retinal cells. *Nature Protocols, 4*, 811–824. https://doi.org/10.1038/nprot.2009.51
26. Sridhar, A., Ohlemacher, S. K., Langer, K. B., & Meyer, J. S. (2016). Robust differentiation of mRNA-reprogrammed human induced pluripotent stem cells toward a retinal lineage. *Stem Cells Translational Medicine, 5*, 417–426. https://doi.org/10.5966/sctm.2015-0093
27. Sridhar, A., Steward, M. M., & Meyer, J. S. (2013). Nonxenogeneic growth and retinal differentiation of human induced pluripotent stem cells. *Stem Cells Translational Medicine, 2*, 255–264. https://doi.org/10.5966/sctm.2012-0101
28. Nakano, T., Ando, S., Takata, N., Kawada, M., Muguruma, K., Sekiguchi, K., et al. (2012). Self-formation of optic cups and storable stratified neural retina from human ESCs. *Cell Stem Cell, 10*, 771–785. https://doi.org/10.1016/j.stem.2012.05.009
29. Phillips, M. J., Wallace, K. A., Dickerson, S. J., Miller, M. J., Verhoeven, A. D., Martin, J. M., et al. (2012). Blood-derived human iPS cells generate optic vesicle-like structures with the

capacity to form retinal laminae and develop synapses. *Investigative Ophthalmology & Visual Science, 53*, 2007–2019. https://doi.org/10.1167/iovs.11-9313
30. Zhong, X., Gutierrez, C., Xue, T., Hampton, C., Vergara, M. N., Cao, L. H., et al. (2014). Generation of three-dimensional retinal tissue with functional photoreceptors from human iPSCs. *Nature Communications, 5*, 4047. https://doi.org/10.1038/ncomms5047
31. Kaewkhaw, R., Kaya, K. D., Brooks, M., Homma, K., Zou, J., Chaitankar, V., et al. (2015). Transcriptome dynamics of developing photoreceptors in three-dimensional retina cultures recapitulates temporal sequence of human cone and rod differentiation revealing cell surface markers and gene networks. *Stem Cells (Dayton, OH), 33*, 3504–3518. https://doi.org/10.1002/stem.2122
32. Kuwahara, A., Ozone, C., Nakano, T., Saito, K., Eiraku, M., & Sasai, Y. (2015). Generation of a ciliary margin-like stem cell niche from self-organizing human retinal tissue. *Nature Communications, 6*, 6286. https://doi.org/10.1038/ncomms7286
33. Lowe, A., Harris, R., Bhansali, P., Cvekl, A., & Liu, W. (2016). Intercellular adhesion-dependent cell survival and ROCK-regulated actomyosin-driven forces mediate self-formation of a retinal organoid. *Stem Cell Reports, 6*, 743–756. https://doi.org/10.1016/j.stemcr.2016.03.011
34. Mellough, C. B., Collin, J., Khazim, M., White, K., Sernagor, E., Steel, D. H., et al. (2015). IGF-1 signaling plays an important role in the formation of three-dimensional laminated neural retina and other ocular structures from human embryonic stem cells. *Stem Cells (Dayton, OH), 33*, 2416–2430. https://doi.org/10.1002/stem.2023
35. Reichman, S., Slembrouck, A., Gagliardi, G., Chaffiol, A., Terray, A., Nanteau, C., et al. (2017). Generation of storable retinal organoids and retinal pigmented epithelium from adherent human iPS cells in xeno-free and feeder-free conditions. *Stem Cells (Dayton, OH)*. https://doi.org/10.1002/stem.2586
36. Singh, R. K., Mallela, R. K., Cornuet, P. K., Reifler, A. N., Chervenak, A. P., West, M. D., et al. (2015). Characterization of three-dimensional retinal tissue derived from human embryonic stem cells in adherent monolayer cultures. *Stem Cells and Development, 24*, 2778–2795. https://doi.org/10.1089/scd.2015.0144
37. Volkner, M., Zschatzsch, M., Rostovskaya, M., Overall, R. W., Busskamp, V., Anastassiadis, K., et al. (2016). Retinal organoids from pluripotent stem cells efficiently recapitulate retinogenesis. *Stem Cell Reports, 6*, 525–538. https://doi.org/10.1016/j.stemcr.2016.03.001
38. Wahlin, K. J., Maruotti, J. A., Sripathi, S. R., Ball, J., Angueyra, J. M., Kim, C., et al. (2017). Photoreceptor outer segment-like structures in long-term 3D retinas from human pluripotent stem cells. *Scientific Reports, 7*, 766. https://doi.org/10.1038/s41598-017-00774-9
39. Wiley, L. A., Burnight, E. R., DeLuca, A. P., Anfinson, K. R., Cranston, C. M., Kaalberg, E. E., et al. (2016). cGMP production of patient-specific iPSCs and photoreceptor precursor cells to treat retinal degenerative blindness. *Scientific Reports, 6*, 30742. https://doi.org/10.1038/srep30742
40. Chow, R. L., & Lang, R. A. (2001). Early eye development in vertebrates. *Annual Review of Cell and Developmental Biology, 17*, 255–296. https://doi.org/10.1146/annurev.cellbio.17.1.255
41. Fuhrmann, S. (2010). Eye morphogenesis and patterning of the optic vesicle. *Current Topics in Developmental Biology, 93*, 61–84. https://doi.org/10.1016/b978-0-12-385044-7.00003-5
42. Kolb, H., Nelson, R., Ahnelt, P., & Cuenca, N. (2001). Cellular organization of the vertebrate retina. *Progress in Brain Research, 131*, 3–26.
43. Neves, G., & Lagnado, L. (1999). The retina. *Current Biology, 9*, R674–R677.
44. Dowling, J. E., & Werblin, F. S. (1971). Synaptic organization of the vertebrate retina. *Vision Research, 3*, 1–15.
45. Reynolds, J., & Lamba, D. A. (2014). Human embryonic stem cell applications for retinal degenerations. *Experimental Eye Research, 123*, 151–160. https://doi.org/10.1016/j.exer.2013.07.010

46. Lamb, T. D., Collin, S. P., & Pugh, E. N. (2007). Evolution of the vertebrate eye: Opsins, photoreceptors, retina and eye cup. *Nature Reviews. Neuroscience, 8*, 960–976. https://doi.org/10.1038/nrn2283
47. Andreazzoli, M. (2009). Molecular regulation of vertebrate retina cell fate. *Birth Defects Research. Part C, Embryo Today, 87*, 284–295. https://doi.org/10.1002/bdrc.20161
48. Cepko, C. L., Austin, C. P., Yang, X., Alexiades, M., & Ezzeddine, D. (1996). Cell fate determination in the vertebrate retina. *Proceedings of the National Academy of Sciences of the United States of America, 93*, 589–595.
49. Livesey, F. J., & Cepko, C. L. (2001). Vertebrate neural cell-fate determination: Lessons from the retina. *Nature Reviews. Neuroscience, 2*, 109–118. https://doi.org/10.1038/35053522
50. Zaghloul, N. A., Yan, B., & Moody, S. A. (2005). Step-wise specification of retinal stem cells during normal embryogenesis. *Biology of the Cell, 97*, 321–337. https://doi.org/10.1042/bc20040521
51. Klimanskaya, I., Hipp, J., Rezai, K. A., West, M., Atala, A., & Lanza, R. (2004). Derivation and comparative assessment of retinal pigment epithelium from human embryonic stem cells using transcriptomics. *Cloning and Stem Cells, 6*, 217–245. https://doi.org/10.1089/clo.2004.6.217
52. Carr, A. J., Vugler, A., Lawrence, J., Chen, L. L., Ahmado, A., Chen, F. K., et al. (2009). Molecular characterization and functional analysis of phagocytosis by human embryonic stem cell-derived RPE cells using a novel human retinal assay. *Molecular Vision, 15*, 283–295. https://doi.org/10.5966/sctm.2012-0163
53. Idelson, M., Alper, R., Obolensky, A., Ben-Shushan, E., Hemo, I., Yachimovich-Cohen, N., et al. (2009). Directed differentiation of human embryonic stem cells into functional retinal pigment epithelium cells. *Cell Stem Cell, 5*, 396–408. https://doi.org/10.1016/j.stem.2009.07.002
54. Buchholz, D. E., Hikita, S. T., Rowland, T. J., Friedrich, A. M., Hinman, C. R., Johnson, L. V., et al. (2009). Derivation of functional retinal pigmented epithelium from induced pluripotent stem cells. *Stem Cells (Dayton, OH), 27*, 2427–2434. https://doi.org/10.1002/stem.189
55. Singh, R., Phillips, M. J., Kuai, D., Meyer, J., Markin J. M., Smith, M. A., et al. (2013). Functional analysis of serially expanded human iPS cell-derived RPE cultures. *Investigative Ophthalmology & Visual Science, 54*, 6767–6778. https://doi.org/10.1093/hmg/dds469
56. Singh, R., Shen, W., Kuai, D., Martin, J. M., Guo, X., Smith, M. A., et al. (2013). iPS cell modeling of best disease: Insights into the pathophysiology of an inherited macular degeneration. *Human Molecular Genetics, 22*, 593–607. https://doi.org/10.1016/j.stem.2016.03.021
57. Tucker, B. A., Mullins, R. F., Streb, L. M., Anfinson, K., Eyestone, M. E., Kaalberg, E., et al. (2013). Patient-specific iPSC-derived photoreceptor precursor cells as a means to investigate retinitis pigmentosa. *eLife, e00824*, 2. https://doi.org/10.1038/mt.2014.100
58. Riazifar, H., Jia, Y., Chen, J., Lynch, G., & Huang, T. (2014). Chemically induced specification of retinal ganglion cells from human embryonic and induced pluripotent stem cells. *Stem Cells Translational Medicine, 3*, 424–432. https://doi.org/10.5966/sctm.2013-0147
59. Ferrer, M., Corneo, B., Davis, J., Wan, Q., Miyagishima, K. J., King, R., et al. (2014). A multiplex high-throughput gene expression assay to simultaneously detect disease and functional markers in induced pluripotent stem cell-derived retinal pigment epithelium. *Stem Cells Translational Medicine, 3*, 911–922. https://doi.org/10.1167/iovs.11-9313
60. Maruotti, J., Sripathi, S. R., Bharti, K., Fuller, J., Wahlin, K. J., Ranganathan, V., et al. (2015). Small-molecule-directed, efficient generation of retinal pigment epithelium from human pluripotent stem cells. *Proceedings of the National Academy of Sciences of the United States of America, 112*, 10950–10955. https://doi.org/10.1073/pnas.1422818112
61. Zhou, S., Flamier, A., Abdouh, M., Tetreault, N., Barabino, A., Wadhwa, S., et al. (2015). Differentiation of human embryonic stem cells into cone photoreceptors through simultaneous inhibition of BMP, TGFbeta and Wnt signaling. *Development, 142*, 3294–3306. https://doi.org/10.1242/dev.125385

62. Sluch, V. M., Davis, C. H., Ranganathan, V., Kerr, J. M., Krick, K., Martin, R., et al. (2015). Differentiation of human ESCs to retinal ganglion cells using a CRISPR engineered reporter cell line. *Scientific Reports, 5*, 16595. https://doi.org/10.1038/srep16595
63. Ohlemacher, S. K., Sridhar, A., Xiao, Y., Hochstetler, A. E., Sarfarazi, M., Cummins, T. R., et al. (2016). Stepwise differentiation of retinal ganglion cells from human pluripotent stem cells enables analysis of glaucomatous neurodegeneration. *Stem Cells (Dayton, OH)*. https://doi.org/10.1002/stem.2356
64. Barnea-Cramer, A. O., Wang, W., Lu, S. J., Singh, M. S., Luo, C., Huo, H., et al. (2016). Function of human pluripotent stem cell-derived photoreceptor progenitors in blind mice. *Scientific Reports, 6*, 29784.
65. Gill, K. P., Hung, S. S., Sharov, A., Lo, C. Y., Needham, K., Lidgerwood, G. E., et al. (2016). Enriched retinal ganglion cells derived from human embryonic stem cells. *Scientific Reports, 6*, 30552. https://doi.org/10.1038/srep30552
66. Sluch, V. M., Chamling, X., Liu, M. M., Berlinicke, C. A., Cheng, J., Mitchell, K. L., et al. (2017). Enhanced stem cell differentiation and immunopurification of genome engineered human retinal ganglion cells. *Stem Cells Translational Medicine, 6*, 1972–1986. https://doi.org/10.1002/sctm.17-0059
67. Langer, K. B., Ohlemacher, S. K., Phillips, M. J., Fligor, C. M., Jiang, P., Gamm, D. M., et al. (2018). Retinal ganglion cell diversity and subtype specification from human pluripotent stem cells. *Stem Cell Reports, 10*, 1282–1293. https://doi.org/10.1016/j.stemcr.2018.02.010
68. Reh, T. A., Lamba, D., & Gust, J. (2010). Directing human embryonic stem cells to a retinal fate. *Methods in Molecular Biology, 636*, 139–153. https://doi.org/10.1007/978-1-60761-691-7_9
69. Osakada, F., Jin, Z. B., Hirami, Y., Ikeda, H., Danjyo, T., Watanabe, K., et al. (2009). In vitro differentiation of retinal cells from human pluripotent stem cells by small-molecule induction. *Journal of Cell Science, 122*, 3169–3179. https://doi.org/10.1242/jcs.050393
70. Bharti, K., Miller, S. S., & Arnheiter, H. (2011). The new paradigm: Retinal pigment epithelium cells generated from embryonic or induced pluripotent stem cells. *Pigment Cell & Melanoma Research, 24*, 21–34. https://doi.org/10.1111/j.1755-148X.2010.00772.x
71. Buchholz, D. E., Pennington, B. O., Croze, R. H., Hinman, C. R., Coffey, P. J., & Clegg, D. O. (2013). Rapid and efficient directed differentiation of human pluripotent stem cells into retinal pigmented epithelium. *Stem Cells Translational Medicine, 2*, 384–393. https://doi.org/10.1038/nbt.3070
72. Capowski, E. E., Simonett, J. M., Clark, E. M., Wright, L. S., Howden S. E., Wallace, K. A., et al. (2014). Loss of MITF expression during human embryonic stem cell differentiation disrupts retinal pigment epithelium development and optic vesicle cell proliferation. *Human Molecular Genetics, 23*, 6332–6344. https://doi.org/10.1007/978-1-4614-3209-8_20
73. Clarke, L., Ballios, B. G., & van der Kooy, D. (2012). Generation and clonal isolation of retinal stem cells from human embryonic stem cells. *The European Journal of Neuroscience, 36*, 1951–1959. https://doi.org/10.1111/j.1460-9568.2012.08123.x
74. Maeda, T., Lee, M. J., Palczewska, G., Marsilli, S., Tesar, P. J., Palczewski, K., et al. (2013). Retinal pigmented epithelial cells obtained from human induced pluripotent stem cells possess functional visual cycle enzymes in vitro and in vivo. *The Journal of Biological Chemistry, 288*, 34484–34493. https://doi.org/10.5966/sctm.2014-0038
75. Okamoto, S., & Takahashi, M. (2011). Induction of retinal pigment epithelial cells from monkey iPS cells. *Investigative Ophthalmology & Visual Science, 52*, 8785–8790. https://doi.org/10.1167/iovs.11-8129
76. Singh, R., Kuai, D., Guziewicz, K. E., Meyer, J., Wilson M., Lu, J., et al. (2015). Pharmacological modulation of photoreceptor outer segment degradation in a human iPS cell model of inherited macular degeneration. *Molecular Therapy, 23*, 1700–1711. https://doi.org/10.1093/hmg/ddu053
77. Vugler, A., Carr, A. J., Lawrence, J., Chen, L. L., Burrell, K., Wright A., et al. (2008). Elucidating the phenomenon of HESC-derived RPE: Anatomy of cell genesis, expansion

and retinal transplantation. *Experimental Neurology, 214*, 347–361. https://doi.org/10.1016/j.expneurol.2008.09.007

78. Boucherie, C., Sowden, J. C., & Ali, R. R. (2011). Induced pluripotent stem cell technology for generating photoreceptors. *Regenerative Medicine, 6*, 469–479. https://doi.org/10.2217/rme.11.37
79. Deng, F., Chen, M., Liu, Y., Hu, H., Xiong, Y., Xu, C., et al. (2016). Stage-specific differentiation of iPSCs toward retinal ganglion cell lineage. *Molecular Vision, 22*, 536–547.
80. Li, K., Zhong, X., Yang, S., Luo, S., Li, K., Liu, Y., et al. (2017). HiPSC-derived retinal ganglion cells grow dendritic arbors and functional axons on a tissue-engineered scaffold. *Acta Biomaterialia*. https://doi.org/10.1016/j.actbio.2017.02.032
81. Tanaka, T., Yokoi, T., Tamalu, F., Watanabe, S., Nishina, S., & Azuma, N. (2015). Generation of retinal ganglion cells with functional axons from human induced pluripotent stem cells. *Scientific Reports, 5*, 8344. https://doi.org/10.1038/srep08344
82. Teotia, P., Chopra, D. A., Dravid, S. M., Van Hook, M. J., Qiu, F., Morrison, J., et al. (2017). Generation of functional human retinal ganglion cells with target specificity from pluripotent stem cells by chemically defined recapitulation of developmental mechanism. *Stem Cells (Dayton, OH), 35*, 572–585. https://doi.org/10.1002/stem.2513
83. Jin, Z. B., & Takahashi, M. (2012). Generation of retinal cells from pluripotent stem cells. *Progress in Brain Research, 201*, 171–181. https://doi.org/10.1016/b978-0-444-59544-7.00008-1
84. Parfitt, D. A., Lane, A., Ramsden C. M., Carr, A. J., Munro, P. M., Jovanovic, K., et al. (2016). Identification and correction of mechanisms underlying inherited blindness in human iPSC-derived optic cups. *Cell Stem Cell, 18*, 769–781.
85. Wiley, L. A., Burnight, E. R., Songstad, A. E., Drack, A. V., Mullins, R. F., Stone, E. M., et al. (2015). Patient-specific induced pluripotent stem cells (iPSCs) for the study and treatment of retinal degenerative diseases. *Progress in Retinal and Eye Research, 44*, 15–35. https://doi.org/10.1016/j.preteyeres.2014.10.002
86. Congdon, N., O'Colmain, B., Klaver, C. C., Klein, R., Muñoz, B., Friedman, D. S., et al. (2004). Causes and prevalence of visual impairment among adults in the United States. *Archives of Ophthalmology, 122*, 477–485. https://doi.org/10.1001/archopht.122.4.477
87. Veleri, S., Lazar, C. H., Chang, B., Sieving, P. A., Banin, E., & Swaroop, A. (2015). Biology and therapy of inherited retinal degenerative disease: Insights from mouse models. *Disease Models & Mechanisms, 8*, 109–129. https://doi.org/10.1242/dmm.017913
88. Fletcher, E. L., Jobling, A. I., Greferath, U., Mills, S. A., Waugh M., Ho, T., et al. (2014). Studying age-related macular degeneration using animal models. *Optometry and Vision Science, 91*, 878–886. https://doi.org/10.1097/opx.0000000000000322
89. Fletcher, E. L., Jobling, A. I., Vessey, K. A., Luu, C., Guymer, R. H., & Baird, P. N. (2011). Animal models of retinal disease. *Progress in Molecular Biology and Translational Science, 100*, 211–286. https://doi.org/10.1016/b978-0-12-384878-9.00006-6
90. Iglesias, A. I., Springelkamp, H., Ramdas, W. D., Klaver, C. C., Willemsen, R., & van Duijn, C. M. (2015). Genes, pathways, and animal models in primary open-angle glaucoma. *Eye (London, England), 29*, 1285–1298. https://doi.org/10.1038/eye.2015.160
91. Jones, M. K., Lu, B., Girman, S., & Wang, S. (2017). Cell-based therapeutic strategies for replacement and preservation in retinal degenerative diseases. *Progress in Retinal and Eye Research*. https://doi.org/10.1016/j.preteyeres.2017.01.004
92. Kostic, C., & Arsenijevic, Y. (2016). Animal modelling for inherited central vision loss. *The Journal of Pathology, 238*, 300–310. https://doi.org/10.1002/path.4641
93. Niwa, M., Aoki, H., Hirata, A., Tomita, H., Green, P. G., & Hara, A. (2016). Retinal cell degeneration in animal models. *International Journal of Molecular Sciences, 17*. https://doi.org/10.3390/ijms17010110
94. Pennesi, M. E., Neuringer, M., & Courtney, R. J. (2012). Animal models of age related macular degeneration. *Molecular Aspects of Medicine, 33*, 487–509. https://doi.org/10.1016/j.mam.2012.06.003

95. Jin, Z. B., Okamoto, S., Osakada, F., Homma, K., Assawachananont, J., Hirami, Y., et al. (2011). Modeling retinal degeneration using patient-specific induced pluripotent stem cells. *PLoS One, 6*, e17084. https://doi.org/10.1371/journal.pone.0017084
96. Tucker, B. A., Scheetz, T. E., Mullins, R. F., DeLuca, A. P., Hoffmann, J. M., Johnston, R. M., et al. (2011). Exome sequencing and analysis of induced pluripotent stem cells identify the cilia-related gene male germ cell-associated kinase (MAK) as a cause of retinitis pigmentosa. *Proceedings of the National Academy of Sciences of the United States of America, 108*, E569–E576. https://doi.org/10.1073/pnas.1108918108
97. Minegishi, Y., Iejima, D., Kobayashi, H., Chi, Z. L., Kawase, K., Yamamoto T., et al. (2013). Enhanced optineurin E50K-TBK1 interaction evokes protein insolubility and initiates familial primary open-angle glaucoma. *Human Molecular Genetics, 22*, 3559–3567. https://doi.org/10.1093/hmg/ddt210
98. Lustremant, C., Habeler, W., Plancheron, A., Goureau, O., Grenot, L., de la Grange, P., et al. (2013). Human induced pluripotent stem cells as a tool to model a form of Leber congenital amaurosis. *Cellular Reprogramming, 15*, 233–246. https://doi.org/10.1089/cell.2012.0076
99. Tucker, B. A., Solivan-Timpe, F., Roos, B. R., Anfinson, K. R., Robin, A. L., Wiley L. A., et al. (2014). Duplication of TBK1 stimulates autophagy in iPSC-derived retinal cells from a patient with normal tension glaucoma. *Journal of Stem Cell Research and Therapy, 3*, 161. https://doi.org/10.4172/2157-7633.1000161
100. Yang, J., Li, Y., Chan, L., Tsai, Y. T., Wu, W. H., Nguyen, H. V., et al. (2014). Validation of genome-wide association study (GWAS)-identified disease risk alleles with patient-specific stem cell lines. *Human Molecular Genetics, 23*, 3445–3455.
101. Cereso, N., Pequignot, M. O., Robert, L., Becker, F., De Luca, V., Nabholz, N., et al. (2014). Proof of concept for AAV2/5-mediated gene therapy in iPSC-derived retinal pigment epithelium of a choroideremia patient. *Molecular Therapy – Methods & Clinical Development, 1*, 14011. https://doi.org/10.1038/mtm.2014.11
102. Yoshida, T., Ozawa, Y., Suzuki, K., Yuki, K., Ohyama, N., Akamatsu, W., et al. (2014). The use of induced pluripotent stem cells to reveal pathogenic gene mutations and explore treatments for retinitis pigmentosa. *Molecular Brain, 7*, 45.
103. Li, Y., Wu, W. H., Hsu, C. W., Nguyen, H. V., Tsai, Y. T., Chan, L., et al. (2014). Gene therapy in patient-specific stem cell lines and a preclinical model of retinitis pigmentosa with membrane frizzled-related protein defects. *Molecular Therapy, 22*, 1688–1697. https://doi.org/10.1186/1756-6606-7-45
104. Schwarz, N., Carr, A. J., Lane, A., Moeller, F., Chen, L. L., Aguilà, M., et al. (2015). Translational read-through of the RP2 Arg120stop mutation in patient iPSC-derived retinal pigment epithelium cells. *Human Molecular Genetics, 24*, 972–986. https://doi.org/10.1371/journal.pone.0017084
105. Moshfegh, Y., Velez, G., Li, Y., Bassuk, A. G., Mahajan, V. B., & Tsang, S. H. (2016). BESTROPHIN1 mutations cause defective chloride conductance in patient stem cell-derived RPE. *Human Molecular Genetics, 25*, 2672–2680. https://doi.org/10.1093/hmg/ddw126
106. Chen, J., Riazifar, H., Guan, M. X., & Huang, T. (2016). Modeling autosomal dominant optic atrophy using induced pluripotent stem cells and identifying potential therapeutic targets. *Stem Cell Research & Therapy, 7*, 2. https://doi.org/10.1186/s13287-015-0264-1
107. Saini, J. S., Corneo, B., Miller, J. D., Kiehl, T. R., Qang, Q., Boles, N. C., et al. (2017). Nicotinamide ameliorates disease phenotypes in a human iPSC model of age-related macular degeneration. *Cell Stem Cell*. https://doi.org/10.1016/j.stem.2016.12.015
108. Ramsden, C. M., Nommiste, B., R Lane, A., Carr, A. F., Powner, M. B., J K Smart, M., et al. (2017). Rescue of the MERTK phagocytic defect in a human iPSC disease model using translational read-through inducing drugs. *Scientific Reports, 7*, 51. https://doi.org/10.1038/s41598-017-00142-7
109. Hallam, D., Collin, J., Bojic, S., Chichagova, V., Buskin, A., Xu, Y., et al. (2017). An induced pluripotent stem cell patient specific model of complement factor H (Y402H) polymorphism displays characteristic features of age-related macular degeneration and indicates a

beneficial role for UV light exposure. *Stem Cells (Dayton, OH), 35*, 2305–2320. https://doi.org/10.1002/stem.2708

110. Comyn, O., Lee, E., & MacLaren, R. E. (2010). Induced pluripotent stem cell therapies for retinal disease. *Current Opinion in Neurology, 23*, 4–9. https://doi.org/10.1097/WCO.0b013e3283352f96
111. Wu, S. M., & Hochedlinger, K. (2011). Harnessing the potential of induced pluripotent stem cells for regenerative medicine. *Nature Cell Biology, 13*, 497–505. https://doi.org/10.1038/ncb0511-497
112. Chichagova, V., Hallam D., Collin, J., Buskin, A., Saretzki, G., Armstrong, L., et al. (2017). Human iPSC disease modelling reveals functional and structural defects in retinal pigment epithelial cells harbouring the m.3243A > G mitochondrial DNA mutation. *Scientific Reports, 7*, 12320. https://doi.org/10.1038/s41598-017-12396-2
113. Croze, R. H., & Clegg, D. O. (2014). Differentiation of pluripotent stem cells into retinal pigmented epithelium. *Developments in Ophthalmology, 53*, 81–96. https://doi.org/10.1159/000357361
114. Du, H., Lim, S. L., Grob, S., & Zhang, K. (2011). Induced pluripotent stem cell therapies for geographic atrophy of age-related macular degeneration. *Seminars in Ophthalmology, 26*, 216–224. https://doi.org/10.3109/08820538.2011.577498
115. Howden, S. E., Gore, A., Li, Z., Fung, H. L., Nisler, B. S., Nie, J., et al. (2011). Genetic correction and analysis of induced pluripotent stem cells from a patient with gyrate atrophy. *Proceedings of the National Academy of Sciences of the United States of America, 108*, 6537–6542. https://doi.org/10.5966/sctm.2012-0163
116. Jin, Z. B., Okamoto, S., Xiang, P., & Takahashi, M. (2012). Integration-free induced pluripotent stem cells derived from retinitis pigmentosa patient for disease modeling. *Stem Cells Translational Medicine, 1*, 503–509. https://doi.org/10.2119/molmed.2012.00242
117. Leach, L. L., Buchholz, D. E., Nadar, V. P., Lowenstein, S. E., & Clegg, D. O. (2015). Canonical/beta-catenin Wnt pathway activation improves retinal pigmented epithelium derivation from human embryonic stem cells. *Investigative Ophthalmology & Visual Science, 56*, 1002–1013. https://doi.org/10.1167/iovs.14-15835
118. Phillips, M. J., Perez, E. T., Martin, J. M., Reshel S. T., Wallace, K. A., Capowski, E. E., et al. (2014). Modeling human retinal development with patient-specific induced pluripotent stem cells reveals multiple roles for visual system homeobox 2. *Stem Cells (Dayton, OH), 32*, 1480–1492. https://doi.org/10.1093/hmg/ddu351
119. Yvon, C., Ramsden, C. M., Lane, A., Powner, M. B., da Cruz, L., Coffey, P. J., et al. (2015). Using stem cells to model diseases of the outer retina. *Computational and Structural Biotechnology Journal, 13*, 382–389. https://doi.org/10.1016/j.csbj.2015.05.001
120. Cedrone, C., Mancino, R., Cerulli, A., Cesareo, M., & Nucci, C. (2008). Epidemiology of primary glaucoma: Prevalence, incidence, and blinding effects. *Progress in Brain Research, 173*, 3–14. https://doi.org/10.1016/s0079-6123(08)01101-1
121. Tham, Y. C., Li, X., Wong, T. Y., Quigley, H. A., Aung, T., & Cheng, C. Y. (2014). Global prevalence of glaucoma and projections of glaucoma burden through 2040: A systematic review and meta-analysis. *Ophthalmology, 121*, 2081–2090. https://doi.org/10.1016/j.ophtha.2014.05.013
122. Maekawa, Y., Onishi, A., Matsushita, K., Koide, N., Mandai, M., Suzuma, K., et al. (2016). Optimized culture system to induce neurite outgrowth from retinal ganglion cells in three-dimensional retinal aggregates differentiated from mouse and human embryonic stem cells. *Current Eye Research, 41*, 558–568. https://doi.org/10.3109/02713683.2015.1038359
123. Bershteyn, M., Nowakowski, T. J., Pollen, A. A., Di Lullo, E., Nene, A., Wynshaw-Boris, A., et al. (2017). Human iPSC-derived cerebral organoids model cellular features of lissencephaly and reveal prolonged mitosis of outer radial glia cell. *Stem Cell, 20*, 435–449 e434. https://doi.org/10.1016/j.stem.2016.12.007
124. Qian, X., Nguyen, H. N., Jacob, F., Song, H., & Ming, G. L. (2017). Using brain organoids to understand Zika virus-induced microcephaly. *Development, 144*, 952–957.

125. Qian, X., Nguyen, H. N., Song, M. M., Hadiono, C., Ogden S. C., Hammack, C., et al. (2016). Brain-region-specific organoids using mini-bioreactors for modeling ZIKV exposure. *Cell, 165*, 1238–1254. https://doi.org/10.1242/dev.140707
126. Grskovic, M., Javaherian, A., Strulovici, B., & Daley, G. Q. (2011). Induced pluripotent stem cells—opportunities for disease modelling and drug discovery. *Nature Reviews. Drug Discovery, 10*, 915–929. https://doi.org/10.1038/nrd3577
127. Gunaseeli, I., Doss, M. X., Antzelevitch, C., Hescheler, J., & Sachinidis, A. (2010). Induced pluripotent stem cells as a model for accelerated patient- and disease-specific drug discovery. *Current Medicinal Chemistry, 17*, 759–766.
128. Wahlin, K. J., Maruotti, J., & Zack, D. J. (2014). Modeling retinal dystrophies using patient-derived induced pluripotent stem cells. *Advances in Experimental Medicine and Biology, 801*, 157–164. https://doi.org/10.1007/978-1-4614-3209-8_20
129. Wright, L. S., Phillips, M. J., Pinilla, I., Hei, D., & Gamm, D. M. (2014). Induced pluripotent stem cells as custom therapeutics for retinal repair: Progress and rationale. *Experimental Eye Research, 123*, 161–172. https://doi.org/10.1016/j.exer.2013.12.001
130. Chang, Y. C., Chang, W. C., Hung, K. H., Yang, D. M., Cheng, Y. H., Liao, Y. W., et al. (2014). The generation of induced pluripotent stem cells for macular degeneration as a drug screening platform: Identification of curcumin as a protective agent for retinal pigment epithelial cells against oxidative stress. *Frontiers in Aging Neuroscience, 6*, 191. https://doi.org/10.3389/fnagi.2014.00191
131. Kokkinaki, M., Sahibzada, N., & Golestaneh, N. (2011). Human induced pluripotent stem-derived retinal pigment epithelium (RPE) cells exhibit ion transport, membrane potential, polarized vascular endothelial growth factor secretion, and gene expression pattern similar to native RPE. *Stem Cells (Dayton, OH), 29*, 825–835. https://doi.org/10.1002/stem.635
132. Santos-Ferreira, T. F., Borsch, O., & Ader, M. (2016). Rebuilding the missing part-A review on photoreceptor transplantation. *Frontiers in Systems Neuroscience, 10*, 105.
133. Streilein, J. W. (2003). Ocular immune privilege: The eye takes a dim but practical view of immunity and inflammation. *Journal of Leukocyte Biology, 74*, 179–185. https://doi.org/10.1083/jcb.201006020
134. Sung, C. H., & Chuang, J. Z. (2010). The cell biology of vision. *The Journal of Cell Biology, 190*, 953–963.
135. Lamba, D. A., Gust, J., & Reh, T. A. (2009). Transplantation of human embryonic stem cell-derived photoreceptors restores some visual function in Crx-deficient mice. *Cell Stem Cell, 4*, 73–79. https://doi.org/10.1016/j.stem.2008.10.015
136. Zhu, J., Cifuentes, H., Reynolds, J., & Lamba, D. A. (2017). Immunosuppression via loss of IL2rgamma enhances long-term functional integration of hESC-derived photoreceptors in the mouse retina cell. *Stem Cell, 20*(374–384), e375. https://doi.org/10.1016/j.stem.2016.11.019
137. Coles, B. L., Angénieux, B., Inoue, T., Del Rio-Tsonis, K., Spence, J. R., McInnes, R. R., et al. (2004). Facile isolation and the characterization of human retinal stem cells. *Proceedings of the National Academy of Sciences of the United States of America, 101*, 15772–15777.
138. Czekaj, M., Haas, J., Gebhardt, M., Müller-Reichert, T., Humphries, P., Farrar, J., et al. (2012). In vitro expanded stem cells from the developing retina fail to generate photoreceptors but differentiate into myelinating oligodendrocytes. *PLoS One, 7*, e41798.
139. Klassen, H. J., Ng, T. F., Kurimoto, Y., Kirov, I., Shatos, M., Coffey, P., et al. (2004). Multipotent retinal progenitors express developmental markers, differentiate into retinal neurons, and preserve light-mediated behavior. *Investigative Ophthalmology & Visual Science, 45*, 4167–4173. https://doi.org/10.1167/iovs.04-0511
140. Eberle, D., Kurth, T., Santos-Ferreira, T., Wilson, J., Corbeil, D., & Ader, M. (2012). Outer segment formation of transplanted photoreceptor precursor cells. *PLoS One, 7*, e46305.
141. Lund, R. D., Wang, S., Klimanskaya, I., Holmes, T., Ramos-Kelsey R., Lu, B., et al. (2006). Human embryonic stem cell-derived cells rescue visual function in dystrophic RCS rats. *Cloning and Stem Cells, 8*, 189–199. https://doi.org/10.1089/clo.2006.8.189

142. Assawachananont, J., Mandai, M., Okamoto, S., Yamada, C. Eiraku, M., Yonemura, S., et al. (2014). Transplantation of embryonic and induced pluripotent stem cell-derived 3D retinal sheets into retinal degenerative mice. *Stem Cell Reports, 2*, 662–674.
143. Chao, J. R., Lamba, D. A., Klesert, T. R., Torre, A., Hoshino, A., Taylor R. J., et al. (2017). Transplantation of human embryonic stem cell-derived retinal cells into the subretinal space of a non-human primate. *Translational Vision Science & Technology, 6*, 4. https://doi.org/10.1167/tvst.6.3.4
144. da Cruz, L., Chen, F. K., Ahmado, A., Greenwood, J., & Coffey, P. (2007). RPE transplantation and its role in retinal disease. *Progress in Retinal and Eye Research, 26*, 598–635. https://doi.org/10.1016/j.preteyeres.2007.07.001
145. Falkner-Radler, C. I., Krebs, I., Glittenberg, C., Povazay, B., Drexler, W., Graf, A., et al. (2011). Human retinal pigment epithelium (RPE) transplantation: Outcome after autologous RPE-choroid sheet and RPE cell-suspension in a randomised clinical study. *The British Journal of Ophthalmology, 95*, 370–375. https://doi.org/10.1136/bjo.2009.176305
146. Kamao, H., Mandai, M., Okamoto, S., Sakai, N., Suga, A., Sugita, S., et al. (2014). Characterization of human induced pluripotent stem cell-derived retinal pigment epithelium cell sheets aiming for clinical application. *Stem Cell Reports, 2*, 205–218. https://doi.org/10.5966/sctm.2014-0038
147. Lakowski, J., Gonzalez-Cordero, A., West, E. L., Han, Y. T., Welby E., Naeem, A., et al. (2015). Transplantation of photoreceptor precursors isolated via a cell surface biomarker panel from embryonic stem cell-derived self-forming retina. *Stem Cells (Dayton, OH), 33*, 2469–2482
148. Mandai, M., Watanabe, A., Kurimoto, Y., Hirami, Y., Morinaga, C., Daimon, T., et al. (2017). Autologous induced stem-cell-derived retinal cells for macular degeneration. *The New England Journal of Medicine, 376*, 1038–1046. https://doi.org/10.1056/NEJMoa1608368
149. Park, U. C., Cho, M. S., Park, J. H., Kim, S. J., Ku, S. Y., Choi, Y. M., et al. (2011). Subretinal transplantation of putative retinal pigment epithelial cells derived from human embryonic stem cells in rat retinal degeneration model. *Clinical and Experimental Reproductive Medicine, 38*, 216–221. https://doi.org/10.5653/cerm.2011.38.4.216
150. Petrus-Reurer, S., Bartuma, H., Aronsson, M., Westman, S., Lanner, F., & Kvanta, A. (2018). Subretinal transplantation of human embryonic stem cell derived-retinal pigment epithelial cells into a large-eyed model of geographic atrophy. *Journal of Visualized Experiments*. https://doi.org/10.3791/56702
151. Radtke, N. D., Aramant, R. B., Petry, H. M., Green, P. T., Pidwell, D. J., & Seiler, M. J. (2008). Vision improvement in retinal degeneration patients by implantation of retina together with retinal pigment epithelium. *American Journal of Ophthalmology, 146*, 172–182. https://doi.org/10.1016/j.ajo.2008.04.009
152. Schwartz, S. D., Regillo, C. D., Lam, B. L., Eliott, D., Rosenfeld, P. J., Gregori, N. Z., et al. (2015). Human embryonic stem cell-derived retinal pigment epithelium in patients with age-related macular degeneration and Stargardt's macular dystrophy: Follow-up of two open-label phase 1/2 studies. *Lancet, 385*, 509–516. https://doi.org/10.1016/S0140-6736(14)61376-3
153. Cooke, J. A., & Meyer, J. S. (2015). Human pluripotent stem cell-derived retinal ganglion cells: Applications for the study and treatment of optic neuropathies. *Current Ophthalmology Reports, 3*, 200–206.
154. Lebrun-Julien, F., & Di Polo, A. (2008). Molecular and cell-based approaches for neuroprotection in glaucoma. *Optometry and Vision Science, 85*, 417–424. https://doi.org/10.1097/OPX.0b013e31817841f7
155. Sluch, V. M., & Zack, D. J. (2014). Stem cells, retinal ganglion cells and glaucoma. *Developments in Ophthalmology, 53*, 111–121.
156. Shirai, H., Mandai, M., Matsushita, K., Kuwahara, A., Yonemura, S., Nakano, T., et al. (2016). Transplantation of human embryonic stem cell-derived retinal tissue in two primate models of retinal degeneration. *Proceedings of the National Academy of Sciences of the United States of America, 113*, E81–E90.

157. Boucherie, C., Mukherjee, S., Henckaerts, E., Thrasher, A. J., Sowden, J. C., & Ali, R. R. (2013). Brief report: Self-organizing neuroepithelium from human pluripotent stem cells facilitates derivation of photoreceptors. *Stem Cells (Dayton, OH), 31*, 408–414. https://doi.org/10.1002/stem.1268
158. Viczian, A. S., Solessio, E. C., Lyou, Y., & Zuber, M. E. (2009). Generation of functional eyes from pluripotent cells. *PLoS Biology, 7*, e1000174. https://doi.org/10.1371/journal.pbio.1000174
159. Lu, B., Malcuit, C., Wang, S., Girman, S., Francis, P., Lemieux, L., et al. (2009). Long-term safety and function of RPE from human embryonic stem cells in preclinical models of macular degeneration. *Stem Cells (Dayton, OH), 27*, 2126–2135. https://doi.org/10.1002/stem.149
160. MacLaren, R. E., Pearson, R. A., MacNeil, A., Douglas, R. H., Salt, T. E., Akimoto, M., et al. (2006). Retinal repair by transplantation of photoreceptor precursors. *Nature, 444*, 203–207. https://doi.org/10.1038/nature05161
161. Song, W. K., Park, K. M., Kim, H. J., Lee, J. H., Choi, J., Chong, S. Y., et al. (2015). Treatment of macular degeneration using embryonic stem cell-derived retinal pigment epithelium: Preliminary results in Asian patients. *Stem Cell Reports, 4*, 860–872. https://doi.org/10.1038/srep12910
162. Rowland, T. J., Blaschke, A. J., Buchholz, D. E., Hikita, S. T., Johnson, L. V., & Clegg, D. O. (2013). Differentiation of human pluripotent stem cells to retinal pigmented epithelium in defined conditions using purified extracellular matrix proteins. *Journal of Tissue Engineering and Regenerative Medicine, 7*, 642–653. https://doi.org/10.1002/term.1458

Chapter 3
Bioengineered and Regenerative Medicine Strategies for Retina Repair

Linyang Yu, Vianney Delplace, Samantha L. Payne, and Molly S. Shoichet

Abstract With the development of biomedical engineering and regenerative medicine, new ways of treating retinal diseases are being explored. We provide a summary of the most recent bioengineering strategies employed for the delivery of therapeutics and cells in the treatment of retinal diseases. We describe current challenges in the field of ocular drug and cell delivery, and the way that innovative materials, advanced protein therapies, and controlled release systems are being used as potential treatments. New implants and nanoparticles are promising for drug and protein delivery, while hydrogels and biomaterial scaffolds are useful for enhanced survival and integration of transplanted retinal cells. This chapter concludes with an outlook for future studies in the field of ophthalmic delivery.

Keywords Regenerative medicine · Drug delivery · Cell delivery · Protein therapeutics · Biomaterials · Implants · Nanoparticles · Hydrogels

---

Linyang Yu and Vianney Delplace contributed equally to this chapter and are considered co-first authors.

This chapter is an updated version of an original review published by Delplace, V., Payne, S., & Shoichet, M. (2015). Delivery strategies for treatment of age-related ocular diseases: From a biological understanding to biomaterial solutions. *Journal of Controlled Release 219*, 652–668.

L. Yu · V. Delplace · S. L. Payne · M. S. Shoichet (✉)
Department of Chemical Engineering and Applied Chemistry, University of Toronto, Toronto, ON, Canada
e-mail: ly.yu@mail.utoronto.ca; vianney.delplace@utoronto.ca; samanthal.payne@mail.utoronto.ca; molly.shoichet@utoronto.ca

B. G. Ballios, M. J. Young (eds.), *Regenerative Medicine and Stem Cell Therapy for the Eye*, Fundamental Biomedical Technologies,
https://doi.org/10.1007/978-3-319-98080-5_3

3.1 Introduction

Millions of people worldwide suffer from retinal degenerative diseases, from partial vision loss to complete blindness. Age-related macular degeneration (AMD) and diabetic retinopathy (DR) are the leading causes of retinal degeneration [1]. AMD affects more than 11 million individuals in the United States (U.S.), and the global prevalence is 170 million [2–4]. With the aging population, the number of patients with AMD is expected to continue to increase. In addition, individuals diagnosed with diabetes represent approximately 8.5% of the world adult population, and one in three diabetics has some degree of DR [5]. With more than six million people affected, glaucoma is another leading cause of irreversible blindness; however, the pathophysiology of glaucoma, which results in optic nerve damage due to increased intraocular pressure, differs significantly from that of AMD and DR. The lack of inherent regenerative capacity of the retina, along with limited treatment options for millions of patients, has led to research into innovative repair strategies.

Biomedical engineering strategies have advanced over the past several decades, resulting in new ways to deliver both biomolecules and cells to the retina. For example, drugs such as ranibizumab and aflibercept have been developed and approved by the FDA to treat AMD, while liposomes and other nanoparticles have been designed to achieve more efficacious drug delivery to the eye. Moreover, with the advances in stem cell biology, their delivery and that of their progenitors have been pursued with the goal of reversing blindness [6]. Biomaterials have been shown to promote cell survival and integration [7, 8]. Here, we focus on both biomolecule delivery to slow the progression of retinal degeneration, and cell delivery to promote retinal regeneration.

3.2 Drug Delivery to the Posterior Segment of the Eye

The eye is considered an immunoprivileged site due to the blood retinal barrier (BRB) and the absence of a lymphatic system [9]. Penetration into the retina by systemically administered drugs is limited by the BRB, necessitating local delivery. The eye has a number of properties that make it well suited for local delivery strategies, including its relative accessibility, its size, and the inner barriers, which limit drug clearance [10]. However, topical drug delivery to the posterior segment of the eye is limited by pre-corneal drainage, the lipoidal nature of the corneal epithelium, and systemic circulation absorption, which leads to low drug bioavailability in the retina—typically less than 5% [11–13]. Therefore, subconjunctival/subtenon injections (between the conjunctiva/tenon and the sclera) or intravitreal injections (into the vitreous humor) are the most common routes of administration for drug delivery to the posterior segment of the eye [14]. Routes of administration, barriers, and transporters have been extensively reviewed elsewhere [15–17] and will not be further discussed here.

3.2.1 Drugs and Growth Factors for Ocular Treatment

3.2.1.1 Neurotrophic Factors

Various neurotrophic factors have been studied as potential candidates to treat retinal degeneration by supporting endogenous cell survival and growth (Table 3.1), including: brain-derived neurotrophic factor (BDNF) [18–20], pigment epithelium-derived factor (PEDF) [21], nerve growth factor (NGF) [22], glial cell line-derived neurotrophic factor (GDNF) [19, 20, 23–26], and ciliary neurotrophic factor (CNTF) [23, 27–29].

BDNF has attracted much attention as it has been shown to limit the progression of retinal dysfunction in a variety of animal models [19, 30, 31]. In rats with light-induced retinal degeneration, intravitreal administration of BDNF significantly slows down the apoptosis of photoreceptors [30]. Another study showed that BDNF increases pERK and c-fos immunoreactivity in Müller glia and ganglion cells, and not in photoreceptors, supporting photoreceptor protection via a secondary mechanism [32]. BDNF was also demonstrated to be effective in treating glaucoma. In a chronic glaucoma model developed from DBA/2J mice, intravitreal injections of BDNF showed improvement in retinal ganglion cell (RGC) survival after 2 weeks [18, 33]; and visual function rescue was observed in the early stage of the degeneration [18].

GDNF, the first member of the GDNF family discovered, has both morphological and functional protective effects on rod photoreceptors, and was shown to be more potent than BDNF for the treatment of irreversible rod degeneration in early stages of retinitis pigmentosa (RP) [34]. Naturally secreted by retinal glial cells (e.g., Müller glia), GDNF has been shown to: significantly attenuate the degeneration of adult rat RGCs [26]; protect the retina from degeneration induced by light damage [35]; and prevent cell death of rod photoreceptors in a RP mouse model [34]. In an in vitro study using dissociated rat rod photoreceptors, high concentrations of GDNF, but not BDNF, resulted in better cell survival and light-responsiveness [23]. In a retinal degeneration (rd) mouse model, subretinal injection of GDNF preserved the normal structure and function of photoreceptors [34]. In addition, increased outer nuclear layer (ONL) thickness, rod photoreceptor survival, and improved ERG signals were observed in transgenic S334ter rhodopsin (Rho) rat (line 4) after recombinant adeno-associated virus (rAAV)-mediated GDNF delivery [36]. GDNF also exhibits a protective effect on RGCs. For example, in D2 mice with spontaneous glaucoma, a single-intravitreal injection of GDNF-loaded PLGA microspheres protected the RGCs from chronic glaucoma effectively for at least 10 weeks [24]. Compared with BDNF, intravitreal injection of GDNF has superior protective effects on RGCs; and a combined delivery of BDNF and GDNF further enhanced RGC survival [26].

Other trophic factors have attracted attention for the treatment of retinal disorders, including: ciliary neurotrophic factor (CNTF), pigment epithelium-derived factor (PEDF), epidermal growth factor (EGF), fibroblast growth factor (FGF2),

Table 3.1 Neurotrophic/trophic factors for the treatment of retinal degenerations

Neurotrophic factor	Delivery method	Disease model	Outcome	Reference
BDNF	Intravitreal injection	Light damage rats	Increase the ONL thickness	[40]
	Intravitreal injections	Mutant mouse/rat with inherited retinal degeneration	Slowing down the degeneration of photoreceptors	[41]
	Subretinal injection	S334ter line 3 rhodopsin retinal degenerate rat	Neurotrophic effect was not significant	[42]
	Intravitreal injection	Normal rat	Improved retinal function	[18]
	Subretinal injection	S334ter line 3 rhodopsin retinal degenerate rat	Better response to optokinetic and visual electrophysiological responses	[43]
GDNF	In vitro explant	rd/rd mouse	Twofold increase in the number of viable photoreceptors	[44]
	Subretinal injection	rd/rd mouse	Improvement of rod photoreceptor survival	[34]
	Subretinal injection	S334ter line 3 rhodopsin retinal degenerate rat	Enhanced responsiveness of retina in both rosette and laminated transplants	[45]
CNTF	Intravitreal injection	Light damage rat	Increase the ONL thickness	[30]
	Intravitreal injection	Mutant mouse/rat with inherited retinal degeneration	Improved cell survival	[23]
	Intravitreal injection	Normal rat	CNTF impaired retinal function	[29]
Axokine (rhCNTF)	Intravitreal injection	Light damage rat	No photoreceptor rescue	[40]
	Subretinal injection	Q344ter mouse	Prolonged photoreceptor survival	[46]
	Intravitreal injection	AD rod-cone feline dystrophy	Prolonged photoreceptor survival	[47]
PEDF	Intravitreal injection	rd/rd mouse	Prolonged photoreceptor survival	[48]
	Intravitreal injection	Mutant mouse/rat with inherited retinal degeneration	Improved retinal function	[21]
	Intravitreal injection	PDT Brown-Norway rat	Increased photoreceptor survival and function	[49]
	Subretinal injection	Ischemia Wistar rat	Increased IRL, improved RGC survival	[50]

(continued)

Table 3.1 (continued)

Neurotrophic factor	Delivery method	Disease model	Outcome	Reference
	Intravitreal injection	Light damage SD rat	Preserved photoreceptor morphology, improved ERG response	[51]
	Intravitreal injection	RCS rat	Photoreceptor rescue, better retinal light sensitivity	[52]
RdCVF	In vitro explant	Chicken embryos	Improved cone survival	[53]
	Subretinal injection	P23H rat	Increase in cone number and ERG	[54]
NGF	Intravitreal injection	Light damage rat	Low degree photoreceptor rescue	[40]
	Intravitreal injection	RCS rat	Enhanced photoreceptor survival	[22]
NT-3	Intravitreal injection	SD rat with nitric oxide damage	INL rescue	[55]
	Intravitreal injection	SD rat with light damage	Low degree photoreceptor rescue	[40]
	Intravitreal injection	Light damage mouse; rd/rd, rds/+, q334ter mutant mouse	No retina protection	[46]
FGF2	Intravitreal injection	Light damage rat	Increase the ONL thickness	[40]
aFGF	Intravitreal injection	Light damage rat	High degree photoreceptor rescue	[40]
IGF-1	Intravitreal injection	Light damage rat	Low degree photoreceptor rescue	[40]
	Subretinal injection	Rd10 rat	Low degree photoreceptor rescue	[56]

insulin-like growth factor (IGF), and transforming growth factor (TGF-β) (Table 3.1). For example, in damaged rat retinas, increased CNTF expression in Müller glia directly activates the JAK/STAT and MAPK pathways [37], and mediates photoreceptor cell survival in both light-induced [38] and neurodegenerative disease models [39]. However, the short half-life of CNTF and the resulting concerns of repeated intraocular injections have limited its translation to the clinic. The mechanism of action and side effects associated with many of these factors remain unclear.

3.2.1.2 Angiogenesis Inhibitors

Both diabetic retinopathy and wet AMD stem from abnormal neovasculature due to vascular endothelial growth factor (VEGF) overexpression. Different categories of drugs have been studied as potential anti-VEGF candidates for intravitreal injections, which decrease either vascular permeability or neovascularization [57].

Intraocular injections of corticosteroids (such as dexamethasone and fluocinolone acetonide) have been shown to inhibit both VEGF and VEGF gene expression, resulting in improved vision in patients with diabetic macular edema (DME) compared to traditional laser photocoagulation. Unfortunately, corticosteroid use is limited by adverse side effects, such as elevated intraocular pressure leading to glaucoma and cataract-related adverse effects [58].

Recent innovations in anti-angiogenic antibodies and nucleotides have generated new opportunities for ocular treatments of age-related diseases. Three monoclonal antibodies, aflibercept (Eylea™) [59], bevacizumab (Avastin™) [60], and ranibizumab (Lucentis™) [61], have shown reduced macular edema and improved vision in the treatment of AMD, with significantly better results than macular laser therapy. Pegaptanib (Macugen™), a PEGylated RNA aptamer designed to treat neovascular ophthalmic diseases, was the first aptamer therapeutic approved for use in humans [62]. Although potentially safer than the two antibodies due to a more selective inhibition, it was ultimately shown to be less effective [63]. Ranibizumab and aflibercept remain the only two FDA-approved VEGF inhibitors for the treatment of wet AMD and DME [59], despite similar results with off-label use of bevacizumab [58, 64]. Aside from the previously mentioned angiogenic inhibitors, a variety of alternative mechanisms of action and gene transfer therapies are currently under investigation, such as targeting vascular endothelial receptor tyrosine kinase [65], overexpressed integrins [66], hypoxia-induced activator [67], or metalloproteinases [68], among others [69]. Notwithstanding the potential of these strategies, only preliminary studies have been conducted and the possible adverse effects on normal choroidal vessels and retinal neurons need to be evaluated [70]. Pursuing the identification of genes involved in neovascular AMD also holds great promise for future treatments [71].

Although rates of adverse ocular effects are generally low, anti-angiogenic treatments require repeated intravitreal injections over prolonged periods of time (months to years), which can result in complications such as ocular inflammation, endophthalmitis (in 1% of patients), subconjunctival hemorrhage, increased intraocular pressure, and retinal detachment [58]. Topical administration via eye drops avoids these complications, yet suffers from low bioavailability, which has restricted this route mainly to the treatment of anterior segment and corneal diseases. These observations underscore the need for innovative delivery strategies.

3.2.2 Limitations of Protein Delivery to the Retina

Conventional ocular administration of therapeutic factors in saline has numerous limitations: rapid inactivation of factors prior to reaching their targets, toxicity from overdose caused by burst release, patient compliance, and convenience. Various ocular barriers also pose more significant challenges compared with drug delivery to other parts of the body, including both static and dynamic barriers. Sclera, choroid and retinal pigment epithelium (RPE) are typical static barriers, while conjunctival blood and lymphatic circulation are the main dynamic barriers [11].

3.2.3 *Delivery Strategy*

The ideal ocular delivery system for age-related diseases should meet the following criteria: (1) sustained and controlled drug release, with the rate and duration optimized to the clinical condition (usually months to years); (2) limited number of surgical interventions (and fewer is better); (3) high specificity to the targeted tissue; (4) limited side effects; and (5) fast clinical translation. While viral vectors for ocular gene therapy are promising [72, 73], they will not be discussed here. Instead, we will focus on implants, nanocarriers, and innovative delivery strategies.

3.2.3.1 Implants

To avoid frequent intravitreal injections, intravitreal rod-shaped implants containing corticosteroids for prolonged drug release were developed [74]. The recently FDA-approved Iluvien™ offers 36-month delivery of fluocinolone acetonide from an injectable silicone adhesive tube, with the drug diffusing through a poly(vinyl alcohol) membrane [75]. The choice to use a nondegradable material is questionable, as it requires surgical removal. This problem has been overcome in Ozurdex™, a similar material that delivers dexamethasone from a degradable matrix of poly(lactic-co-glycolic acid) (PLGA) [76]. Most recently, the safety and efficacy of the Ozurdex™ implant was confirmed in a clinical trial. However, in the previous study, the efficacious duration is shorter than 6 months and consequently, reinjection of Ozurdex™ at 6 months may be suggested for maximum benefit [77].

Implants reduce the risk of vision loss and improve the speed of visual recovery while avoiding side effects associated with multiple injections. However, the use of implants can increase intraocular pressure and cataract progression due to both traction exerted on the vitreous [78] and the side effects of corticosteroid therapy. As an example, the results of a 3-year randomized phase II/III clinical trial on another implant containing fluocinolone acetonide, Retisert™, showed reduced recurrence rate of noninfectious posterior uveitis, but also an overall elevated intraocular pressure (IOP) incidence of 68–71% (vs. 19–24% for non-implanted eyes). Moreover, nearly all (94%) of the patients with a phakic-implanted eye—that is an eye with a native crystalline lens implanted with Retisert™—required subsequent cataract surgery (vs. 30% for phakic non-implanted eyes) [79, 80].

Another remarkable implant-based strategy consists of transplanting a capsule that contains genetically engineered mammalian cells able to sustainably produce therapeutic proteins [81]. The NT-501 implant contains a human RPE cell line transfected to release CNTF and encapsulated in a poly(ethylene terephthalate)

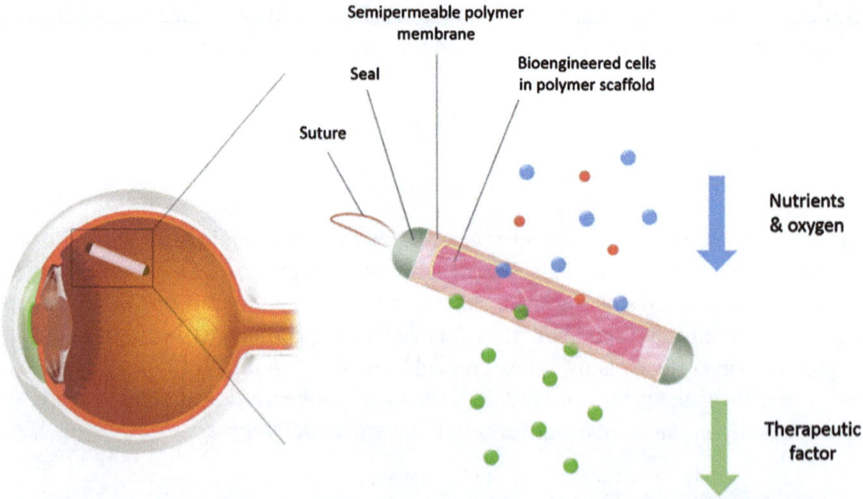

Fig. 3.1 Encapsulated cell therapy (ECT) for ocular drug delivery. Reproduced with permission [87]

scaffold (see Fig. 3.1) [82]. Encouraging results, such as long-term photoreceptor protection, were obtained using several animal models (rats, rabbits, and dogs) [39, 83], and led to advanced clinical trials [84]. The NT-501 is currently undergoing a Phase 2 trial as a 12-month treatment for patients with macular telangiectasia and glaucoma [85]. It should be emphasized that no significant improvement over controls has been observed in retinitis pigmentosa patients necessitating further examination to definitively assess implant efficacy [86]. While encapsulated cell therapy has been investigated for over two decades, and this approach shows some promise, it also underscores the complex interactions of the disease and cell therapy, even when the cells are physically isolated from the system to deliver bioactive proteins. Nevertheless, this approach may inspire future innovations in therapeutic delivery.

3.2.3.2 Ocular Nanotherapy

Drug encapsulation in nanoparticles has been widely investigated for ocular delivery, as described in numerous review articles. For example, Kompella et al. [88] extensively reviewed the field of ocular nanotherapies, Bochot et al. [89] presented a complete review of state of the art liposomes for intravitreal drug delivery, and Kang-Mieler et al. [90] summarized the delivery systems currently in clinical trials. Here, we highlight the most recent developments in nanomedicine concerning age-related ocular diseases and discuss the emerging use of transscleral administration and peptide-targeted therapies.

Nanotherapy requires high drug loading, a defined size between 30 and 200 nm, stability of the delivered biomolecules in the intraocular space, material biocompatibility, and bioresorbability. A variety of systems have been proposed, based on solid lipid nanoparticles [91, 92], polymeric nanoparticles [93, 94], liposomes [95], micelles [96], and dendrimers [97, 98]. When injected in the vitreous, these nanotherapies allow prolonged drug release [99, 100], which reduces the number of injections required. However, although reduced, similar potential complications to traditional intravitreal drug delivery may be expected, including ocular inflammation and elevated intraocular pressure. Therefore, transscleral delivery for topical applications has attracted attention because it constitutes a noninvasive route of administration to deliver drugs to the posterior segment of the eye. For example, polymeric micelles of dexamethasone applied topically provide a promising alternative to intravitreal injections. Using materials such as poly(hydroxyethyl-aspartamide) or crosslinked poly(N-isopropylacrylamide-co-vinyl pyrrolidone) result in greater drug bioavailability and significantly reduced inflammation in animal models [101, 102], Furthermore, eye drops containing dexamethasone-cyclodextrin microparticles have demonstrated some clinical success for the treatment of diabetic macular edema [103, 104] and cystoid macular edema [105, 106]. Thus, despite a lingering skepticism about the appropriateness of topical treatments for the posterior segment, these results suggest that this administration route is promising for retinal disease.

Liposomes, which are the most studied system for any ocular treatment [107], have recently been advanced by grafting specific binding peptides for active tissue targeting. For example, PEGylated 1,2-distearoyl-sn-glycero-3-phosphoethanolamine (DSPE-PEG) has been modified with a series of cell-adhesive peptides to promote binding to receptors expressed in neovascular tissue: the integrin binding RGD tripeptide [108, 109], the angiogenic vessel-specific APRPG pentapeptide [110], and an ephrin mimetic dodecapeptide (YSAYPDSVPMMS) [111]. In animal models, these studies showed enhanced drug delivery, downregulated VEGF expression, and/or reduced choroidal neovascularization. Despite the appeal of such advanced strategies and more than 20 years of research on liposomes for ocular delivery [112], no liposomal formulation for intravitreal or topical treatment of retinal disorders has reached the market to date [89]. This highlights the major limitations of the two main routes of administration—side effects after intravitreal injections and low bioavailability after transscleral administration. Conversely, systemic administration of liposomes has been successful in the treatment of other diseases [113]. Interestingly, the only marketed liposomal treatment for age-related retinopathy, Visudyne®, uses systemic administration to deliver a photosensitizing agent (Verteporfin) for photodynamic therapy of neovascularized retina [114]. Systemic administration also prevents some blurring effects induced by the presence of particles in the humor, which is a well-known side effect of intravitreal micro/nanotherapies and a source of visual discomfort [89]. An inspiring alternative is intravenous delivery of RGD-functionalized PLGA nanoparticles that are designed to deliver a recombinant Flt23k intracellular receptor, an antagonist of VEGF, and have been shown to restore 40% of vision loss in a monkey choroidal neovascularization model [115, 116].

The benefits of anti-angiogenic therapy have already resulted in the translation of antibodies to the market for ocular therapies: bevacizumab (Avastin™) and ranibizumab (Lucentis™). However, advanced delivery strategies have been ineffective at increasing their efficacy, suggesting opportunities for further innovation. For example, bevacizumab-loaded, peptide-conjugated liposomes showed enhanced drug delivery and prolonged intravitreal residency in animal models, yet efficacy was not demonstrated [99]. Similarly, bevacizumab delivery from a thermosensitive, biodegradable, and biocompatible intravitreal hydrogel revealed no significant improvement over controls [117]. This emphasizes the challenge and the opportunity: local, sustained delivery of anti-angiogenic factors remains a fertile area for strategic investment. In that regard, inspiration may be found from the recent development of "VEGF sticky-traps", where VEGF-receptors binding domains (similar to aflibercept) are recombinantly synthesized together with heparin-binding domains to allow prolonged, local binding of the drug to the heparin sulfate proteoglycans of the ECM, therefore improving VEGF inhibition and reducing side effects [118].

3.2.3.3 Advanced Synthetic Materials

Advanced materials play an increasingly important role in ocular drug delivery and are one of the key features to versatile, tunable, and controlled delivery systems. For example, crosslinked hydrogels allow sustained, controlled release by tuning the crosslinking and functionalization. A crosslinked PNIPAAm-based hydrogel was injected into the vitreous with no indication of unwanted side effects [119] while a thiolated poly(aspartic acid) demonstrated in situ gelation for use as a potential ocular mucoadhesive drug delivery system, with a sustained, 24 h drug bioavailability [120]. Nanoparticle-hydrogel composite systems have the potential to further reduce burst and sustain release for local delivery. For example, the encapsulation of ranibizumab or aflibercept in PLGA nanoparticles prior to entrapment in a PNIPAAm hydrogel resulted in an extended drug release for up to ~200 days [121]. When this treatment was applied in vivo, using a laser-induced model of choroidal neovascularization, it showed significantly reduced lesion areas in the treated animals compared with the non-treated group. However, no significant functional improvement was observed by ERG [122].

Triggered release from stimuli-responsive materials is an interesting strategy for ocular treatment as it offers on-demand, spatiotemporally controlled drug bioactivity. Using the unique characteristic of transparency of the eye, a light-degradable polymeric nanocarrier has been used for minimally invasive release of a tyrosine-kinase inhibitor, nintedanib, under low-power UV exposure. Although UV irradiation may damage retinal cells and should be used carefully [123, 124], this system demonstrated release upon irradiation up to 30 weeks after intravitreal injection and long-term suppression of choroidal neovasculature in rats.

Advanced materials have begun to be pursued to provide new solutions to the challenge of targeted delivery to the eye. For example, while there is considerable enthusiasm for gene therapy, an appropriate delivery vehicle is needed to effectively target the tissue of interest. Recently, cationic lipid-based systems, which are known to facilitate endocytosis and endolysosomal escape, were applied to ocular delivery and resulted in higher gene transfection efficiency, underlining the importance of understanding cell trafficking in the field of drug delivery [125, 126]. This should, however, be carefully considered, as cationic surface charge may induce a strong inflammatory cell response [127]. Finally, as a proof-of-concept study for ocular imaging, an amphiphilic vinyl block copolymer modified with a single-strand collagen mimetic peptide (CMP) was co-nanoprecipitated with a semiconducting polymer with good fluorescent properties. Stable nanoparticles of 40 nm were obtained and demonstrated selective binding to collagen in histology sections of mouse cornea tissue. This paves the way toward ocular "theranostic" approaches where the diagnostic and therapeutic agents can be combined into one system to better target and image the cells [128–130].

3.2.3.4 Innovative Delivery Concepts

Among the new mechanisms of action that have been explored as treatment strategies for age-related retinal degeneration, the prevention of excessive reactive oxygen species (ROS) formation should be highlighted. ROS formation is known to trigger oxidative stress which in turn damages retinal tissue and is thus a good therapeutic target [131]. Two systems have been designed to overcome the tissue damage associated with ROS by intravitreal injection of either human serum albumin nanoparticles containing a plasmid encoding a superoxide dismutase (SOD1) gene [132], and nanoceria crystallites composed of cerium oxide. Human serum albumin nanoparticles coated with hyaluronic acid were administered intravitreally, followed by ultrasound treatment via the transscleral route, leading to higher nanoparticle motility and increased retinal tissue penetration. This method caused minimum damage to the retinal tissue and demonstrated a better delivery profile [133]. The latter strategy is particularly exciting because no excipients are required for delivery and it was efficacious in a rat model; angiogenesis-associated pathologies were prevented, including reduced levels of retinal ROS and VEGF, vascular lesions, subretinal neovascular tufts, light damage, and blindness [134, 135]. In a separate study, a peptide was delivered via a modified chitosan nanocarrier with the goal of regulating RPE phagocytosis, which is known to be a key factor in photoreceptor survival [136]. Another study reports the inhibitory effect of gold nanoparticles on retinal neovascularization in a mouse model of retinopathy of prematurity (ROP) [137].

Beyond improved traditional implants and nanotherapies, novel concepts have been developed. Contact lenses as drug carriers for sustained release have been evaluated for the delivery of anti-glaucoma treatments. This complex system, in which coated nano-diamond clusters were successively embedded in chitosan spheres and a poly(hydroxyethylmethacrylate) (PHEMA) hydrogel matrix, showed controlled and sustained release of encapsulated timolol maleate and prolonged drug activity on primary human trabecular meshwork cells from the cornea (see Fig. 3.2) [138]. Another interesting study for glaucoma treatment used a molecular-

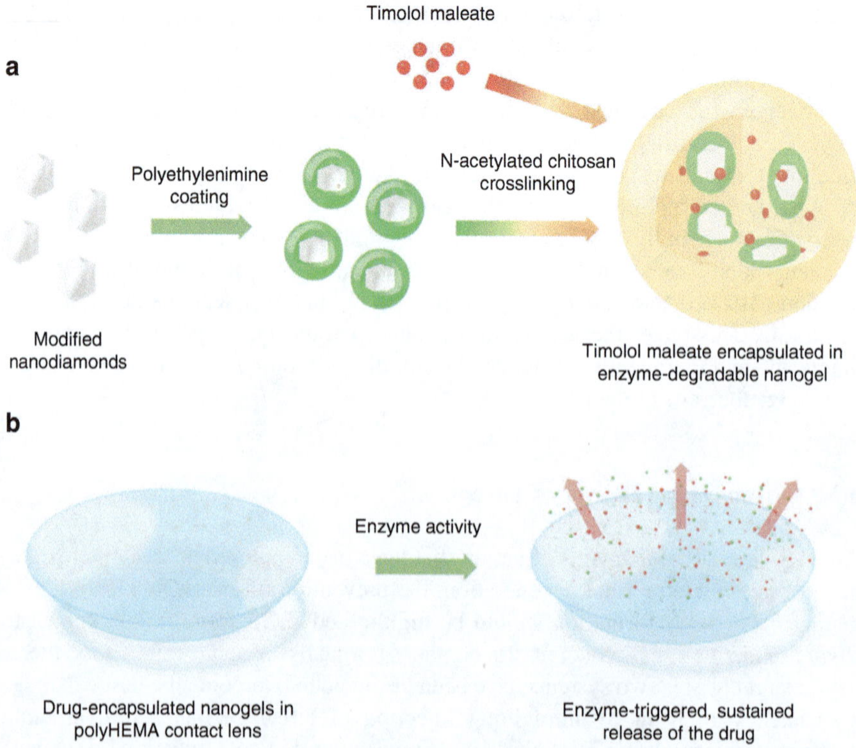

Fig. 3.2 Contact lenses as drug carriers for sustained delivery of anti-glaucoma treatment. (**a**) Chemically modified nano-diamonds are successively coated with polyethylenimine (PEI) and co-encapsulated with an anti-glaucoma drug (timolol maleate) in crosslinked chitosan nanogels. (**b**) Drug-encapsulated nanogels are then embedded in poly(hydroxyethylmethacrylate) (poly-HEMA) hydrogel matrix and cast into contact lenses for enzyme-triggered delivery to the eye. Reproduced with permission [87]

imprinting technique to combine the PHEMA contact lens with carbonic anhydrase inhibitors [139]. This study demonstrated a better-controlled release profile. However, the long-term biocompatibility and physical properties of this delivery system after drug encapsulation need further investigation [140]. Similarly, the development of microfabrication techniques now allows the design of unusual delivery systems, such as micropatterned planar microdevices for drug delivery across the RPE [141] or hollow microneedles for suprachoroidal particle injection [142, 143]. For example, when nanoparticle-stabilized emulsion droplets were injected in the suprachoroidal space using hollow micro-needles, at least 50% of the nanoparticles were found at the back of rabbit eyes [144]. A recent study demonstrated the use of an electroforming cobalt-nickel microtube designed to be loaded with drugs by capillary action before minimally invasive intravitreal injection and wireless magnetic positioning within the eye. The so-called swiveling tubular magnetic microrobot is a good example of what could be the future of ocular delivery [145].

Despite major discoveries and progress in the treatment of age-related ocular diseases, ocular drug delivery systems remain somewhat invasive with limited duration of drug bioavailability in the posterior segment of the eye. Moreover, current therapies only slow the progression of vision loss rather than reversing the disease. Current regenerative strategies require cell delivery and here, innovative combinations of cells, biomaterials, and delivery vehicles are being pursued.

3.3 Cell Delivery Strategies

3.3.1 Introduction

Cell therapy promises to limit and reverse retinal degeneration. There are two approaches being pursued for cell therapy: (1) transplantation of cells to replace damaged tissue; and (2) transplantation of cells to locally produce therapeutic molecules to promote endogenous repair [147]. Endogenous stem cell stimulation is a third option; however, this approach is rendered more difficult by the quiescent nature of the endogenous retinal stem cells [148]. For successful transplantation and replacement of endogenous tissue, cells must survive and integrate into the host system. This requires migration to the target tissue, differentiation into the correct cell type for integration into the existing circuitry, and restoration of long-term function, all while being exposed to the hostile conditions of the aging or degenerating retina. Although age-related degenerative diseases have different origins, they all culminate in the death of specific cell types of the retina, whether photoreceptors, RPE cells, or retinal ganglion cells (RGCs) [149]. To address this, cell delivery strategies include the transplantation of healthy photoreceptors [150], photoreceptor precursor cells [151], RGCs [6, 152, 153], RPE cells [154, 155], or somatic cell types that secrete trophic factors [156, 157].

3.3.2 Challenges with Cell Transplantation

Despite encouraging results with retinal transplantation, there remain a number of challenges that must be overcome for long-term functional recovery. The method of delivery must be carefully considered as the degree of trauma caused by the delivery of cells can affect the inflammatory response and thus the success of the transplantation [158, 159]. Currently there are two locations in the eye used for delivery: the subretinal space, which is more technically challenging and potentially disruptive, but is the location where the photoreceptors and/or RPE cells are lost; and the vitreous, which is less invasive and not as technically challenging, but requires cells to survive in the vitreous and then migrate to the retina. Ideally, cells injected as a suspension into the subretinal space should migrate and distribute along the retina as a monolayer. However, following a bolus injection of cells in saline, aggregation is often observed, resulting in cell death and limited (if any) host tissue integration [160, 161]. Introducing the use of materials to support cells and provide proper distribution in the subretinal space increases cell survival following transplantation [7, 162, 163].

Another major concern for cell transplantation is the survival and integration of cells. It is unclear how many cells survive in many cases, as quantification of cell survival is not often reported, but it appears that the majority of cells undergo apoptosis [160, 164, 165]. One reason for this is the problem of anoikis, an apoptotic cascade initiated by lack of cell anchorage to an extracellular matrix [165, 166]. This is supported by the observation that RPE cells transplanted as a whole sheet survive better than as a suspension [167, 168]. An additional contributing factor to cell death is the hostility of the microenvironment, which includes the presence of apoptotic signals, inflammatory cells, and damage to the surrounding architecture [153, 169, 170]. To compound this issue, cells that fail to integrate and undergo apoptosis secrete pro-apoptotic signals and create debris, causing further damage to the surrounding tissue [158, 170]. Beside survival, integration, and functional recovery, another challenge comes from the method used to assess cell migration and integration. Previously, transplantation of GFP^+ donor cells was considered convenient for imaging [171]. However, in recent studies, an unexpected mismatch in nuclear morphology between donor and integrated photoreceptors was reported, indicating that reporter-positive cells in the recipient retina may not reflect the integration of donor photoreceptors. Researchers have reported cellular material, including RNA and/or proteins transfer between donor cells and host photoreceptors, which results in the presence of a variety of donor-derived proteins in the host photoreceptors. The results also indicate that this material transfer process occurs in a majority of host cells; however, the cellular mechanisms of this process remain to be determined [172, 173].

3.3.3 Biomaterials for Cell Transplantation

3.3.3.1 Biomaterial Scaffolds

Scaffolds for cell delivery have been designed to promote cell survival and integration (see Table 3.2) [8, 174], and protect cells from the hostile degenerating environment [175, 176]. Scaffolds can support uniform cell distribution and direct the differentiation of transplanted stem cell progeny [177], and be designed to mimic natural ECM properties (e.g., stiffness, composition) [178, 179]. To rescue aging ocular tissues, RPE cells have been transplanted on synthetic polymer sheets meant

Table 3.2 In vivo studies of biomaterial-supported cell/stem-cell delivery to the posterior segment of the eye

Cell type	Material	Disease model	Outcome	Reference
Retinal pigmented epithelium (RPE) cells				
Adult and fetal human RPE stem cells	Polyester matrix membrane	Wild-type rabbit	Cell survival up to 4 weeks with polarity markers maintained; no retinal scarring	[181]
Human ESC[a]-derived RPE cells	Parylene plate	Royal college of surgeons rat	Successful implantation of intact synthetic monolayer seeded with cells; loss of less than 2 % of cells after 7 days	[182]
Human ESC[a]-derived RPE cells	PET[b] or P(LA-co-CL)[c]	Wild-type rabbit	Subretinal biocompatibility over 14 days, some migration of native RPE cells into scaffold	[181]
Retinal progenitor cells (RPCs)				
Primary mouse RPCs	Nanowire PCL[d] scaffold	rhodopsin null mouse	Supported cell growth in vitro, some migration and differentiation in vivo after 30 days	[183]
Primary mouse RPCs	Thin-film PCL[d] scaffold	Rhodopsin-null C57Bl6 mouse	Localized to ONL and expressed photoreceptor markers in vivo	[184]
Primary mouse RPCs	PGS[e] scaffold	Rhodopsin-null and wild-type C57Bl6 mice	Transplanted cell migration into retina and maturation, cells survived for 1 month	[185]
Mouse RPCs	Hyaluronic acid hydrogel	Rhodopsin-null mice	No damage during injection; cell distributed evenly in subretinal space; cell survival up to 3 weeks	[186]
Retinal stem cell-derived rods	Hyaluronic acid/ methylcellulose hydrogel	Adult albino CD10 and TKO[f] mice	Cell survival and migration significantly greater in gel than saline; improvement in pupillary light response	[162]
Retinal stem cells	Hyaluronic acid/ methylcellulose hydrogel	CD10/ Gnat2$^{-/-}$ mice	Cell survival for 4 weeks in vivo, superior cell distribution in subretinal space	[7]

[a]Embryonic stem cell
[b]Polyethylene terephthalate
[c]Poly(L-lactide-co-ε-caprolactone)
[d]Polycaprolactone
[e]Poly(glycerol sebacate)
[f]Triple knock-out

to mimic the Bruch's membrane, which naturally support the RPE cells in the eye; and injectable hydrogels can be used to deliver photoreceptors to the subretinal space [180].

3.3.3.1.1 Synthetic Bruch's Membrane for RPE Delivery

As the RPE lacks the capacity to regenerate, age-related diseases such as AMD and diabetic retinopathy result in degeneration of the RPE and blindness. RPE degeneration is accompanied by pathological changes to Bruch's membrane, such as thickening, collagen crosslinking, calcification, and drusen deposition [187, 188]. Transplantation of RPE cells to the aging posterior segment aims to reverse disease progression and rescue damaged tissues. Success is measured by tissue repair where transplanted RPE cells express mature markers, such as CRALBP, RPE65 and bestrophin, and functional repair with regained vision.

Direct RPE transplantation onto aged Bruch's membrane results in poor cell survival, adhesion, and organization [169, 189, 190]. Given the highly organized cell structure of the RPE, successful RPE cell implantation appears to require replacement of the damaged Bruch's membrane as well [187]. For this reason, RPE cells are transplanted as a pre-cultured sheet on a support membrane and transplanted as a monolayer. This necessitates a complicated surgery and poses a biomaterial challenge in the design of an optimal artificial Bruch's membrane onto which RPE cells are cultured and then transplanted. To overcome the difficulty in transplantation of an ultra-thin membrane, a platform device made of parylene, which has a "U" shape with barriers along the edge, can be loaded with a graft containing the desired cells for transplantation; the space inside the barriers works as the loading chamber [182].

Recognizing the importance of the porous nature of Bruch's membrane, various materials have been studied for the culture and delivery of RPE cells. For example, a porous polycaprolactone (PCL) thin-film (vs. polyester transwell and a nonporous PCL thin-film) enhanced maturation of fetal human RPE (fhRPE) monolayers as demonstrated by improved tight junction localization and cellular density, and expression of RPE-associated genes such as RPE65, RLBP1, and BEST1 [191]. Interestingly, a porous polyimide membrane coated with cell-adhesive proteins, such as laminin and collagen I, promoted the adhesion of hESC-derived RPE [192]. By simply using porous collagen membranes [193], a functional RPE was created as demonstrated by the RPE phagocytosing the outer segments of photoreceptors in vitro [194]. Preliminary tests for potential subconjunctival or subretinal transplantation of a similar collagen-based scaffold were performed in rabbits and showed no inflammatory or immune response [195]. Post-transplantation cell survival has also been demonstrated with the use of a polyester membrane that has properties similar to the native Bruch's membrane to transplant human RPE stem cells into the rabbit retina [181].

Recently, materials have been developed that closely mimic the native membrane. Although the Bruch's membrane has a complex pentalaminar structure, it is the inner collagenous layer that is targeted for mimicry. It consists of a porous, mesh-like architecture made of collagen fibers with diameters of 60 nm and a packing density of 48%, which allows nutrient and oxygen transfer [187]. Using electrospinning, various nanofibrillar delivery systems have been designed to mimic Bruch's membrane for RPE delivery. Different materials have been investigated:

silk and PCL [196], polyimide [197], a combination of silk and poly(ethylene glycol) (PEG) [198], and an RGD-functionalized poly(methyl methacrylate-co-poly(ethylene glycol) methacrylate) (P(MMA-co-PEGMA)) [199]. RPE cells seeded on 200–300 nm nanofibrous scaffolds formed the expected poly(hexagonal) structure with a striking resemblance to native RPE, and expressed typical RPE markers [187]. Surprisingly, the key distinguishing material property was the nanofibrous property regardless of the material used. This was demonstrated independently in two studies comparing different nanofibrillar supports, one comparing PLGA and collagen, and the other polyethylene terephthalate and poly(L-lactide-co-ε-caprolactone) [181, 187]. While the nanofiber structure enables RPE survival and organization, implantation requires the use of a relatively rigid backing [181]. Notwithstanding the complex material design and consequent surgery, the nanofibrous strategy holds great promise, yet the influence of restored RPE on aging ocular tissues remains to be evaluated in these strategies.

3.3.3.1.2 Photoreceptor and Retinal Progenitor Cell Delivery

Regenerative strategies that include photoreceptor replacement aim to overcome vision loss due to RPE degradation and photoreceptor death. Successful transplantation of photoreceptors relies on their integration into the neural circuitry of the retina. Biomaterial scaffolds designed for photoreceptor delivery must enhance both their survival and migration out of the scaffold to enable cell integration. Biodegradable (or bioresorbable) scaffolds with high porosity (permeability) contribute to photoreceptor survival by regulating nutrient and oxygen diffusion [181] and have shown the most success to date [184, 200].

The first generation of degradable scaffolds for RPC delivery consisted of porous membranes of PLA/PLGA blends, the porosity of which was adjusted using different PLA/PLGA ratios and phase inversion/separation techniques [201]. In vitro, cells migrated into the porous scaffolds, attached therein, and showed downregulation of immature markers and upregulation of differentiation markers [202]. Cell attachment to the material was correlated with cell differentiation. In vivo, RPCs transplanted on these degradable scaffolds into the mouse subretinal space showed greater survival than RPCs transplanted as a single-cell suspension. Importantly, grafted RPCs migrated into the host retina and expressed several mature markers (neurofilament 200, glial fibrillary acidic protein, protein kinase C-α, recoverin, and rhodopsin), demonstrating in vivo differentiation [160]. A similar observation was made when RPCs were transplanted into a porcine retina [203].

Notwithstanding the positive results attained with the transplantation of cells on these PLA/PLGA scaffolds, there is a mismatch in modulus and flexibility between the scaffold and the subretinal space, resulting in tissue damage [180]. Moreover, the acidic degradation products of PLA- and PLGA-based supports may lead to a chronic inflammatory response in the confined subretinal space [181, 204]. Consequently, both slow-degrading porous membranes [183] and more flexible systems [185] were developed.

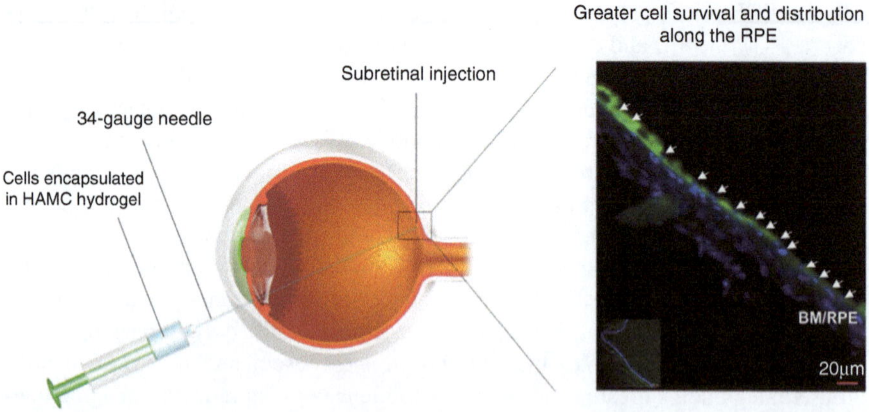

Fig. 3.3 Delivery of retinal progenitor cells (RPCs) to the subretinal space of a damaged retina, using hyaluronan/methylcellulose (HAMC) hydrogel as biomaterial support. HAMC allows in situ rapid gelation and results in greater cell survival and distribution. Reproduced with permission [87]

Bioresorbable/biodegradable hydrogels facilitate cell injection into the subretinal space. The tunability of injectable hydrogels, which allows the viscosity, composition, and degradation rate to be adjusted, enables minimally invasive surgery and results in limited tissue damage [7, 162]. Hydrogels provide a protective environment and an even distribution of RPCs within the retinal tissue [186]. An injectable hydrogel composed of hyaluronan/methylcellulose (HAMC) promoted greater survival and distribution of both transplanted RPCs and retinal stem cell (RSC)-derived photoreceptor rods than the identical cells delivered in conventional saline to the subretinal space of mice (see Fig. 3.3) [7, 8, 162]. The HAMC hydrogel is particularly well suited for delivery into the subretinal space as it is minimally swelling, bioresorbed within 1 week and gels rapidly on injection [177, 205, 206]. Interestingly, hyaluronan also promotes the survival of primary rod photoreceptors in vitro by an mTOR-mediated pathway [207]. Fibrin has also been shown to be useful for cell delivery as it promotes cell adhesion, via an integrin-RGD-mediated mechanism [208]. These pioneering studies on hydrogel-based systems pave the way for future ocular regenerative therapies.

3.3.3.1.3 Retinal Ganglion Cell Delivery

Unlike RPE and photoreceptor transplantation, retinal ganglion cell (RGC) transplantation has had limited attention to date [209, 210] because regeneration of optic nerve cells likely requires a more complicated strategy. For example, the arrangement of the transplant should recapitulate the radial organization of the native nerve fibers in order to facilitate proper impulse conduction. An electrospun scaffold composed of densely packed, radially aligned PLA nanofibers was coated with laminin, and resulted in enhanced in vitro cell adhesion, survival, and preserved RGC electrical properties within a radial pattern of RGC axons [209]. After in vivo transplantation, RGCs survived in the recipient retina, and integrated into the ganglion cell layer [211].

3.3.3.2 Co-delivery of Growth Factors and Retinal Cells

In previous studies, transplantation of a variety of cell types to the adult retina showed limited donor cell integration [212] or functional recovery [164]. In order to improve the survival and integration of transplanted cells and encourage functional recovery after transplantation, co-delivery of neurotrophic factors with transplanted cells has been investigated. For example, BDNF was co-delivered with RPCs by encapsulating it in microspheres embedded in sheets of RPCs, and tested in rat models of blindness. Functional repair was tested by measuring the response to low light stimuli, and it was reported that 80% of rats treated with both BDNF and RPCs showed some functional repair whereas only 50% of rats treated with RPCs alone showed repair [42]. The effect of co-delivering GDNF or BDNF with retinal sheet transplants to restore vision in a RP mouse model was investigated by Yang et al. [20]. In this study, pre-cultured E19 retinal cell sheets were transplanted with no neurotrophic factor, GDNF microparticles, or BDNF microparticles. The neurotrophic factor-treated cell sheets were transplanted in three rat models of rapid degeneration. The histology of the transplanted retinal cell sheets was analyzed and classified into two groups: laminate and rosette (more disordered form). BDNF improved the integration of cell sheets when they were in the laminated form. GDNF enhanced integration and functional recovery in both laminated and rosetted cell sheets, suggesting that GDNF and BDNF may act through different pro-survival pathways. Overall, both GDNF and BDNF improved the integration and function of the recipient retina, highlighting the benefit of co-delivering trophic factors and cells [20].

While nanoparticles have been used to deliver proteins for decades, protein encapsulation is inherently limited by the formulation process, resulting in typically <0.1% of protein encapsulated per polymeric nanoparticle by mass. This has led to affinity release systems where proteins are incorporated into scaffolds designed to weakly bind the proteins and prolong their release without further encapsulation. For example, the native interaction between Src Homology 3 (SH3) and its binding peptides has been incorporated into injectable hydrogels where the protein of interest is expressed as a fusion protein with SH3 and the binding peptide is covalently bound to the injectable hydrogel [213–216]. Recently, RPE cells cultured in the presence of affinity released IGF-1 showed greater cell viability in an injectable hyaluronan-methylcellulose (HAMC) hydrogel than controls without IGF-1 [217]. The combined delivery of trophic factors and retinal cells is promising, yet mostly unexplored to date, opening opportunities for future research.

3.4 Outlook

Despite the promising results of anti-angiogenic (anti-VEGF) therapies, such as bevacizumab (Avastin™) and ranibizumab (Lucentis™), the progression of vision loss is slowed, but not reversed. Given the safety concerns of current repeated intravitreal injections, there is great opportunity for local delivery strategies to be developed to achieve sustained release. Nanoparticulate systems, which have been long studied for cancer therapies and diagnosis, may be a rich source of inspiration [218–220]. As a promising example, nanocarriers loaded with common anticancer drugs, such as paclitaxel [221, 222] or doxorubicin [111, 223], have been recently reported for the treatment of choroidal neovasculature. However, the BRB may limit the use of intravenous delivery of nanocarriers, and intravitreal injections may result in undesirable complications. Thus, topical administration of eye drops for drug delivery to the posterior segment of the eye constitutes a noninvasive approach that may provide significant benefit, but remains challenging due to multiple drainage and biological barriers, including blinking and rapid dispersion of eye drops. To counteract the fast clearing of eye drops, nanoparticles have been modified on the surface with phenylboronic acid and shown to sustain the release of cyclosporin-A for up to 5 days [224]. Solutions might also come from annexin-functionalized nanocarriers which have demonstrated significant uptake and transcytosis of liposomal drug carriers across corneal epithelial barriers [117]. In addition, emerging techniques are currently being evaluated for better penetration of the nanoparticles through the sclera and to the retina. For example, transscleral delivery can be enhanced with either iontophoresis where penetration is enhanced with a low electric current [225, 226] or ultrasound [227, 228]. Implants are an interesting alternative to nanoparticles for prolonged drug release. Currently under investigation, transscleral, refillable implants may provide an innovative strategy for sustained drug release [229, 230].

Cell delivery constitutes a more ambitious, yet still emerging, approach for ocular treatments. The need for artificial Bruch's membranes to support RPE cell transplantation has prompted the development of novel nanofiber-based, degradable scaffolds, which have already demonstrated beneficial properties in vitro and a good biocompatibility in vivo [187, 196]. However, the nanofibrillar system necessitates the use of an additional support material with appropriate permeability, flexibility, bioresorption, and biocompatibility. A micropatterned porous thin-film co-carrier [191] coated with nanofibers may meet the requirements; however, in vivo efficacy of transplanted RPE has yet to be demonstrated clinically. RPC transplantation is an alternate strategy, which obviates the need for a supportive membrane yet requires injection into the subretinal space, followed by in situ differentiation and tissue integration.

Injectable, bioresorbable hydrogel delivery vehicles are useful for cell delivery to the retina, as the surgical strategy is less invasive than implantation of a solid scaffold. Moreover, the hydrogel can be designed to have a modulus that matches that of the retina, and biochemical properties that enhance cell survival/integration after transplantation [8, 175, 231] and direct stem-cell fate in situ [176, 177]. While side effects and complications from subretinal injection need to be assessed, advanced hydrogel design for RPC or precursor cell delivery provides tremendous opportunity for successful ocular cell delivery.

Whether for drug or cell delivery, the choice of material constitutes one of the major challenges for the future. Since regulatory approval is a long, laborious process, most of the materials presented herein have a long regulatory history, including natural lipids (mainly for liposomes), polysaccharides (such as chitosan, hyaluronan, and cyclodextrins), or synthetic polymers (typically PLA, PLGA, PCL). The confined space in which ocular delivery systems are administered may complicate biocompatibility as the acidic degradation products of synthetic polyesters may be harmful to the tissue [6, 181, 183]. The tremendous progress in synthetic materials should provide advanced properties to these delivery systems. In particular, functional polymers for drug/peptide grafting and delivery may lead to innovations in controlled and targeted drug release to the posterior segment of the eye. Well-documented strategies and advanced designs of stimuli-responsive and functional polymer-based nanocarriers for other biomedical applications may serve as examples [232, 233]. However, for these materials to be useful, they (and their degradation products) must be biocompatible and bioresorbable [146]. Therefore, intraocular toxicity of any innovative material should be investigated early.

Another exciting potential strategy to explore for vision repair is the combination of delivery techniques, such as nanoparticle-encapsulated hydrogels, for sustained release of multiple therapeutics, co-delivery of cells and drugs, or multifunctional delivery systems [161, 234]. For example, PLGA particles are often used in the study of controlled release of therapeutic proteins. However, organic solvents and high shear forces used during the formulation process can denature the proteins, resulting in low bioavailability [235]. To overcome encapsulation, two new strategies have been developed for bioactive protein delivery: affinity-controlled release and electrostatic-controlled release.

Affinity release has been achieved typically with a heparin-based biomaterial and heparin-binding factors [236]. The biomaterial can also mimic natural interactions between delivered proteins and their natural extracellular matrix, and have a sequential multi-therapeutic delivery, which may improve ocular drug delivery [237]. There is a balance between the extracellular matrix and the protein that regulates the concentration and activity of the protein in the microenvironment. Proteins can also interact with other glycosaminoglycans (GAGs), and a recent study showed that the release profile of a protein released from GAG-modified hydrogels can be controlled by the GAG concentration, binding affinity, and the crosslinking density of the matrix [238]. The use of fusion proteins of SH3 with the growth factor of interest and the biomaterial modified with an SH3-binding peptide provides a more controlled strategy for protein release. In electrostatic release, negatively charged PLGA nanoparticles were shown to interact with positively charged growth factors and to demonstrate the same release profile and bioactivity of encapsulated proteins [2]. This overcomes the low loading and bioavailability of encapsulated proteins.

With these studies, sustaining the therapeutic release in a mild condition becomes possible for ocular drug delivery.

Improvements of in vitro and in vivo models will be required for better translation to the clinic. Indeed, creating an accurate animal model of retinal degeneration is challenging. Diseases such as AMD and diabetic retinopathy are the culmination of complex genetic and environmental factors and, while current models can mimic several pathological characteristics of degeneration, none of these fully recapitulate all anatomical features. Furthermore, because of this variation in models, it is difficult to translate results from one study to another; an investigation of six different genetic mouse models of degeneration found that photoreceptor transplantation success varied widely depending on the model used [153]. This underlines the importance of testing strategies in different animal models of disease and ideally in different laboratories, similar to the way a clinical trial would be executed at multiple sites.

Ultimately, to overcome the challenges associated with age-related or degenerative ocular diseases, collaboration between scientists, engineers, and clinicians in academia and industry is required for successful translation of innovative strategies to the clinic. Advances in each discipline—cell biology, drug discovery and delivery, and surgical strategies—are required to overcome these daunting challenges to repair and restore vision.

Glossary

AMD Age-related macular degeneration.
BDNF Brain-derived neurotrophic factor.
BRB Blood retinal barrier.
CMP Collagen mimetic peptide.
CNTF Ciliary neurotrophic factor.
DME Diabetic macular edema.
DR Diabetic retinopathy.
EGF Epidermal growth factor.
ERG Electroretinography.
FDA US Food and Drug Administration.
FGF2 Fibroblast growth factor.
GAG Glycosaminoglycan.
GDNF Glial cell line-derived neurotrophic factor.
IGF Insulin-like growth factor.
NGF Nerve growth factor.
NT-3 Neurotrophin-3.
PCL Polycaprolactone.
PEDF Pigment epithelium-derived factor.
PEG Poly(ethylene glycol).
PLGA Poly(lactic-co-glycolic acid).

rAAV Recombinant adeno-associated virus.
RGC Retinal ganglion cells.
ROP Retinopathy of prematurity.
ROS Reactive oxygen species.
RPE Retinal pigment epithelium.
RSC Retinal stem cell.
SH3 Src homology 3.
SOD1 Superoxide dismutase.
TGF-β Transforming growth factor beta.
VEGF Vascular endothelial growth factor.

Acknowledgments We are grateful to the Canada First Research Excellence Fund for funding through Medicine by Design and to The Shoichet Lab at the University of Toronto, and especially Nick Mitrousis, for thoughtful review of this chapter.

References

1. Thapa, R., Bajimaya, S., Paudyal, G., Khanal, S., Tan, S., Thapa, S. S., et al. (2015). Population awareness of diabetic eye disease and age related macular degeneration in Nepal: The Bhaktapur Retina Study. *BMC Ophthalmology, 15*(1), 188.
2. Pakulska, M. M., Donaghue, I. E., Obermeyer, J. M., Tuladhar, A., McLaughlin, C. K., Shendruk, T. N., et al. (2016). Encapsulation-free controlled release: Electrostatic adsorption eliminates the need for protein encapsulation in PLGA nanoparticles. *Science Advances, 2*, e1600519.
3. Wong, W. L., Su, X., Li, X., Cheung, C. M. G., Klein, R., Cheng, C. Y., et al. (2014). Global prevalence of age-related macular degeneration and disease burden projection for 2020 and 2040: A systematic review and meta-analysis. *The Lancet Global Health, 2*(2), 106–116.
4. Pennington, K. L., & DeAngelis, M. M. (2016). Epidemiology of age-related macular degeneration (AMD): Associations with cardiovascular disease phenotypes and lipid factors. *Eye and Vision, 3*(1), 34.
5. Yau, J. W., Rogers, S. L., Kawasaki, R., Lamoureux, E. L., Kowalski, J. W., Bek, T., et al. (2012). Global prevalence and major risk factors of diabetic retinopathy. *Diabetes Care, 35*(3), 56–64.
6. Gonzalez-Cordero, A., West, E. L., Pearson, R. A., Duran, Y., Carvalho, L. S., Chu, C. J., et al. (2013). Photoreceptor precursors derived from three-dimensional embryonic stem cell cultures integrate and mature within adult degenerate retina. *Nature Biotechnology, 31*, 741–747.
7. Ballios, B. G., Cooke, M. J., van der Kooy, D., & Shoichet, M. S. (2010). A hydrogel-based stem cell delivery system to treat retinal degenerative diseases. *Biomaterials, 31*(9), 2555–2564.
8. Tam, R. Y., Fuehrmann, T., Mitrousis, N., & Shoichet, M. S. (2014). Regenerative therapies for central nervous system diseases: A biomaterials approach. *Neuropsychopharmacology, 39*, 169–188.
9. Nissen, M. H., & Röpke, C. (2005). Innate and adaptive immunity of the eye. *Advances in Organ Biology, 10*, 291–305.
10. Hosoya, K. I., Tomi, M., & Tachikawa, M. (2011). Strategies for therapy of retinal diseases using systemic drug delivery: Relevance of transporters at the blood–retinal barrier. *Expert Opinion on Drug Delivery, 8*(12), 1571–1587.

11. Gaudana, R., Ananthula, H. K., Parenky, A., & Mitra, A. K. (2010). Ocular drug delivery. *The AAPS Journal, 12*(3), 348–360.
12. Mishra, G. P., Bagui, M., Tamboli, V., & Mitra, A. K. (2011). Recent applications of liposomes in ophthalmic drug delivery. *Journal of Drug Delivery, 2011*, 863734.
13. Urtti, A. (2006). Challenges and obstacles of ocular pharmacokinetics and drug delivery. *Advanced Drug Delivery Reviews, 58*, 1131–1135.
14. Kim, S. H., Lutz, R. J., Wang, N. S., & Robinson, M. R. (2007). Transport barriers in transscleral drug delivery for retinal diseases. *Ophthalmic Research, 39*, 244–254.
15. Barar, J., Javadzadeh, A. R., & Omidi, Y. (2008). Ocular novel drug delivery: Impacts of membranes and barriers. *Expert Opinion on Drug Delivery, 5*, 567–558.
16. Tomi, M., & Hosoya, K. I. (2010). The role of blood–ocular barrier transporters in retinal drug disposition: An overview. *Expert Opinion on Drug Metabolism & Toxicology, 6*, 1111–1124.
17. Jordán, J., & Ruíz-Moreno, J. M. (2013). Advances in the understanding of retinal drug disposition and the role of blood–ocular barrier transporters. *Expert Opinion on Drug Metabolism & Toxicology, 9*, 1181–1192.
18. Domenici, L., Origlia, N., Falsini, B., Cerri, E., Barloscio, D., Fabiani, C., et al. (2014). Rescue of retinal function by BDNF in a mouse model of glaucoma. *PLoS One, 9*, 115579.
19. Hu, Z. L., Li, N., Wei, X., Tang, L., Wang, T. H., & Chen, X. M. (2017). Neuroprotective effects of BDNF and GDNF in intravitreally transplanted mesenchymal stem cells after optic nerve crush in mice. *International Journal of Ophthalmology, 10*(1), 35.
20. Yang, P. B., Seiler, M. J., Aramant, R. B., Yan, F., Mahoney, M. J., Kitzes, L. M., et al. (2010). Trophic factors GDNF and BDNF improve function of retinal sheet transplants. *Experimental Eye Research, 91*(5), 727–738.
21. Tombran-Tink, J., & Barnstable, C. J. (2003). PEDF: A multifaceted neurotrophic factor. *Nature Reviews Neuroscience, 4*(8), 628–636.
22. Lenzi, L., Coassin, M., Lambiase, A., Bonini, S., Amendola, T., & Aloe, L. (2005). Effect of exogenous administration of nerve growth factor in the retina of rats with inherited retinitis pigmentosa. *Vision Research, 45*(12), 1491–1500.
23. Carwile, M. E., Culbert, R. B., Sturdivant, R. L., & Kraft, T. W. (1998). Rod outer segment maintenance is enhanced in the presence of bFGF, CNTF and GDNF. *Experimental Eye Research, 66*(6), 791–805.
24. Checa-Casalengua, P., Jiang, C., Bravo-Osuna, I., Tucker, B. A., Molina-Martínez, I. T., Young, M. J., et al. (2011). Retinal ganglion cells survival in a glaucoma model by GDNF/Vit E PLGA microspheres prepared according to a novel microencapsulation procedure. *Journal of Controlled Release, 156*(1), 92–100.
25. Lingor, P., Unsicker, K., & Krieglstein, K. (2000). GDNF and NT-4 protect midbrain dopaminergic neurons from toxic damage by iron and nitric oxide. *Experimental Neurology, 163*(1), 55–62.
26. Yan, Q., Wang, J., Matheson, C. R., & Urich, J. L. (1999). Glial cell line–derived neurotrophic factor (GDNF) promotes the survival of axotomized retinal ganglion cells in adult rats: Comparison to and combination with brain-derived neurotrophic factor (BDNF). *Journal of Neurobiology, 38*(3), 382–390.
27. Cao, W., Li, F., Steinberg, R. H., & Lavail, M. M. (2001). Development of normal and injury-induced gene expression of aFGF, bFGF, CNTF, BDNF, GFAP and IGF-I in the rat retina. *Experimental Eye Research, 72*(5), 591–604.
28. Liang, F. Q., Aleman, T. S., Dejneka, N. S., Dudus, L., Fisher, K. J., Maguire, A. M., et al. (2001). Long-term protection of retinal structure but not function using RAAV. CNTF in animal models of retinitis pigmentosa. *Molecular Therapy, 4*(5), 461–472.
29. McGill, T. J., Prusky, G. T., Douglas, R. M., Yasumura, D., Matthes, M. T., Nune, G., et al. (2007). Intraocular CNTF reduces vision in normal rats in a dose-dependent manner. *Investigative Ophthalmology & Visual Science, 48*(12), 5756–5766.
30. LaVail, M. M., Unoki, K., Yasumura, D., Matthes, M. T., Yancopoulos, G. D., & Steinberg, R. H. (1992). Multiple growth factors, cytokines, and neurotrophins rescue photorecep-

tors from the damaging effects of constant light. *Proceedings of the National Academy of Sciences, 89*(23), 11249–11253.
31. Mysona, B. A., Zhao, J., & Bollinger, K. E. (2017). Role of BDNF/TrkB pathway in the visual system: Therapeutic implications for glaucoma. *Expert Review of Ophthalmology, 12*(1), 69–81.
32. Wahlin, K. J., Campochiaro, P. A., Zack, D. J., & Adler, R. (2000). Neurotrophic factors cause activation of intracellular signaling pathways in Muller cells and other cells of the inner retina, but not photoreceptors. *Investigative Ophthalmology & Visual Science, 41*, 927–936.
33. Fu, Q. L., Li, X., Yip, H. K., Shao, Z., Wu, W., Mi, S., et al. (2009). Combined effect of brain-derived neurotrophic factor and LINGO-1 fusion protein on long-term survival of retinal ganglion cells in chronic glaucoma. *Neuroscience, 162*, 375–382.
34. Frasson, M., Picaud, S., Léveillard, T., Simonutti, M., Mohand-Said, S., Dreyfus, H., et al. (1999). Glial cell line–derived neurotrophic factor induces histologic and functional protection of rod photoreceptors in the RD/RD mouse. *Investigative Ophthalmology & Visual Science, 40*(11), 2724–2734.
35. Dong, A., Shen, J. K., Krause, M., Hackett, S. F., & Campochiaro, P. A. (2007). Increased expression of glial cell line-derived neurotrophic factor protects against oxidative damage-induced retinal degeneration. *Journal of Neurochemistry, 103*(3), 1041–1052.
36. Sanftner, L. H. M., Abel, H., Hauswirth, W. W., & Flannery, J. G. (2001). Glial cell line derived neurotrophic factor delays photoreceptor degeneration in a transgenic rat model of retinitis pigmentosa. *Molecular Therapy, 4*(6), 622–629.
37. Kassen, S. C., Thummel, R., Campochiaro, L. A., Harding, M. J., Bennett, N. A., & Hyde, D. R. (2009). CNTF induces photoreceptor neuroprotection and Müller glial cell proliferation through two different signaling pathways in the adult zebrafish retina. *Experimental Eye Research, 88*(6), 1051–1064.
38. Joly, S., Pernet, V., Chemtob, S., Di Polo, A., & Lachapelle, P. (2007). Neuroprotection in the juvenile rat model of light-induced retinopathy: Evidence suggesting a role for FGF-2 and CNTF. *Investigative Ophthalmology & Visual Science, 48*(5), 2311–2320.
39. Sieving, P. A., Caruso, R. C., Tao, W., Coleman, H. R., Thompson, D. J., Fullmer, K. R., et al. (2006). Ciliary neurotrophic factor (CNTF) for human retinal degeneration: Phase I trial of CNTF delivered by encapsulated cell intraocular implants. *Proceedings of the National Academy of Sciences of the United States of America, 103*(10), 3896–3901.
40. LaVail, M. M., et al. (1992). Multiple growth factors, cytokines, and neurotrophins rescue photoreceptors from the damaging effects of constant light. *National Academy of Sciences, 89*(23), 11249–11253.
41. Azadi, S., Johnson, L. E., Paquet-Durand, F., Perez, M. T. R., Zhang, Y., Ekström, P. A., et al. (2007). CNTF+ BDNF treatment and neuroprotective pathways in the rd1 mouse retina. *Brain Research, 1129*, 116–129.
42. Seiler, M. J., Thomas, B. B., Chen, Z., Arai, S., Chadalavada, S., Mahoney, M. J., et al. (2008). BDNF-treated retinal progenitor sheets transplanted to degenerate rats: Improved restoration of visual function. *Experimental Eye Research, 86*(1), 92–104.
43. Thomas, B. B., Aramant, R. B., Qiu, G. T., Arai, S., Chen, Z., Sadda, S. R., et al. (2004). BDNF microsphere treatment increases functional effects of retinal transplants. *Investigative Ophthalmology & Visual Science, 45*(13), 5184–5184.
44. Ogilvie, J. M., Speck, J. D., & Lett, J. M. (2000). Growth factors in combination, but not individually, rescue rd mouse photoreceptors in organ culture. *Experimental Neurology, 161*(2), 676–685.
45. Dalkara, D., Kolstad, K. D., Guerin, K. I., Hoffmann, N. V., Visel, M., Klimczak, R. R., et al. (2011). AAV mediated GDNF secretion from retinal glia slows down retinal degeneration in a rat model of retinitis pigmentosa. *Molecular Therapy, 19*(9), 1602–1608.
46. LaVail, M. M., Yasumura, D., Matthes, M. T., Lau-Villacorta, C., Unoki, K., Sung, C. H., et al. (1998). Protection of mouse photoreceptors by survival factors in retinal degenerations. *Investigative Ophthalmology & Visual Science, 39*(3), 592–602.

47. Wenzel, A., Grimm, C., Samardzija, M., & Remé, C. E. (2005). Molecular mechanisms of light-induced photoreceptor apoptosis and neuroprotection for retinal degeneration. *Progress in Retinal and Eye Research, 24*(2), 275–306.
48. Cayouette, M., Smith, S. B., Becerra, S. P., & Gravel, C. (1999). Pigment epithelium-derived factor delays the death of photoreceptors in mouse models of inherited retinal degenerations. *Neurobiology of Disease, 6*(6), 523–532.
49. Gao, G., Li, Y., Fant, J., Crosson, C. E., Becerra, S. P., & Ma, J. X. (2002). Difference in ischemic regulation of vascular endothelial growth factor and pigment epithelium-derived factor in brown norway and sprague dawley rats contributing to different susceptibilities to retinal neovascularization. *Diabetes, 51*(4), 1218–1225.
50. Ogata, N., Wang, L., Jo, N., Tombran-Tink, J., Takahashi, K., Mrazek, D., et al. (2001). Pigment epithelium derived factor as a neuroprotective agent against ischemic retinal injury. *Current Eye Research, 22*(4), 245–252.
51. Cao, W., Tombran-Tink, J., Elias, R., Sezate, S., Mrazek, D., & McGinnis, J. F. (2001). In vivo protection of photoreceptors from light damage by pigment epithelium-derived factor. *Investigative Ophthalmology & Visual Science, 42*(7), 1646–1652.
52. Akiyama, G., Sakai, T., Kuno, N., Kimura, E., Okano, K., Kohno, H., et al. (2012). Photoreceptor rescue of pigment epithelium-derived factor-impregnated nanoparticles in Royal College of Surgeons rats. *Molecular Vision, 18*, 3079–3086.
53. Léveillard, T., Mohand-Saïd, S., Lorentz, O., Hicks, D., Fintz, A. C., Clérin, E., et al. (2004). Identification and characterization of rod-derived cone viability factor (RdCVF). *Nature Genetics, 36*(7), 755–759.
54. Yang, Y., Mohand-Said, S., Danan, A., Simonutti, M., Fontaine, V., Clerin, E., et al. (2009). Functional cone rescue by RdCVF protein in a dominant model of retinitis pigmentosa. *Molecular Therapy, 17*(5), 787–795.
55. Peinado-Ramon, P., Salvador, M., Villegas-Perez, M. P., & Vidal-Sanz, M. (1996). Effects of axotomy and intraocular administration of NT-4, NT-3, and brain-derived neurotrophic factor on the survival of adult rat retinal ganglion cells. A quantitative in vivo study. *Investigative Ophthalmology & Visual Science, 37*(4), 489–500.
56. Kolomeyer, A. M., & Zarbin, M. A. (2014). Trophic factors in the pathogenesis and therapy for retinal degenerative diseases. *Survey of Ophthalmology, 59*(2), 134–165.
57. Stewart, M. W. (2012). The expanding role of vascular endothelial growth factor inhibitors in ophthalmology. *Mayo Clinic Proceedings, 87*, 77–88.
58. Boyer, D. S., Hopkins, J. J., Sorof, J., & Ehrlich, J. S. (2013). Anti-vascular endothelial growth factor therapy for diabetic macular edema. *Therapeutic Advances in Endocrinology and Metabolism, 4*, 151–169.
59. Stewart, M., Grippon, S., & Kirkpatrick, P. (2012). Aflibercept. *Nature Reviews Drug Discovery, 11*(4), 269–270.
60. Summers, J., Cohen, M. H., Keegan, P., & Pazdur, R. (2010). FDA drug approval summary: Bevacizumab plus interferon for advanced renal cell carcinoma. *The Oncologist, 15*(1), 104–111.
61. Smith, A. G., & Kaiser, P. K. (2016). Therapeutic monoclonal antibodies and fragments: Ranibizumab. *Retinal Pharmacotherapeutics, 55*, 246–251.
62. Ng, E. W., Shima, D. T., Calias, P., Cunningham Jr., E. T., Guyer, D. R., & Adamis, A. P. (2006). A targeted anti-VEGF aptamer for ocular vascular disease. *Nature Reviews. Drug Discovery, 5*, 123–132.
63. Colquitt, J. L., Jones, J., Tan, S. C., Takeda, A. L., Clegg, A. J., & Price, A. (2008). Ranibizumab and pegaptanib for the treatment of age-related macular degeneration: A systematic review and economic evaluation. *Health Technology Assessment, 12*, 9–201.
64. Stewart, M. W. (2013). Critical appraisal of ranibizumab in the treatment of diabetic macular edema. *Clinical Ophthalmology, 7*, 1257–1267.
65. Moss, A. (2013). The angiopoietin:tie 2 interaction: A potential target for future therapies in human vascular disease. *Cytokine & Growth Factor Reviews, 24*, 579–592.

66. Santulli, R. J., Kinney, W. A., Ghosh, S., DeCorte, B. L., Liu, L., Tuman, R. W., et al. (2008). Studies with an orally bioavailable alpha V integrin antagonist in animal models of ocular vasculopathy: Retinal neovascularization in mice and retinal vascular permeability in diabetic rats. *The Journal of Pharmacology and Experimental Therapeutics, 324*, 894–901.
67. Nguyen, Q. D., Schachar, R. A., Nduaka, C. I., Sperling, M., Basile, A. S., Klamerus, K. J., et al. (2012). Dose-ranging evaluation of intravitreal siRNA PF-04523655 for diabetic macular edema (the DEGAS study). *Investigative Ophthalmology & Visual Science, 53*, 7666–7674.
68. Takahashi, T., Nakamura, T., Hayashi, A., Kamei, M., Nakabayashi, M., Okada, A. A., et al. (2000). Inhibition of experimental choroidal neovascularization by overexpression of tissue inhibitor of metalloproteinases-3 in retinal pigment epithelium cells. *American Journal of Ophthalmology, 130*, 774–781.
69. Vadlapatla, R. K., Vadlapudi, A. D., Pal, D., Mukherji, M., & Mitra, A. K. (2014). Ritonavir inhibits HIF-1alpha-mediated VEGF expression in retinal pigment epithelial cells in vitro. *Eye (London, England), 28*, 93–101.
70. Prea, S. M., Chan, E. C., Dusting, G. J., Vingrys, A. J., Bui, B. V., & Liu, G. S. (2015). Gene therapy with endogenous inhibitors of angiogenesis for neovascular age-related macular degeneration: Beyond anti-VEGF therapy. *Journal of Ophthalmology, 2015*, 201726.
71. Ma, L., Brelen, M. E., Tsujikawa, M., Chen, H., Chu, W. K., Lai, T. Y., et al. (2017). Identification of ANGPT2 as a new gene for neovascular age-related macular degeneration and polypoidal choroidal vasculopathy in the chinese and japanese populations ANGPT2 in AMD and PCV. *Investigative Ophthalmology & Visual Science, 58*(2), 1076–1083.
72. Thomas, C. E., Ehrhardt, A., & Kay, M. A. (2003). Progress and problems with the use of viral vectors for gene therapy. *Nature Reviews Genetics, 4*(5), 346–358.
73. Garoon, R. B., & Stout, J. T. (2016). Update on ocular gene therapy and advances in treatment of inherited retinal diseases and exudative macular degeneration. *Current Opinion in Ophthalmology, 27*(3), 268–273.
74. Haghjou, N., Soheilian, M., & Abdekhodaie, M. J. (2011). Sustained release intraocular drug delivery devices for treatment of uveitis. *Journal of Ophthalmic & Vision Research, 6*, 317–329.
75. Kane, F. E., Burdan, J., Cutino, A., & Green, K. E. (2008). Iluvien: A new sustained delivery technology for posterior eye disease. *Expert Opinion on Drug Delivery, 5*, 1039–1046.
76. Allergan Inc. (2014). *OZURDEX® (dexamethasone intravitreal implant) for intravitreal injection*. Dublin: Allergan Inc..
77. Fong, A. H., Lau, T. W., & Luk, F. O. (2016). Safety and efficacy of dexamethasone intravitreal implant injection for macular edema associated with retinal vein occlusion. *Hong Kong Journal of Ophthalmology, 20*(2), 78–83.
78. Bakri, S. J., & Omar, A. F. (2012). Evolution of vitreomacular traction following the use of the dexamethasone intravitreal implant (Ozurdex) in the treatment of macular edema secondary to central retinal vein occlusion. *Journal of Ocular Pharmacology and Therapeutics, 28*, 547–549.
79. Jaffe, G. J., Martin, D., Callanan, D., Pearson, P. A., Levy, B., Comstock, T., et al. (2006). Fluocinolone acetonide implant (Retisert) for noninfectious posterior uveitis: Thirty-four-week results of a multicenter randomized clinical study. *Ophthalmology, 113*, 1020–1027.
80. Sangwan, V. S., Pearson, P. A., Paul, H., & Comstock, T. L. (2015). Use of the fluocinolone acetonide intravitreal implant for the treatment of noninfectious posterior uveitis: 3-Year results of a randomized clinical trial in a predominantly Asian population. *Ophthalmology and Therapy, 4*, 1–19.
81. Zanin, M. P., Pettingill, L. N., Harvey, A. R., Emerich, D. F., Thanos, C. G., & Shepherd, R. K. (2012). The development of encapsulated cell technologies as therapies for neurological and sensory diseases. *Journal of Controlled Release, 160*, 3–13.
82. Tao, W., Wen, R., Goddard, M. B., Sherman, S. D., O'Rourke, P. J., Stabila, P. F., et al. (2002). Encapsulated cell-based delivery of CNTF reduces photoreceptor degeneration in

animal models of retinitis pigmentosa. *Investigative Ophthalmology & Visual Science, 43*, 3292–3298.
83. Thanos, C. G., Bell, W. J., O'Rourke, P., Kauper, K., Sherman, S., Stabila, P., et al. (2004). Sustained secretion of ciliary neurotrophic factor to the vitreous, using the encapsulated cell therapy-based NT-501 intraocular device. *Tissue Engineering, 10*, 1617–1622.
84. Kauper, K., McGovern, C., Sherman, S., Heatherton, P., Rapoza, R., Stabila, P., et al. (2012). Two-year intraocular delivery of ciliary neurotrophic factor by encapsulated cell technology implants in patients with chronic retinal degenerative diseases. *Investigative Ophthalmology & Visual Science, 53*, 7484–7491.
85. Wong, F. S., Tsang, K. K., & Lo, A. C. (2017). Delivery of therapeutics to posterior eye segment: Cell-encapsulating systems. *Neural Regeneration Research, 12*(4), 576.
86. Birch, D. G., Weleber, R. G., Duncan, J. L., Jaffe, G. J., Tao, W., & Ciliary Neurotrophic Factor Retinitis Pigmentosa Study Groups. (2013). Randomized trial of ciliary neurotrophic factor delivered by encapsulated cell intraocular implants for retinitis pigmentosa. *American Journal of Ophthalmology, 156*, 283–292.
87. Delplace, V., Payne, S., & Shoichet, M. (2015). Delivery strategies for treatment of age-related ocular diseases: From a biological understanding to biomaterial solutions. *Journal of Controlled Release, 219*, 652–668.
88. Kompella, U. B., Amrite, A. C., Ravi, R. P., & Durazo, S. A. (2013). Nanomedicines for back of the eye drug delivery, gene delivery, and imaging. *Progress in Retinal and Eye Research, 36*, 172–198.
89. Bochot, A., & Fattal, E. (2012). Liposomes for intravitreal drug delivery: A state of the art. *Journal of Controlled Release, 161*, 628–634.
90. Kang-Mieler, J. J., Osswald, C. R., & Mieler, W. F. (2014). Advances in ocular drug delivery: Emphasis on the posterior segment. *Expert Opinion on Drug Delivery, 11*, 1647–1660.
91. Araújo, J., Nikolic, S., Egea, M. A., Souto, E. B., & Garcia, M. L. (2011). Nanostructured lipid carriers for triamcinolone acetonide delivery to the posterior segment of the eye. *Colloids and Surfaces B: Biointerfaces, 88*, 150–157.
92. Araújo, J., Garcia, M. L., Mallandrich, M., Souto, E. B., & Calpena, A. C. (2012). Release profile and transscleral permeation of triamcinolone acetonide loaded nanostructured lipid carriers (TA-NLC): In vitro and ex vivo studies. *Nanomedicine, 8*, 1034–1041.
93. Park, K., Chen, Y., Hu, Y., Mayo, A. S., Kompella, U. B., Longeras, R., et al. (2009). Nanoparticle-mediated expression of an angiogenic inhibitor ameliorates ischemia-induced retinal neovascularization and diabetes-induced retinal vascular leakage. *Diabetes, 58*, 1902–1913.
94. Jin, J., Zhou, K. K., Park, K., Hu, Y., Xu, X., Zheng, Z., et al. (2011). Anti-inflammatory and antiangiogenic effects of nanoparticle-mediated delivery of a natural angiogenic inhibitor. *Investigative Ophthalmology & Visual Science, 52*, 6230–6237.
95. Liu, H. A., Liu, Y. L., Ma, Z. Z., Wang, J. C., & Zhang, Q. (2011). A lipid nanoparticle system improves siRNA efficacy in RPE cells and a laser-induced murine CNV model. *Investigative Ophthalmology & Visual Science, 52*, 4789–4794.
96. Pepić, I., Hafner, A., Lovrić, J., Pirkić, B., & Filipović-Grčić, J. (2010). A nonionic surfactant/chitosan micelle system in an innovative eye drop formulation. *Journal of Pharmaceutical Sciences, 99*, 4317–4325.
97. Marano, R. J., Toth, I., Wimmer, N., Brankov, M., & Rakoczy, P. E. (2005). Dendrimer delivery of an anti-VEGF oligonucleotide into the eye: A long-term study into inhibition of laser-induced CNV, distribution, uptake and toxicity. *Gene Therapy, 12*, 1544–1550.
98. Marano, R. J., Wimmer, N., Kearns, P. S., Thomas, B. G., Toth, I., Brankov, M., et al. (2004). Inhibition of in vitro VEGF expression and choroidal neovascularization by synthetic dendrimer peptide mediated delivery of a sense oligonucleotide. *Experimental Eye Research, 79*, 525–535.
99. Abrishami, M., Ganavati, S. Z., Soroush, D., Rouhbakhsh, M., Jaafari, M. R., & Malaekeh-Nikouei, B. (2009). Preparation, characterization, and in vivo evaluation of nanoliposomes-encapsulated bevacizumab (avastin) for intravitreal administration. *Retina, 29*, 699–703.

100. Panda, J. J., Yandrapu, S., Kadam, R. S., Chauhan, V. S., & Kompella, U. B. (2013). Self-assembled phenylalanine-alpha,beta-dehydrophenylalanine nanotubes for sustained intravitreal delivery of a multi-targeted tyrosine kinase inhibitor. *Journal of Controlled Release, 172*, 1151–1160.
101. Civiale, C., Licciardi, M., Cavallaro, G., Giammona, G., & Mazzone, M. G. (2009). Polyhydroxyethylaspartamide-based micelles for ocular drug delivery. *International Journal of Pharmaceutics, 378*, 177–186.
102. Rafie, F., Javadzadeh, Y., Javadzadeh, A. R., Ghavidel, L. A., Jafari, B., Moogooee, M., et al. (2010). In vivo evaluation of novel nanoparticles containing dexamethasone for ocular drug delivery on rabbit eye. *Current Eye Research, 35*, 1081–1089.
103. Tanito, M., Hara, K., Takai, Y., Matsuoka, Y., Nishimura, N., Jansook, P., et al. (2011). Topical dexamethasone-cyclodextrin microparticle eye drops for diabetic macular edema. *Investigative Ophthalmology & Visual Science, 52*, 7944–7948.
104. Ohira, A., Hara, K., Jóhannesson, G., Tanito, M., Ásgrímsdóttir, G. M., Lund, S. H., et al. (2015). Topical dexamethasone γ-cyclodextrin nanoparticle eye drops increase visual acuity and decrease macular thickness in diabetic macular oedema. *Acta Ophthalmologica, 93*, 610–615.
105. Krag, S., & Hessellund, A. (2014). Topical dexamethasone-cyclodextrin microparticle eye drops for uveitic macular oedema. *Acta Ophthalmologica, 92*, 689–690.
106. Shulman, S., Jóhannesson, G., Stefánsson, E., Loewenstein, A., Rosenblatt, A., & Habot-Wilner, Z. (2015). Topical dexamethasone-cyclodextrin nanoparticle eye drops for non-infectious Uveitic macular oedema and vitritis – a pilot study. *Acta Ophthalmologica, 93*, 411–415.
107. Shimazaki, H., Hironaka, K., Fujisawa, T., Tsuruma, K., Tozuka, Y., Shimazawa, M., et al. (2011). Edaravone-loaded liposome eyedrops protect against light-induced retinal damage in mice. *Investigative Ophthalmology & Visual Science, 52*, 7289–7297.
108. Chen, C. W., Lu, D. W., Yeh, M. K., Shiau, C. Y., & Chiang, C. H. (2011). Novel RGD-lipid conjugate-modified liposomes for enhancing siRNA delivery in human retinal pigment epithelial cells. *International Journal of Nanomedicine, 6*, 2567–2580.
109. Chen, C. W., Yeh, M. K., Shiau, C. Y., Chiang, C. H., & Lu, D. W. (2013). Efficient downregulation of VEGF in retinal pigment epithelial cells by integrin ligand-labeled liposome-mediated siRNA delivery. *International Journal of Nanomedicine, 8*, 2613–2627.
110. Honda, M., Asai, T., Umemoto, T., Araki, Y., Oku, N., & Tanaka, M. (2011). Suppression of choroidal neovascularization by intravitreal injection of liposomal SU5416. *Archives of Ophthalmology, 129*, 317–321.
111. Wang, J. L., Liu, Y. L., Li, Y., Dai, W. B., Guo, Z. M., Wang, Z. H., et al. (2012). EphA2 targeted doxorubicin stealth liposomes as a therapy system for choroidal neovascularization in rats. *Investigative Ophthalmology & Visual Science, 53*, 7348–7357.
112. Meisner, D., & Mezei, M. (1995). Liposome ocular delivery systems. *Advanced Drug Delivery Reviews, 16*, 75–93.
113. Allen, T. M., & Cullis, P. R. (2013). Liposomal drug delivery systems: From concept to clinical applications. *Advanced Drug Delivery Reviews, 65*, 36–48.
114. Verteporfin in Photodynamic Therapy Study Group. (2001). Verteporfin therapy of subfoveal choroidal neovascularization in age-related macular degeneration: Two-year results of a randomized clinical trial including lesions with occult with no classic choroidal neovascularization–verteporfin in photodynamic therapy report 2. *American Journal of Ophthalmology, 131*, 541–560.
115. Singh, S. R., Grossniklaus, H. E., Kang, S. J., Edelhauser, H. F., Ambati, B. K., & Kompella, U. B. (2009). Intravenous transferrin, RGD peptide and dual-targeted nanoparticles enhance anti-VEGF intraceptor gene delivery to laser-induced CNV. *Gene Therapy, 16*, 645–659.
116. Luo, L., Zhang, X., Hirano, Y., Tyagi, P., Barabás, P., Uehara, H., et al. (2013). Targeted intraceptor nanoparticle therapy reduces angiogenesis and fibrosis in primate and murine macular degeneration. *ACS Nano, 7*, 3264–3275.

117. Davis, B. M., Normando, E. M., Guo, L., Turner, L. A., Nizari, S., O'Shea, P., et al. (2014). Topical delivery of Avastin to the posterior segment of the eye in vivo using annexin A5-associated liposomes. *Small, 10*, 1575–1584.
118. Michael, I. P., Westenskow, P. D., Hacibekiroglu, S., Greenwald, A. C., Ballios, B. G., Kurihara, T., et al. (2014). Local acting sticky-trap inhibits VEGF dependent pathological angiogenesis in the eye. *EMBO Molecular Medicine, 6*, 604–623.
119. Turturro, S. B., Guthrie, M. J., Appel, A. A., Drapala, P. W., Brey, E. M., Pérez-Luna, V. H., et al. (2011). The effects of cross-linked thermo-responsive PNIPAAm-based hydrogel injection on retinal function. *Biomaterials, 32*, 3620–3626.
120. Horvát, G., Gyarmati, B., Berkó, S., Szabó-Révész, P., Szilágyi, B. Á., Szilágyi, A., et al. (2015). Thiolated poly(aspartic acid) as potential in situ gelling, ocular mucoadhesive drug delivery system. *European Journal of Pharmaceutical Sciences, 67*, 1–11.
121. Osswald, C. R., & Kang-Mieler, J. J. (2016). Controlled and extended in vitro release of bioactive anti-vascular endothelial growth factors from a microsphere-hydrogel drug delivery system. *Current Eye Research, 14*, 1–7.
122. Osswald, C. R., Guthrie, M. J., Avila, A., Valio Jr., J. A., Mieler, W. F., & Kang-Mieler, J. J. (2017). In vivo efficacy of an injectable microsphere-hydrogel ocular drug delivery system. *Current Eye Research, 42*, 1293–1301.
123. Glickman, R. D. (2011). Ultraviolet phototoxicity to the retina. *Eye & Contact Lens, 37*, 196–205.
124. Huu, V. A. N., Luo, J., Zhu, J., Zhu, J., Patel, S., Boone, A., et al. (2015). Light-responsive nanoparticle depot to control release of a small molecule angiogenesis inhibitor in the posterior segment of the eye. *Journal of Controlled Release, 200*, 71–77.
125. Jiang, M., Gan, L., Zhu, C., Dong, Y., Liu, J., & Gan, Y. (2012). Cationic core-shell liponanoparticles for ocular gene delivery. *Biomaterials, 33*, 7621–7630.
126. Puras, G., Mashal, M., Zárate, J., Agirre, M., Ojeda, E., Grijalvo, S., et al. (2014). A novel cationic niosome formulation for gene delivery to the retina. *Journal of Controlled Release, 174*, 27–36.
127. Omri, S., Behar-Cohen, F., de Kozak, Y., Sennlaub, F., Verissimo, L. M., Jonet, L., et al. (2011). Microglia/macrophages migrate through retinal epithelium barrier by a transcellular route in diabetic retinopathy: Role of PKCζ in the goto Kakizaki rat model. *The American Journal of Pathology, 179*, 942–953.
128. Santos, J. L., Li, Y., Culver, H. R., Michael, S. Y., & Herrera-Alonso, M. (2014). Conducting polymer nanoparticles decorated with collagen mimetic peptides for collagen targeting. *Chemical Communications, 50*, 15045–15048.
129. Liu, Y., Shi, M., Xu, M., Yang, H., & Wu, C. (2012). Multifunctional nanoparticles of Fe(3) O(4)•SiO(2)(FITC)/PAH conjugated the recombinant plasmid of pIRSE2-EGFP/ VEGF(165) with dual functions for gene delivery and cellular imaging. *Expert Opinion on Drug Delivery, 9*, 1197–1207.
130. Wahyudi, H., Reynolds, A. A., Li, Y., Owen, S. C., & Yu, S. M. (2016). Targeting collagen for diagnostic imaging and therapeutic delivery. *Journal of Controlled Release, 240*, 323–331.
131. Subrizi, A., Toropainen, E., Ramsay, E., Airaksinen, A. J., Kaarniranta, K., & Urtti, A. (2015). Oxidative stress protection by exogenous delivery of rhHsp70 chaperone to the retinal pigment epithelium (RPE), a possible therapeutic strategy against RPE degeneration. *Pharmaceutical Research, 32*, 211–221.
132. Mo, Y., Barnett, M. E., Takemoto, D., Davidson, H., & Kompella, U. B. (2007). Human serum albumin nanoparticles for efficient delivery of Cu, Zn superoxide dismutase gene. *Molecular Vision, 13*, 746–757.
133. Huang, D., Chen, Y. S., Thakur, S. S., & Rupenthal, I. D. (2017). Ultrasound-mediated nanoparticle delivery across ex vivo bovine retina after intravitreal injection. *European Journal of Pharmaceutics and Biopharmaceutics, 119*, 125–136.
134. Chen, J., Patil, S., Seal, S., & McGinnis, J. F. (2006). Rare earth nanoparticles prevent retinal degeneration induced by intracellular peroxides. *Nature Nanotechnology, 1*, 142–150.

135. Zhou, X., Wong, L. L., Karakoti, A. S., Seal, S., & McGinnis, J. F. (2011). Nanoceria inhibit the development and promote the regression of pathologic retinal neovascularization in the vldlr knockout mouse. *PLoS One, 6*(2), e16733.
136. Jayaraman, M. S., Bharali, D. J., Sudha, T., & Mousa, S. A. (2012). Nano chitosan peptide as a potential therapeutic carrier for retinal delivery to treat age-related macular degeneration. *Molecular Vision, 18*, 2300–2308.
137. Kim, J. H., Kim, M. H., Jo, D. H., Yu, Y. S., Lee, T. G., & Kim, J. H. (2011). The inhibition of retinal neovascularization by gold nanoparticles via suppression of VEGFR-2 activation. *Biomaterials, 32*, 1865–1871.
138. Kim, H. J., Zhang, K., Moore, L., & Ho, D. (2014). Diamond nanogel-embedded contact lenses mediate lysozyme-dependent therapeutic release. *ACS Nano, 8*, 2998–3005.
139. Ribeiro, A., Veiga, F., Santos, D., Torres-Labandeira, J. J., Concheiro, A., & Alvarez-Lorenzo, C. (2011). Bioinspired imprinted PHEMAhydrogels for ocular delivery of carbonic anhydrase inhibitor drugs. *Biomacromolecules, 12*(3), 701–709.
140. Maulvi, F. A., Lakdawala, D. H., Shaikh, A. A., Desai, A. R., Choksi, H. H., Vaidya, R. J., et al. (2016). In vitro and in vivo evaluation of novel implantation technology in hydrogel contact lenses for controlled drug delivery. *Journal of Controlled Release, 226*, 47–56.
141. Wade, J. S., & Desai, T. A. (2014). Planar microdevices enhance transport of large molecular weight molecules across retinal pigment epithelial cells. *Biomedical Microdevices, 16*, 629–638.
142. Patel, S. R., Lin, A. S., Edelhauser, H. F., & Prausnitz, M. R. (2011). Suprachoroidal drug delivery to the back of the eye using hollow microneedles. *Pharmaceutical Research, 28*, 166–176.
143. Patel, S. R., Berezovsky, D. E., BE, M. C., Zarnitsyn, V., Edelhauser, H. F., & Prausnitz, M. R. (2012). Targeted administration into the suprachoroidal space using a microneedle for drug delivery to the posterior segment of the eye. *Investigative Ophthalmology & Visual Science, 53*, 4433–4441.
144. Kim, Y. C., Edelhauser, H., & Prausnitz, M. R. (2014). Particle-stabilized emulsion droplets for gravity-mediated targeting in the posterior segment of the eye. *Advanced Healthcare Materials, 3*(8), 1272–1282.
145. Chatzipirpiridis, G., Ergeneman, O., Pokki, J., Ullrich, F., Fusco, S., Ortega, J. A., et al. (2015). Electroforming of implantable tubular magnetic microrobots for wireless ophthalmologic applications. *Advanced Healthcare Materials, 4*, 209–214.
146. Delplace, V., & Nicolas, J. (2015). Degradable vinyl polymers for biomedical applications. *Nature Chemistry, 7*, 771–784.
147. Atala, A., Lanza, R., Thomson, J. A., & Nerem, R. (Eds.). (2010). *Principles of regenerative medicine*. Cambridge, MA: Academic.
148. Balenci, L., & van der Kooy, D. (2014). Notch signaling induces retinal stem-like properties in perinatal neural retina progenitors and promotes symmetrical divisions in adult retinal stem cells. *Stem Cells and Development Stem Cells and Devlopment, 23*(3), 230–244.
149. Kinnunen, K., Petrovski, G., Moe, M. C., Berta, A., & Kaarniranta, K. (2012). Molecular mechanisms of retinal pigment epithelium damage and development of age-related macular degeneration. *Acta Ophthalmologica, 90*, 299–309.
150. Reh, T. A. (2016). Photoreceptor transplantation in late stage retinal degenerationphotoreceptor transplantation in late stage RD. *Investigative Ophthalmology & Visual Science, 57*(5), 1–7.
151. Singh, M. S., Balmer, J., Barnard, A. R., Aslam, S. A., Moralli, D., Green, C. M., et al. (2016). Transplanted photoreceptor precursors transfer proteins to host photoreceptors by a mechanism of cytoplasmic fusion. *Nature Communications, 7*, 13537.
152. Pearson, R. A., Barber, A. C., Rizzi, M., Hippert, C., Xue, T., West, E. L., et al. (2012). Restoration of vision after transplantation of photoreceptors. *Nature, 485*, 99–103.
153. Barber, A. C., Hippert, C., Duran, Y., West, E. L., Bainbridge, J. W., Warre-Cornish, K., et al. (2013). Repair of the degenerate retina by photoreceptor transplantation. *Proceedings of the National Academy of Sciences of the United States of America, 110*, 354–359.

154. Lu, B., Malcuit, C., Wang, S., Girman, S., Francis, P., Lemieux, L., et al. (2009). Long-term safety and function of RPE from human embryonic stem cells in preclinical models of macular degeneration. *Stem Cells, 27*, 2126–2135.
155. Kamao, H., Mandai, M., Okamoto, S., Sakai, N., Suga, A., Sugita, S., et al. (2014). Characterization of human induced pluripotent stem cell-derived retinal pigment epithelium cell sheets aiming for clinical application. *Stem Cell Reports, 2*, 205–218.
156. Mead, B., Logan, A., Berry, M., Leadbeater, W., & Scheven, B. A. (2013). Intravitreally transplanted dental pulp stem cells promote neuroprotection and axon regeneration of retinal ganglion cells after optic nerve injury. *Investigative Ophthalmology & Visual Science, 54*, 7544–7556.
157. Mathivanan, I., Trepp, C., Brunold, C., Baerlocher, G., & Enzmann, V. (2015). Retinal differentiation of human bone marrow-derived stem cells by co-culture with retinal pigment epithelium in vitro. *Experimental Cell Research, 333*, 11–20.
158. Pearson, R. A. (2014). Advances in repairing the degenerate retina by rod photoreceptor transplantation. *Biotechnology Advances, 32*, 485–491.
159. Tomita, M., Lavik, E., Klassen, H., Zahir, T., Langer, R., & Young, M. J. (2005). Biodegradable polymer composite grafts promote the survival and differentiation of retinal progenitor cells. *Stem Cells, 23*, 1579–1588.
160. Yao, J., Tucker, B. A., Zhang, X., Checa-Casalengua, P., Herrero-Vanrell, R., & Young, M. J. (2011). Robust cell integration from co-transplantation of biodegradable MMP2-PLGA microspheres with retinal progenitor cells. *Biomaterials, 32*, 1041–1050.
161. Ballios, B. G., Cooke, M. J., Donaldson, L., Coles, B. L., Morshead, C. M., van der Kooy, D., et al. (2015). A hyaluronan-based injectable hydrogel improves the survival and integration of stem cell progeny following transplantation. *Stem Cell Reports, 4*, 1031–1045.
162. Klassen, H. J., Ng, T. F., Kurimoto, Y., Kirov, I., Shatos, M., Coffey, P., et al. (2004). Multipotent retinal progenitors express developmental markers, differentiate into retinal neurons, and preserve light-mediated behavior. *Investigative Ophthalmology & Visual Science, 45*, 4167–4173.
163. MacLaren, R. E., Pearson, R. A., MacNeil, A., Douglas, R. H., Salt, T. E., Akimoto, M., et al. (2006). Retinal repair by transplantation of photoreceptor precursors. *Nature, 444*, 203–207.
164. Taddei, M. L., Giannoni, E., Fiaschi, T., & Chiarugi, P. (2012). Anoikis: An emerging hallmark in health and diseases. *The Journal of Pathology, 226*, 380–393.
165. Seiler, M. J., & Aramant, R. B. (2012). Cell replacement and visual restoration by retinal sheet transplants. *Progress in Retinal and Eye Research, 31*, 661–687.
166. Diniz, B., Thomas, P., Thomas, B., Ribeiro, R., Hu, Y., Brant, R., et al. (2013). Subretinal implantation of retinal pigment epithelial cells derived from human embryonic stem cells: Improved survival when implanted as a monolayer. *Investigative Ophthalmology & Visual Science, 54*, 5087–5096.
167. Ghosh, F., Rauer, O., & Arnér, K. (2008). Neuroretinal xenotransplantation to immunocompetent hosts in a discordant species combination. *Neuroscience, 152*, 526–533.
168. Gullapalli, V. K., Sugino, I. K., Van Patten, Y., Shah, S., & Zarbin, M. A. (2005). Impaired RPE survival on aged submacular human Bruch's membrane. *Experimental Eye Research, 80*, 235–248.
169. West, E. L., Pearson, R. A., Barker, S. E., Luhmann, U. F., Maclaren, R. E., Barber, A. C., et al. (2010). Long-term survival of photoreceptors transplanted into the adult murine neural retina requires immune modulation. *Stem Cells, 28*, 1997–2007.
170. Smiley, S., Nickerson, P. E., Comanita, L., Daftarian, N., El-Sehemy, A., Tsai, E. L. S., et al. (2016). Establishment of a cone photoreceptor transplantation platform based on a novel cone-GFP reporter mouse line. *Scientific Reports, 6*, 22867.
171. Ortin-Martinez, A., et al. (2017). A reinterpretation of cell transplantation: GFP transfer from donor to host photoreceptors. *Stem Cells, 35*(4), 932–939.
172. Pearson, R. A., et al. (2016). Donor and host photoreceptors engage in material transfer following transplantation of post-mitotic photoreceptor precursors. *Nature Communications, 7*, 13029.

173. Yao, J., Tao, S. L., & Young, M. J. (2011). Synthetic polymer scaffolds for stem cell transplantation in retinal tissue engineering. *Polymer, 3*, 899.
174. Inatani, M., Tanihara, H., Oohira, A., Honjo, M., Kido, N., & Honda, Y. (2000). Upregulated expression of neurocan, a nervous tissue specific proteoglycan, in transient retinal ischemia. *Investigative Ophthalmology & Visual Science, 41*, 2748–2754.
175. Krishnamoorthy, R., Agarwal, N., & Chaitin, M. H. (2000). Upregulation of CD44 expression in the retina during the rds degeneration. *Molecular Brain Research, 77*, 125–130.
176. Kim, H., Cooke, M. J., & Shoichet, M. S. (2012). Creating permissive microenvironments for stem cell transplantation into the central nervous system. *Trends in Biotechnology, 30*, 55–63.
177. Hugar, D. L., & Ivanisevic, A. (2013). Materials characterization and mechanobiology of the eye. *Materials Science and Engineering: C, 33*(4), 1867–1875.
178. Ishikawa, M., Sawada, Y., & Yoshitomi, T. (2015). Structure and function of the interphotoreceptor matrix surrounding retinal photoreceptor cells. *Experimental Eye Research, 133*, 3–18.
179. Hynes, S. R., & Lavik, E. B. (2010). A tissue-engineered approach towards retinal repair: Scaffolds for cell transplantation to the subretinal space. *Graefe's Archive for Clinical and Experimental Ophthalmology, 248*, 763–778.
180. Liu, Z., Yu, N., Holz, F. G., Yang, F., & Stanzel, B. V. (2014). Enhancement of retinal pigment epithelial culture characteristics and subretinal space tolerance of scaffolds with 200 nm fiber topography. *Biomaterials, 35*, 2837–2850.
181. Hu, Y., Liu, L., Lu, B., Zhu, D., Ribeiro, R., Diniz, B., et al. (2012). A novel approach for subretinal implantation of ultrathin substrates containing stem cell-derived retinal pigment epithelium monolayer. *Ophthalmic Research, 48*, 186–191.
182. Redenti, S., Tao, S., Yang, J., Gu, P., Klassen, H., Saigal, S., et al. (2008). Retinal tissue engineering using mouse retinal progenitor cells and a novel biodegradable, thin-film poly(e-caprolactone) nanowire scaffold. *Journal of Ocular Biology, Diseases, and Informatics, 1*, 19–29.
183. Yao, J., Ko, C. W., Baranov, P. Y., Regatieri, C. V., Redenti, S., Tucker, B. A., et al. (2015). Enhanced differentiation and delivery of mouse retinal progenitor cells using a micropatterned biodegradable thin-film polycaprolactone scaffold. *Tissue Engineering Part A, 21*, 1247–1260.
184. Redenti, S., Neeley, W. L., Rompani, S., Saigal, S., Yang, J., Klassen, H., et al. (2009). Engineering retinal progenitor cell and scrollable poly(glycerol-sebacate) composites for expansion and subretinal transplantation. *Biomaterials, 30*, 3405–3414.
185. Liu, Y., Wang, R., Zarembinski, T. I., Doty, N., Jiang, C., Regatieri, C., et al. (2013). The application of hyaluronic acid hydrogels to retinal progenitor cell transplantation. *Tissue Engineering Part A, 19*, 135–142.
186. Warnke, P. H., Alamein, M., Skabo, S., Stephens, S., Bourke, R., Heiner, P., et al. (2013). Primordium of an artificial Bruch's membrane made of nanofibers for engineering of retinal pigment epithelium cell monolayers. *Acta Biomaterialia, 9*, 9414–9422.
187. Okubo, A., Rosa, R. H., Bunce, C. V., Alexander, R. A., Fan, J. T., Bird, A. C., et al. (1999). The relationships of age changes in retinal pigment epithelium and Bruch's membrane. *Investigative Ophthalmology & Visual Science, 40*, 443–449.
188. Cai, H., & Del Priore, L. V. (2006). Bruch membrane aging alters the gene expression profile of human retinal pigment epithelium. *Current Eye Research, 31*, 181–189.
189. Sun, K., Cai, H., Tezel, T. H., Paik, D., Gaillard, E. R., & Del Priore, L. V. (2007). Bruch's membrane aging decreases phagocytosis of outer segments by retinal pigment epithelium. *Molecular Vision, 13*, 2310–2319.
190. McHugh, K. J., Tao, S. L., & Saint-Geniez, M. (2014). Porous poly(ε-caprolactone) scaffolds for retinal pigment epithelium transplantation. *Investigative Ophthalmology & Visual Science, 55*, 1754–1762.
191. Subrizi, A., Hiidenmaa, H., Ilmarinen, T., Nymark, S., Dubruel, P., Uusitalo, H., et al. (2012). Generation of hESC-derived retinal pigment epithelium on biopolymer coated polyimide membranes. *Biomaterials, 33*, 8047–8054.

192. Lu, J. T., Lee, C. J., Bent, S. F., Fishman, H. A., & Sabelman, E. E. (2007). Thin collagen film scaffolds for retinal epithelial cell culture. *Biomaterials, 28*, 1486–1494.
193. Sorkio, A. E., Vuorimaa-Laukkanen, E. P., Hakola, H. M., Liang, H., Ujula, T. A., Valle-Delgado, J. J., et al. (2015). Biomimetic collagen I and IV double layer Langmuir–Schaefer films as microenvironment for human pluripotent stem cell derived retinal pigment epithelial cells. *Biomaterials, 51*, 257–269.
194. Thumann, G., Viethen, A., Gaebler, A., Walter, P., Kaempf, S., Johnen, S., et al. (2009). The in vitro and in vivo behaviour of retinal pigment epithelial cells cultured on ultra-thin collagen membranes. *Biomaterials, 30*, 287–294.
195. Xiang, P., Wu, K. C., Zhu, Y., Xiang, L., Li, C., Chen, D. L., et al. (2014). A novel Bruch's membrane-mimetic electrospun substrate scaffold for human retinal pigment epithelium cells. *Biomaterials, 35*, 9777–9788.
196. Thieltges, F., Stanzel, B. V., Liu, Z., & Holz, F. G. (2011). A nanofibrillar surface promotes superior growth characteristics in cultured human retinal pigment epithelium. *Ophthalmic Research, 46*, 133–140.
197. Shadforth, A. M., George, K. A., Kwan, A. S., Chirila, T. V., & Harkin, D. G. (2012). The cultivation of human retinal pigment epithelial cells on Bombyx mori silk fibroin. *Biomaterials, 33*, 4110–4117.
198. Treharne, A. J., Thomson, H. A., Grossel, M. C., & Lotery, A. J. (2012). Developing methacrylate-based copolymers as an artificial Bruch's membrane substitute. *Journal of Biomedical Materials Research Part A, 100A*, 2358–2364.
199. Tao, S., Young, C., Redenti, S., Zhang, Y., Klassen, H., Desai, T., et al. (2007). Survival, migration and differentiation of retinal progenitor cells transplanted on micro-machined poly(methyl methacrylate) scaffolds to the subretinal space. *Lab on a Chip, 7*, 695–701.
200. Eberle, D., Kurth, T., Santos-Ferreira, T., Wilson, J., Corbeil, D., & Ader, M. (2012). Outer segment formation of transplanted photoreceptor precursor cells. *PLoS One, 7*, 28.
201. Lavik, E. B., Klassen, H., Warfvinge, K., Langer, R., & Young, M. J. (2005). Fabrication of degradable polymer scaffolds to direct the integration and differentiation of retinal progenitors. *Biomaterials, 26*, 3187–3196.
202. Warfvinge, K., Kiilgaard, J. F., Lavik, E. B., Scherfig, E., Langer, R., Klassen, H. J., et al. (2005). Retinal progenitor cell xenografts to the pig retina: Morphologic integration and cytochemical differentiation. *Archives of Ophthalmology, 123*, 1385–1393.
203. Albertsson, A. C., & Varma, I. K. (2003). Recent developments in ring opening polymerization of lactones for biomedical applications. *Biomacromolecules, 4*, 1466–1486.
204. Gupta, D., Tator, C. H., & Shoichet, M. S. (2006). Fast-gelling injectable blend of hyaluronan and methylcellulose for intrathecal, localized delivery to the injured spinal cord. *Biomaterials, 27*, 2370–2379.
205. Baumann, M. D., Kang, C. E., Tator, C. H., & Shoichet, M. S. (2010). Intrathecal delivery of a polymeric nanocomposite hydrogel after spinal cord injury. *Biomaterials, 31*, 7631–7639.
206. Mitrousis, N., Tam, R. Y., Baker, A. E., van der Kooy, D., & Shoichet, M. S. (2016). Hyaluronic acid-based hydrogels enable rod photoreceptor survival and maturation in vitro through activation of the mTOR pathway. *Advanced Functional Materials, 26*(12), 1975–1985.
207. Ahmed, T. A., Ringuette, R., Wallace, V. A., & Griffith, M. (2015). Autologous fibrin glue as an encapsulating scaffold for delivery of retinal progenitor cells. *Frontiers in Bioengineering and Biotechnology, 2*, 85.
208. Kador, K. E., Montero, R. B., Venugopalan, P., Hertz, J., Zindell, A. N., Valenzuela, D. A., et al. (2013). Tissue engineering the retinal ganglion cell nerve fiber layer. *Biomaterials, 34*, 4242–4250.
209. Kador, K. E., Alsehli, H. S., Zindell, A. N., Lau, L. W., Andreopoulos, F. M., Watson, B. D., et al. (2014). Retinal ganglion cell polarization using immobilized guidance cues on a tissue-engineered scaffold. *Acta Biomaterialia, 10*, 4939–4946.
210. Venugopalan, P., Wang, Y., Nguyen, T., Huang, A., Muller, K. J., & Goldberg, J. L. (2016). Transplanted neurons integrate into adult retinas and respond to light. *Nature Communications, 7*, 10472.

211. Joussen, A. (2004). Cell transplantation in age related macular degeneration: Current concepts and future hopes. *Graefe's Archive for Clinical and Experimental Ophthalmology, 242*(1), 1–2.
212. Vulic, K., Pakulska, M. M., Sonthalia, R., Ramachandran, A., & Shoichet, M. S. (2015). Mathematical model accurately predicts protein release from an affinity-based delivery system. *Journal of Controlled Release, 197*, 69–77.
213. Pakulska, M. M., Vulic, K., Tam, R. Y., & Shoichet, M. S. (2015). Hybrid crosslinked methylcellulose hydrogel: A predictable and tunable platform for local drug delivery. *Advanced Materials, 27*, 5002–5008.
214. Pakulska, M. M., Vulic, K., & Shoichet, M. S. (2013). Affinity-based release of chondroitinase ABC from a modified methylcellulose hydrogel. *Journal of Controlled Release, 171*(1), 11–16.
215. Vulic, K., & Shoichet, M. S. (2011). Tunable growth factor delivery from injectable hydrogels for tissue engineering. *Journal of the American Chemical Society, 134*(2), 882–885.
216. Parker, J., Mitrousis, N., & Shoichet, M. S. (2016). Hydrogel for simultaneous tunable growth factor delivery and enhanced viability of encapsulated cells in vitro. *Biomacromolecules, 17*(2), 476–484.
217. Peer, D., Karp, J. M., Hong, S., Farokhzad, O. C., Margalit, R., & Langer, R. (2007). Nanocarriers as an emerging platform for cancer therapy. *Nature Nanotechnology, 2*, 751–760.
218. Brigger, I., Dubernet, C., & Couvreur, P. (2002). Nanoparticles in cancer therapy and diagnosis. *Advanced Drug Delivery Reviews, 54*, 631–651.
219. Godse, R., Singh, K., Shrivastava, A., & Shinde, U. (2016). Polymeric nanoparticulate systems: A potential approach for ocular drug delivery. In Y. Pathak, V. Sutariya, & A. A. Hirani (Eds.), *Nano-biomaterials for ophthalmic drug delivery* (pp. 351–387). New York, NY: Springer.
220. Gross, N., Ranjbar, M., Evers, C., Hua, J., Martin, G., Schulze, B., et al. (2013). Choroidal neovascularization reduced by targeted drug delivery with cationic liposome-encapsulated paclitaxel or targeted photodynamic therapy with verteporfin encapsulated in cationic liposomes. *Molecular Vision, 19*, 54–61.
221. Paciotti, G. F., Zhao, J., Cao, S., Brodie, P. J., Tamarkin, L., Huhta, M., et al. (2016). Synthesis and evaluation of paclitaxel-loaded gold nanoparticles for tumor-targeted drug delivery. *Bioconjugate Chemistry, 27*(11), 2646–2657.
222. Ukawa, M., Ando, H., Shimizu, T., & Ishida, T. (2016). Pharmaceutics of nanoparticles. *Nanomaterials in Pharmacology, 2016*, 219–238.
223. Liu, S., Chang, C. N., Verma, M. S., Hileeto, D., Muntz, A., Stahl, U., et al. (2015). Phenylboronic acid modified mucoadhesive nanoparticle drug carriers facilitate weekly treatment of experimentallyinduced dry eye syndrome. *Nano Research, 8*, 621–635.
224. Eljarrat-Binstock, E., & Domb, A. J. (2006). Iontophoresis: A non-invasive ocular drug delivery. *Journal of Controlled Release, 110*, 479–489.
225. Eljarrat-Binstock, E., Orucov, F., Aldouby, Y., Frucht-Pery, J., & Domb, A. J. (2008). Charged nanoparticles delivery to the eye using hydrogel iontophoresis. *Journal of Controlled Release, 126*, 156–161.
226. Huang, D., Wang, L., Dong, Y., Pan, X., Li, G., & Wu, C. (2014). A novel technology using transscleral ultrasound to deliver protein loaded nanoparticles. *European Journal of Pharmaceutics and Biopharmaceutics, 88*, 104–115.
227. Park, J., Zhang, Y., Vykhodtseva, N., Akula, J. D., & McDannold, N. J. (2012). Targeted and reversible blood-retinal barrier disruption via focused ultrasound and microbubbles. *PLoS One, 7*, 13.
228. Gooch, N., Burr, R. M., Holt, D. J., Gale, B., & Ambati, B. (2013). Design and in vitro biocompatibility of a novel ocular drug delivery device. *Journal of Functional Biomaterials, 4*, 14–26.
229. Rubio, R. G. (2014). Long-acting anti-VEGF delivery. *Retina Today, 2014*, 78–80.

230. van Zeeburg, E. J., Maaijwee, K. J., Missotten, T. O., Heimann, H., & van Meurs, J. C. (2012). A free retinal pigment epithelium-choroid graft in patients with exudative age-related macular degeneration: Results up to 7 years. *American Journal of Ophthalmology, 153*, 120–127.
231. Nicolas, J., Mura, S., Brambilla, D., Mackiewicz, N., & Couvreur, P. (2013). Design, functionalization strategies and biomedical applications of targeted biodegradable/biocompatible polymer-based nanocarriers for drug delivery. *Chemical Society Reviews, 42*, 1147–1235.
232. Mura, S., Nicolas, J., & Couvreur, P. (2013). Stimuli-responsive nanocarriers for drug delivery. *Nature Materials, 12*, 991–1003.
233. Ma, J., Kabiel, M., Tucker, B. A., Ge, J., & Young, M. J. (2011). Combining chondroitinase ABC and growth factors promotes the integration of murine retinal progenitor cells transplanted into Rho(−/−) mice. *Molecular Vision, 17*, 1759–1770.
234. Ye, M., Kim, S., & Park, K. (2010). Issues in long-term protein delivery using biodegradable microparticles. *Journal of Controlled Release, 146*(2), 241–260.
235. Sakiyama-Elbert, S. E., & Hubbell, J. A. (2000). Development of fibrin derivatives for controlled release of heparin-binding growth factors. *Journal of Controlled Release, 65*(3), 389–402.
236. Delplace, V., Obermeyer, J., & Shoichet, M. S. (2016). Local affinity release. *ACS Nano, 10*(7), 6433–6436.
237. Wang, N. X., Sieg, S. F., Lederman, M. M., Offord, R. E., Hartley, O., & Von Recum, H. A. (2013). Using glycosaminoglycan/chemokine interactions for the long-term delivery of 5P12-RANTES in HIV prevention. *Molecular Pharmaceutics, 10*(10), 3564–3573.

Chapter 4
Cell and Animal Models used for Retinal Stem Cell Research

Michael J. Young and Jea Young Park

Abstract There are at least 100 million photoreceptors in the human retina and many more cells relaying visual information and providing metabolic support. The pathophysiology of degenerative diseases of the retina such as age-related macular degeneration (AMD), retinitis pigmentosa (RP), and glaucoma involves progressive loss of cell types necessary for vision (photoreceptors, ganglion cells) as well as inadequate trophic support from neighbor cell types (retinal pigmented epithelial cells (RPE), endothelial cells). A cell delivery approach has promise for restoring vision in patients who otherwise have a poor prognosis. Recent strides in the field allowed retinal cell differentiation from a multitude of sources, including but not limited to embryonic stem cells (ESC), adult tissue-derived stem cells (SC), induced pluripotent stem cells (iPSC) [1], retinal progenitor cells (RPC), and Muller cells [2], with excitement for gene-edited autologous stem cell therapy in the near future [3]. Mature RPE cells, photoreceptors, and ganglion cells have also been used in both animals and humans with very promising results. This chapter will provide a brief overview on the types of cells used in retinal cell therapy research and a summary of recent accomplishments.

Keywords Animal models · Retinal degeneration · Inherited retinal dystrophies · Cell therapy · Retinal stem cells

M. J. Young (✉)
Schepens Eye Research Institute, Harvard Medical School, Boston, MA, USA
e-mail: michael_young@meei.harvard.edu

J. Y. Park (✉)
Department of Ophthalmology, Yale School of Medicine, New Haven, CT, USA

B. G. Ballios, M. J. Young (eds.), *Regenerative Medicine and Stem Cell Therapy for the Eye*, Fundamental Biomedical Technologies,
https://doi.org/10.1007/978-3-319-98080-5_4

4.1 Overview of Retinal Cells for Transplant in Animal Models

There are at least 100 million photoreceptors in the human retina and many more cells relaying visual information and providing metabolic support. The pathophysiology of degenerative diseases of the retina such as age-related macular degeneration (AMD), retinitis pigmentosa (RP), and glaucoma involves progressive loss of cell types necessary for vision (photoreceptors, ganglion cells) as well as inadequate trophic support from neighbor cell types (retinal pigmented epithelial cells (RPE), endothelial cells). A cell delivery approach has promise for restoring vision in patients who otherwise have a poor prognosis. Recent strides in the field allowed retinal cell differentiation from a multitude of sources, including but not limited to embryonic stem cells (ESC), adult tissue-derived stem cells (SC), induced pluripotent stem cells (iPSC) [1], retinal progenitor cells (RPC), and Muller cells [2], with excitement for gene-edited autologous stem cell therapy in the near future [3]. Mature RPE cells, photoreceptors, and ganglion cells have also been used in both animals and humans with very promising results. This chapter will provide a brief overview on the types of cells used in retinal cell therapy research and a summary of recent accomplishments.

4.1.1 RPE: A Critical Regulator of Retina Function

One of the earliest work with retinal cell transplantation on animal models was done by Li and Turner [4] with exogenous RPE rat cells. Cell showed survival of graft of up to 3 months post-transplant. The RPE is a crucial supporter of normal photoreceptor function by forming the blood-retinal barrier and phagocytosing photoreceptor debris, and plays a central role in the pathogenesis of both dry and wet forms of AMD. Recent studies have proven that grafted RPE layers derived from human ESCs [5] and iPSCs [6] are able to perform such critical functions in model animals. RPE is also the earliest of retinal cell types to be involved in clinical trials leading to promising results with improved visual acuity and comparable results to anti-VEGF therapy, as well as evidence of safety [7].

4.1.2 Photoreceptor Cells and Their Precursors

Neural cells derived from the retinal progenitor cell line such as photoreceptors and ganglion cells are directly responsible for vision, culminating in the optic nerve. Given that damage to photoreceptors is the direct cause of decreased vision in most degenerative retinal disorders regardless of underlying etiology, more recent efforts have focused on creating suspensions or grafts of photoreceptor cells for transplantation. While results from early clinical trials with mature rod photoreceptors left much to be desired [8], it was soon discovered that precursor rod photoreceptors have been shown to have vastly superior integration into the outer nuclear layer (ONL) to their mature counterparts, successfully differentiate into mature receptors [9], and form triad synaptic connections with bipolar/horizontal cells [10], with positive visual improvements in animal models [11]. In contrast to rods, one of the biggest challenges of cone transplantation is procuring a pure and large enough sample of cones for transplantation. Some of these issues have been mitigated by cone-enriched mouse lines [12], and fluorescence [13] or magnetism-associated cell sorting [14], with very promising results.

4.1.3 Retinal Ganglion Cells

Retinal ganglion cells (RGC) form the innermost layer of the retina as well as the optic nerve and send signals into brain areas such as the lateral geniculate nucleus (LGN) and superior colliculus (SC) for visual processing. These cells are not only implicated in late stages of degenerative diseases but also are the first to be damaged in acute and chronic forms of glaucoma. Previously thought as difficult targets for cell transplantation given complex structure and connections with other cells, the last few years of research have been characterized by important breakthroughs. Intravitreal transplant of RPCs [15] or RGC precursors [16], into mouse models with RGC depletion leads to partial restoration of RGC function despite modest efficiency of invasion around 10%, and a suboptimal axonal growth into the inner nuclear layer (INL). A later study with polymer-based scaffold designed to have radial extensions to resemble the optic nerve head showed that axonal growth can be promoted in vivo [17], which can be combined with embedded neurotrophic factors [18] or 3d-printing [19]. Most recently, mature RGCs from GFP mice have been successfully transplanted intravitreally into uninjured mature rats, which fused to the optic nerve layer with extensions to the LGN and SC in one subject [20].

4.2 Differences in Ocular Anatomy Among Animal Species

Retinal cell transplantation studies have been conducted in a wide variety of model animal species worldwide. The most common animal models are that of rodents and nonhuman primates, as well as larger mammals such as cats, dogs, sheep, and cow who comprise the non-primate, non-rodent mammal group. Zebrafish models [21, 22] and avian models [23] are rather rare in cell transplantation research, and will not be the main focus in subsequent chapters.

As it is well known, key ocular anatomic differences exist between rodents and primates. First, rodent retina is largely devoid of cone photoreceptors as well as a macula, which brings to question whether rodent models of diseases such as age-related macular degeneration and Stargardt's disease are truly models of macular damage. Primates, who have a fovea, or certain mammals with a high concentration of cones such as canines and pigs, may suit better as models of macular disorders [24]. Second, rodents lack a lamina cribosa, which is critically implicated in development of POAG [25]. Surgical and transgenic methods have still shown to be able to create rodent models with a POAG phenotype, but whether subsequent pressure-induced retinal damage is equivalent to that of humans can be of question. Lastly, blood supply to the retina differs greatly between rodents and other species [26]. For example, in primates, the retina receives a secondary main blood supply via the posterior ciliary arteries (the primary supply being the central retinal artery) [27]; however, in rodents, the central retinal artery is one of the branches of the posterior ciliary artery itself [28]. Therefore, an insult to the retinal artery will cause a larger defect in the retina compared to primates.

Interestingly, there is not always a correlation between time and severity of ocular symptoms versus the size of animals, despite rodents experiencing a much shorter life span with rapid development compared to other mammals. The various different methods of model creation as well as underlying disease condition are therefore important considerations for choosing model animal species.

4.3 Overview of Methods for Animal Model Creation

The first in vivo step in retinal stem cell research is transplantation of stem cells into animal models. Different options of model creation provide researchers a careful trade-off between faithful emulation of pathogenesis and quick generation of results. Current methods of model creation can be roughly categorized to transgenic/genetic and non-transgenic methods.

4.3.1 Non-transgenic Methods

Non-transgenic methods of animal model creation can be divided into surgical and chemical methods. A few examples of surgical methods are laser-induced choroidal neovascularization (CNV) [29], optic nerve crush [30], and episcleral vein occlusion for glaucoma [31]. Many of these surgical methods have been verified in a variety of animal species. However, results can vary depending on the skill of the surgeon, and smaller animals such as rodents demand greater surgical finesse while consuming more time to create.

Examples of chemical methods include intravenous sodium iodate injection for RPE and photoreceptor injury [32], intravitreal NMDA injection for RGC ablation [33], subretinal matrigel injection for experimental CNV [34], and steroid-induced intraocular elevation [35]. Ischemic methods such as IOP-induced ischemia/reperfusion for optic nerve damage [36], and autoimmune models such as experimental autoimmune encephalomyelitis (EAE) [37] can also be considered special subtypes of chemical methods. Chemical methods are both easy to perform and can be used in various animal species. In addition, many of these chemical agents cause damage to specific types of retinal cells. However, one should be aware of any systemic side effects when using potentially dangerous chemicals.

4.3.2 *Transgenic/Genetic Methods*

Transgenic/genetic methods include identification of phenotype in an established inbred strain, or genetic engineering of animals via pronuclear injection [38], somatic cell nuclear transfer [39], Cre/lox [40], or adenovirus-associated viral vectors (AAV) [41]. Quintessential inbred strains for retinal research include Royal College of Science (RCS) rats [42] and *rd1* mice [43] for photoreceptor degeneration, as well as DBA/2J mice for glaucoma [44]. These inbred strains are commonly used due to low cost and high availability, but often have various symptoms outside the retina, which can be undesirable. In contrast, genetically engineered animals often show very specific ocular pathology that resembles its human counterpart. The downsides of transgenic animal models include ocular anatomic differences between model species and humans, as well as shorter lifespan and higher cost compared to inbred strains.

4.4 Models of Outer Retina Diseases

In this section, focus will be on animal models showing abnormal retinal angiogenesis or damage to the retinal pigment epithelium (RPE), which are two main pathological findings commonly seen in outer retinal diseases such as age-related macular degeneration (AMD) and diabetic retinopathy (DR), two of the most common causes of blindness in the developed world (Table 4.1). Models of retinitis pigmentosa (RP), whose pathogenesis depends on RPE as well as rod photoreceptor dysfunction, will be discussed in Sect. 4.5.

4.4.1 *Models of Abnormal Angiogenesis*

Pathological angiogenesis in the retina is associated with diverse retinal conditions, such as wet-type AMD, DR, retinopathy of prematurity (ROP), von Hippel-Lindau disease (vHL), and many more. In these diseases, the abnormal vasculature causes dysfunctional nutrient delivery as well as exudative fluid leakage, which results in damage to a variety of neuronal and non-neuronal retinal cells. Various animal models have greatly advanced our knowledge of these diseases [45]. Transplantation research has repeatedly shown possible visual improvement in animal models that represent these diseases.

Table 4.1 Common animal models of outer retinal diseases for stem cell research

Method	Commonly used species	Category	Characteristics	Previous stem cell research
Models of abnormal angiogenesis				
Laser-induced CNV model	Cat [46] Monkey [29] Rat [47] Mouse [48]	Surgical	Acute damage to Bruch's membrane, current standard of CNV model	RPE, [55] MSC [56]
Subretinal matrigel injection	Mouse [34] Rabbit [59] Rat [60]	Chemical	Gradual and prolonged development of CNV	N/A
Adenovirus-delivered VEGF expression	Rats [41]	Transgenic	Gradual and prolonged development of CNV	N/A
VLDR germline knockout	Mouse [90]	Transgenic	Model of mixed dry- and wet-AMD forms with dense angiogenesis	N/A
Oxygen-induced retinopathy model	Rat [67] Mouse [68] Dog [69] Monkey [70] Zebrafish [71]	Ischemic	Easy to produce; neovascularization develops in few days-weeks	N/A
Streptozocin-induced pancreatic ablation	Mouse [72, 113] Pig [76] Zebrafish [77]	Chemical	Model of Type I diabetes; not all animals develop retinopathy. Shows long-term disease process	ESC-derived perivascular stem cell [114] BMSC [115]
Sub-RPE accumulation models				
Carboxyethylpyrrole-modified mouse albumin model	Mouse [112]	Autoimmune	Induced autoimmunity to oxidized RPE protein	N/A
Complement factor H point mutation	Mouse [38]	Transgenic	Slow development of drusen-like deposits	N/A
Human APOEe4 substitution	Mouse [87]	Transgenic	Copious drusen as well as neovascularization	N/A

4.4.1.1 Laser-Induced Choroidal Neovascularization (CNV) Models

CNV is a severe and common complication of many different retinopathies. Development of CNV typically starts from a break in Bruch's membrane, leading to detachment and atrophy of RPE, neovascularization, and eventual degradation of vision. As early as the 1960s [46] it has been known that laser-induced damage in retinal vessels leads to Bruch's membrane breaks in model animals, as it does in humans, and has since been characterized in various animals [29, 47, 48]. Typically, the eye of the animal is dilated, and then argon or krypton laser is applied through a cover glass acting as a contact lens. The laser is typically focused on several points near the posterior pole of the globe, more than 1–2 optic disc diameters from the disc margin. The retina is then examined for the formation of a central bubble in the laser site, which indicates Bruch's membrane separation. Neovascularization typically occurs 1–2 weeks posttreatment [48, 49].

The laser-induced model causes minimal distress to the animals and is relatively easy to create. Therefore, it has served as the definitive model of CNV for the elucidation of angiogenesis pathways [50, 51] and identification of new therapeutic targets and interventions [52, 53]. Results created by using these CNV models should be interpreted with a caveat that the model does not necessarily recapitulate the aging aspect of the AMD.

Cell transplantation research using CNV models has focused on mature RPEs, because the RPE is the central secretor of vascular endothelial growth factor (VEGF) [54]. Exogenous reprogrammed RPEs have been transplanted to rescue CNV process after laser treatment in rats [55], which may be extended to autologous RPE reprogramming in the future. A neuroprotective approach of cell therapy rather than replacement of damaged cells has been studied for CNV as well—for example, mesenchymal stem cells have been investigated as a slow-releasing vehicle for anti-angiogenic factor delivery [56].

4.4.1.2 Growth Factor-Induced Models of CNV

After key components of CNV pathogenesis have been identified from the laser model, injection of said components has been discovered to induce the CNV process in animals, allowing creation of models closer to the natural progression of disease. VEGF was one of the first factors identified. Because simple intravitreal injections of VEGF lead to neovascularization outside the retina [57], delivery methods such as VEGF-eluting scleral beads for rabbits [58] and subretinal co-injection with Matrigel in mice [34] and rabbits [59] have been explored.

Matrigel is a mixture of various extracellular matrix (ECM) components and growth factors, and it alone has been shown to induce CNV in rats [60]. The subretinal Matrigel injection model is easy to perform and can be customized to elute different angiogenic factors in a controlled fashion. It is an interesting option for stem cell research, especially given that Matrigel itself is commonly used for culture of numerous types of retinal cell lines including RPEs [61].

Transgenic methods have been developed as well: adenoviral vectors of VEGF for rats [41] have allowed a gradual development of CNV from 5 weeks up to 10 months, unlike many other models of CNV that cause an acute insult. Transgenic mice with photoreceptor-specific (rhodopsin) [62, 63] and RPE-specific (VMD2) [64] promoters attached to the VEGF gene have also been created. In both of these models, the reverse tetracycline transactivator (rtTA) promoter can be inserted to install a doxycycline trigger to VEGF expression. This prevents activation of neovascularization too early in the mouse life cycle. It should be noted that in the VMD2/VEGF model, a secondary insult to the RPE [64, 65] or even injection of cultured RPE [66] itself often worsens neovascularization due to physical trauma, which has interesting implications to cell therapy for patients with CNV who have a sensitized RPE layer.

4.4.1.3 Oxygen-Induced Retinopathy (OIR) Model

This particular model is faithful to the pathogenesis of retinopathy of prematurity (ROP), a common complication of oxygen therapy in premature infants. Namely, young animals are exposed to alternating hypoxic and hyperoxic conditions of various durations, and subsequently assessed for neovascularization. Rat [67] models are among the most commonly used, but other models include mice [68], dogs [69], monkeys [70], and zebrafish [71]. Most models use room air control, but the monkey model uses transient occlusion of all branch retinal veins, while zebrafish models use water tanks of varying oxygen concentration. Because of the ease of induction, this model has played an essential role for development of anti-VEGF therapy [68, 70]. In addition, the OIR model is commonly used as a substitute for models of diabetic retinopathy (DR), because it shares similar retinal findings such as microaneurysms with severe forms of DR while being easy to create at the same time [45].

4.4.1.4 Models of Diabetic Retinopathy

Experimental diabetic retinopathy is typically induced in two major ways: ablation of pancreatic beta cells by chemicals such as streptozocin [72] and alloxan [73], or development of various transgenic mice such as $Ins2^{Akita}$ [74] and nonobese diabetic (NOD) [75] mice. Chemical models are the current standard for experimental DR and can be induced in a variety of animals, including pig [76] and even zebrafish [77]. The dosage regimen of streptozocin varies greatly depending on animal species, but all develop hyperglycemia in a few days. Retinopathy then follows in a few weeks for rodents, and months to years in larger mammals [78]. Chemical methods are currently limited to Type 1 diabetic (T1DM) forms; a type 2 diabetes (T2DM)-like presentation can be seen with high-calorie diet animals [79], but they are not favored in retina research due to arduous maintenance and follow-up until they become symptomatic.

Transgenic models can be either T1DM or T2DM like, and T2DM models can be further subdivided into obese and nonobese. The $Ins2^{Akita}$ [74] mouse is a T1DM model based on a point mutation on the Insulin 2 gene, and it considered a model of early and mild DR. It can be used to create even more abnormal angiogenesis by crossing with VEGF-overexpressing mice [62, 80]. The Zucker Diabetic Fatty rat [81] is an obese T2DM model with a relatively fast induction of retinopathy at approximately 6 months. These models allow study of survival and integration of transplanted cells inside host retina under an ongoing disease process, but a selection process among a batch of animals for development of ample neovascularization may be required.

4.4.2 Models of Sub-RPE Accumulation

Dry-type AMD is characterized by drusen, which are sub-RPE depositions of lipids and macrophages causing oxidative stress. Disruption of RPE by drusen leads to focal and geographic forms of atrophy, which involves irreversible loss of surrounding RPE and photoreceptor cells. Ever since the main risk factors and components of drusen formation have been characterized, animal models based on each of these factors have been developed. In this chapter, focus will be on the numerous transgenic rodent models that emulate pathogenesis of AMD via mutation of putative contributory genes. Nonhuman primate models are very useful in terms of anatomy because their retinas contain a macula, and recently a model created by inbreeding has been characterized [82], but these models only represent a minority among animal models for AMD.

4.4.2.1 Models of Lipid Accumulation

As early as the 1990s it has been known that drusen consist largely of lipids and lipoproteins, which gives their distinct color in retina exam and fluorescence angiography [83]. Some of the most notable contributors of lipid accumulation in the retina are Apolipoprotein E (ApoE). Originally developed as a model of hypercholesterolemia and atherosclerosis [84], ApoE-deficient mice have been brought to spotlight in retinal research since discovery of a genetic association between ApoE and AMD [85]. The ApoE knockout mice develop particle accumulations in Bruch's membrane, without needing to trigger a higher blood cholesterol level with a high-fat rodent diet [86]. Subsequent models were created by substituting mouse ApoE genes with *human* ApoE subtypes. These models show various levels lipid accumulation—among the ApoE subtypes, APOEe4 models develop most severe forms of RPE/photoreceptor damage as well as neovascularization [87]. With a similar approach, mouse models expressing human ApoB-100 were likewise developed with findings of lipid accumulation on electron microscopy [88, 89].

While the apolipoprotein models show a dry-type presentation, knockout of lipoprotein receptors such as in Very-low density lipoprotein receptor knockout (Vldlr−/−) mice [90] show a wet-type AMD presentation. Copious neovascularization develops by day 15–18, without high levels of cholesterol. The Vldlr knockout mice are useful for the understanding of molecular pathways of AMD development [91, 92], but they are not as commonly used as a model of neovascularization due to the prevalence of laser-induced models.

4.4.2.2 Models of Complement Pathway Activation and Immune Cell Recruitment

Genetic studies of AMD have also found a correlation of disease and complement protein mutations [93], and drusen have been discovered to have a high immunoreactivity against complement protein 3 (C3) and other associated enzymes [94]. C3 is central to complement pathway activation, whose end result is recruitment of leukocytes and triggering of an immune response. Therefore, mutations on inhibitory factors of C3 such as complement factor H (CFH) [93, 95] that are associated with AMD presumably allow a stronger inflammatory response and result in worse disease presentation. Indeed, aged Cfh knockout mice at 2 years of life show deposition of C3 protein in the retina, with decreased visual function and ERG signals compared to age-matched controls [96]. The model captures the aging aspect of AMD as well as RPE dysfunction in the form of rod outer segment disarrangement, but the animals' Bruch membrane thins rather than thickening as seen in typical cases of AMD. On the other hand, Cfh point mutation (Y402H) mice were reported to have drusen-like deposits and a thicker Bruch's membrane [38].

A related group of animal models focuses on the role of macrophages in AMD. The retina is typically devoid of any immune cells due to its immunoprivileged status, but drusen has been shown to contain macrophages [97, 98], and models with deficient mobilization of macrophages due to knockout of chemokines such as Ccl2 [99] show drusenoid deposits. A genetic association between another chemokine, CX3CR1, and human AMD has been established [100], and subsequent knockout mouse models have been developed [101, 102]. So far, the chemokine models with Ccl2 and Cx3cr1 knockout have demonstrated choroidal neovascularization, unlike the complement models that have a drier presentation [103]. CX3CR1 knockout models have a faster development of drusen and neovascularization compared to Ccl2 models, but this sacrifices the aging aspect of AMD at the same time.

Although these mice are mainly used for confirmation of findings from genetic studies, they can potentially serve as a model of long-term AMD development—specifically, to characterize integration of transplanted stem cells under ongoing inflammatory damage.

4.4.2.3 Models of Oxidative Stress in the Retina

The deposition of lipid as well as recruitment of immune cells contributes to oxidative stress that is critically involved with pathogenesis of AMD. Decreasing oxidative stress via antioxidant vitamins (A, C, and E) is a highly recommended component of dry-type AMD management today [104], and its importance has been proven by animal models of oxidative stress in the retina. Transgenic models of this category include superoxide dismutase (SOD) 1 knockout [105], SOD2 knockdown [106], and DJ-1 knockout [107, 108]. The antioxidant effects of SOD can be knocked out specifically in the RPE as well by using RPE-specific cre recombinases [109], but this model did not show drusenoid deposits, only autofluorescent material.

Oxidation can be induced without the use of genetic techniques. One important example is a group of animal models focusing on oxidized proteins that are naturally found in drusen. Proteomic analysis has revealed that carboxyethylpyrrole (CEP)-mediated modification of proteins is highly specific to the RPE [110], and CEP-modified albumin can be found abundantly in patients with AMD [111].

Hollyfield et al. created a model by injecting CEP-modified mouse albumin in mice [112], in order to cause an inflammatory process against any oxidative process occurring in the RPE. Injected mice developed autoantibodies against CEP-albumin, sub-RPE deposits of complement protein, and geographic atrophy-like lesions, but no evidence of choroidal neovascularization was seen [112]. This model is created with 2–3 intravenous injections over a 2–3 month period, which is simpler than many other surgical approaches to the retina.

4.5 Models of Photoreceptor Damage

This chapter will focus on animal models of retinitis pigmentosa (RP) and various inherited central vision disorders. The debilitating nature of these diseases has sparked a critical interest in photoreceptor transplantation therapy. It should be noted that these animal models are a target for transplantation of early photoreceptors *or* RPE cells, given that the RPE is responsible for protection of photoreceptor function.

4.5.1 Royal College of Science (RCS) Rats

The RCS rat is one of the oldest and most commonly used animal models of retinal dystrophy [42]. Previously thought to be an inbred rat with an unknown mechanism of retinal degeneration, later evidence has shown a failure of rod outer segment phagocytosis by RPE [116, 117] due to mutation in the *Mertk* gene in the *rdy* (retinal dystrophy) locus, which codes for a receptor tyrosine kinase [118]. The pathogenesis is therefore remarkably similar to retinitis pigmentosa in humans. Typically, damage to the photoreceptors can be seen by 14–16 days of life with a layer of outer segment debris being detected by 20 days [119], and degree of photoreceptor degeneration can be exacerbated by continuous light exposure [120]. RCS rats are the current standard for retinal progenitor cell (RPC) transplantation [121, 122] in rodent models.

RCS rats develop a multitude of ocular symptoms outside of retinal dystrophy. The degeneration of photoreceptors eventually leads to abnormal angiogenesis [123], as evidenced by fluorescin angiography [124]. Due to its early development of ample neovascularization, it has been used as a substitute model for models of abnormal retina angiogenesis, for diseases such as AMD and DR. Although no significant underlying RPE pathology is typically seen [125], human RPEs [126] and ESC-derived RPE cells [127, 128] have been transplanted with positive results of integration and restoration of visual function. Other symptoms that RCS rats show include cataracts and increased IOP [129], but other inbred lines such as DBA/2J [44] are typically used for this purpose.

4.5.2 Chemical Models of Photoreceptor Damage

Chemical-induced models can deplete photoreceptors very quickly, faster than the standard RCS rat. Commonly used cytotoxic molecules in the retina include sodium iodate (NaIO3) and *N*-methyl-*N*-nitrosurea (MNU). NaIO3 treated mice cause non-specific damage of both RPE and photoreceptor cells via oxidative stress [130], but there are definitive advantages over other models of photoreceptor damage. First, induction time is almost instantaneous with RPE damage [32] observed within 1 h post-injection. Second, the extent of damage can be readily customized—photoreceptor loss is dose-dependent [131, 132], with damage to cells seen only after 20 mg/kg of injection [133]. Third, different methods of administration have been explored—traditionally, NaIO3 is injected intravenously, but intraperitoneal [130] and retrobulbar venous injection [134] has been studied as well. Lastly, NaIO3 injection can be combined with genetic models with increased photoreceptor susceptibility to induce great ablation of outer retinal cells [126].

MNU causes cell damage via cell cycle inhibition, unlike NaIO3, which causes oxidative stress. MNU injection has all the advantages of NaIO3, but damage is more specific to photoreceptor cells [135]. MNU can be injected intraperitoneally [136] and intravitreally (to avoid systemic effects) [135], and has been characterized in many different mammalian species [137]. As with many other chemical models of retinal diseases in general, NaIO3 and MNU-induced retinal damage represents acute loss of cells, rather than following pathogenesis of underlying disease.

4.5.3 Genetic Models of Retinitis Pigmentosa

Retinitis pigentosa (RP) is a heterogeneous disease entity with more than 60 genes currently identified, and can be associated with multi-organ disorders such as Usher syndrome and Bardet-Biedl syndrome [138]. The Pde6b^{rd1} mouse (rd1 mouse) is a commonly studied model of rod degeneration, related to autosomal recessive variants of RP [139]. The loss of photoreceptors develops quickly within 3 weeks for rods [43], followed by progressive loss of the entire photoreceptor layer throughout the life of the mouse [140]. *Rd1* mutations are very common in mice, and many experimental mouse strains carry the allele. Due to cheaper cost and fast induction of photoreceptor-specific blindness, *rd1* has been the standard model for transplantation studies for various stages of differentiated photoreceptors [141–143].

Another commonly studied RP mutation is that of Rhodopsin (Rho), which is believed to account for most cases of autosomal dominant RP. Minipig models with Rho mutation at P23H [39] show a severe decline in function of rods in 30 days of age with an absent scotopic response. Cone populations develop normally until day 60 of life, but decrease in viability afterward, likely due to absence of trophic support from rods and choroid [144]. Cell replacement therapy has yet to be actively researched with this particular model.

4.5.4 Genetic Models of Other Photoreceptor Anomalies

The most commonly used models of photoreceptor damage such as RCS and RD1 rat often do not represent pathogenesis of other photoreceptor diseases to its full extent. Numerous transgenic models for congenital impairments such as Leber's congenital amaurosis (LCA) [145–147], maculopathies such as Stargardt's disease [148, 149], and various cone-rod dystrophies [146, 150, 151] are currently available. Many of these diseases share common causative genes, all involved in the visual cycle, and thus different mutations of the same gene can account for different phenotypes.

Traditionally, human mutations associated with these disorders have been transduced in larger mammals such as pig and dogs, which have a higher density of cones at their central streak and fovea-like regions, respectively [152]. More recently, however, established mouse models are often bred with Nrl−/− mice [12], which lack a crucial protein that determines cone vs. rod photoreceptor fate. This results in a cone-enriched mouse with the cone genetic defect of choice [153, 154]. This breakthrough has led to easier creation of cone disorder models for cone photoreceptor transplantation (Table 4.2).

Table 4.2 Common animal models of photoreceptor diseases for stem cell research

Method	Commonly used species	Category	Characteristics	Previous stem cell research
Royal College of Science Rats	Rats [42]	Inbred mouse strain	Model of RP and neovascularization	hRPCs [121], hMSCs and hRPCs [122] RPEs [126–128]
Sodium iodate-induced model	Rat [32] Rabbit [155]	Chemical	Acute model of RPE and photoreceptor injury	ESCs [156]
N-methyl-N-nitrosurea-induced model	Mouse [135] Rabbit [157] Monkey [137]	Chemical	Moe specific photoreceptor damage, acute model of injury	N/A
Pde6b^{rd1} mouse	Mouse [43]	Inbred mouse strain	Model of RP; fast degradation of rod, and then cone. Very common mutation in experimental mice	mESC/iPSC derived retinal sheet [141] hESC/PSC derived PhRP [142] Precursor and mature rods [143]
Nrl−/− mouse	Mouse [12]	Transgenic	Model of enhanced S-cone syndrome; often bred with models of cone dystrophies	Often a source of cone photoreceptors for retinal research

4.6 Models of Inner Retina Diseases

4.6.1 Models of RGC Apoptosis and Optic Nerve Injury

The inner retina diseases such as glaucoma and various forms of optic neuropathy (ischemic, traumatic, compressive) have common characteristics of retinal ganglion cell (RGC) loss. Animal models that directly focus on RGC damage do not necessarily replicate the pathogenesis of individual conditions, but have been used extensively due to their ability to show a drastic change in RGC function in many different types of animals.

4.6.1.1 Axotomy and Optic Nerve Crush

RGCs have long axonal connections that extend through the optic nerve toward the lateral geniculate and superior colliculi of the brain, which can be directly manipulated via axotomy or crush. This method has been used as early as in the 1980s [158] as a model for optic nerve axonal degeneration. Both axotomy and crush is typically done via intraorbital approach than intracranial due to easier approach [159]. Once the optic nerve is approached, it is incised while maintaining blood supply and meningeal continuity for axotomy [160], or injury is inflicted by a forceps or balloon [161] for a few seconds for optic nerve crush.

Any animal with an optic nerve-like structure can have their optic nerves manipulated; however, this approach has most extensively been in use in rats. Rats are less expensive compared to primates but easier to surgically manipulate compared to mice. Moreover, in rats, RGC cell population drops abruptly in 1–2 weeks post-axotomy [30], whereas in monkeys, significant loss takes 1–2 months [162]. It should be noted that in certain non-mammalian models such as goldfish RGCs regenerate after initial injury [163], which makes them less of an ideal candidate for intervention.

The fast and marked degeneration of RGCs in optic nerve-crushed rats provides reproducible results for researchers interested in RGC regeneration. Recent work has also shown that RGC loss in optic nerve crush and axotomy is largely mediated by apoptosis via activation of various initiator [164] and effector caspases [165], which recapitulates RGC loss in inner retinal diseases. Therefore, recent research has focused on using neurotropic factors such as BDNF [166], antiexcitotoxic molecules such as brimonidine [167], immune modulators [168], and many more potential drugs [169] to delay or salvage RGC damage after optic nerve crush. Because the effect of these molecules is transitory, neuroprotective cell therapy has the promise of allowing long-term neurotrophic benefits via growth factor production and immunomodulation. Optic nerve-crush was also one of the earliest models of cell replacement therapy experiments with RGCs, with positive results of vision restoration in mice [170].

It should be noted that an incomplete transection/crush of the optic nerve can yield false positive results of optic nerve regeneration [171], which may require more stringent survey of RGCs and their axons before and after treatment. Like many other models of RGC damage, optic nerve crush does not capitulate the disease process that causes the RGC damage, i.e., glaucoma, and results using optic nerve crush models should be considered with this caveat.

4.6.1.2 *N*-Methyl-D-Aspartate (NMDA) Receptor-Induced Injury

Glutamate excitotoxicity is a critical component of the selective death of neuronal cells in ischemic retinopathy [172, 173] as well as glaucoma. This has classically been associated with overstimulation of NMDA receptors [174] in retinal ganglion cells, which leads to intracellular influx of calcium ions, leading to activation of downstream apoptotic cascade. Thus animal models utilizing glutamate excitotoxicity have focused on stimulation of NMDA receptors, over other types of glumate receptors such as AMPA/kainate channels.

There are several advantages of using NMDA-induced models for RGC injury: first, intravitreal injection of NMDA results in dose-dependent RGC loss, ranging from no effect to more than 50% loss in 5 days in mice [33]. This implies that the model can be customized to represent "mild" and "severe" stages of inner retina diseases. Second, intravitreous injection of a single compound is an easier method to administer in small animals such as mice, compared to a more complex surgical method. Last and most importantly, damage via NMDA injection is very specific to RGC cells—recent studies showed that bipolar cells and rhodopsin-positive photoreceptors have remained largely intact after 2 weeks since injection, whereas RGCs took the largest hit, followed by some amacrine cells [175]. Because of this specificity the model has been used extensively for stem cell therapy approaches that attempt to replace lost RGCs, examples include RGCs differentiated from embryonic stem cells (ESCs) [176, 177] and Muller cells [178].

It should be noted that newer evidence suggests desensitization of AMPA receptors via kainate [179], or preconditioning AMPA stimulation with glucose or oxygen deprivation [180]. As of current, AMPA stimulated mouse models have not been extensively used for stem cell research.

4.6.1.3 Transient Ischemia Models

The transient ischemia models are a group of models that induce RGC loss via occlusion or functional inhibition of blood vessels supporting the optic nerve and retina, often followed by reperfusion of said vessels—thus the name "transient" ischemia. These models can be grouped into arterial models and venous models.

Different strategies are used for direct ligation of arteries in large and small animals. Because the central retinal artery (CRA) is a narrower continuation of the ophthalmic artery stemming from the internal cartorid, ligation of the CRA is more feasible in larger animals such as monkeys [181]. On the other hand, larger vessel operations such as filamentous middle cerebral artery occlusion (fMCAO) [182] are more commonly used in small animals such as rodents. Because the outsprouting of the posterior ciliary artery from the internal carotid is close to where the internal carotid becomes the MCA, occlusion via intraluminal suture at this point has been shown to result in retinal ischemia, as evidenced by suppressed ERG and increased GFAP induction in Muller cells [182]. With fMCAO, the duration and severity of the ischemia can be directly controlled by the duration of occlusion. However, RGC loss typically occurs when there is severe insult to the brain as well [183], and it can be debatable whether fMCAO is a primary retinal model.

Other methods of inhibiting blood flow in the central retinal artery of small animals such as rodents include (1) intraocular pressure (IOP) elevation and (2) induction of thrombosis. IOP elevation typically involves connecting the animal's eyes to a sterile saline container, and then artificially increasing intraocular pressure (IOP) above the systolic pressure of the CRA [36, 184, 185]. In this method, reperfusion is commonly accomplished by detaching the saline container and then performing a fundus exam for blood flow. Extensive loss of RGCs is typically seen within 60 min [36]. IOP elevation has been widely used for neuroprotective pharmacologic interventional studies [186, 187] and as well as stem cell research [188, 189], due to its relatively easy procedure and ability to be induced in many different animals. The IOP-induced model causes mechanical damage as well as ischemic damage in RGCs, and so whether it is a specific model for ischemic retinal disease can be debatable.

Thrombotic ischemia/reperfusion often involves a light-sensitive dye (such as the Rose Bengal dye) that induces thrombosis via laser irradiation, followed by tPA administration [190, 191]. The major strength of this method is that it is minimally invasive compared to other methods of transient ischemia. However, while the extent of damage can be controlled by concentration of the dye and the intensity of laser, occlusion occurs end-arterially, which can cause high variability in regards to restoration of blood flow by tPA [192].

Similar strategies are used for retinal venous occlusion as in arterial ischemic models. Some examples include thermal coagulation [193] and photochemical-induced thrombosis [194, 195]. Although central venous-occluded models are not as commonly used compared to episcleral venous occlusion for RGC regenerative stem cell therapy research, it is important to note that they model two different diseases—namely central/branch venous occlusion and primary open-angle glaucoma, respectively.

4.6.1.4 Transgenic Models of RGC Loss

Although more difficult to maintain and breed compared to traditional models, transgenic mouse models using the Cre-Lox system can provide a purely RGC-depleted animal for stem cell research. One example is the *Pou4f2 (Brn3b)* knockout mouse, which utilizes a fused Cre recombinase to the estrogen receptor

(CAGG-Cre-ER transgene) [40]. *Pou4f2* is a crucial gene for RGC differentiation and is expressed throughout life. Because the Cre recombinase is triggered by tamoxifen injection in these mice, the timing of *Pou4f2* knockout can be precisely controlled in the animal's life cycle. RPCs isolated from embryonic retinas have been shown to be able to integrate into the depleted RGC layer in this model [15].

4.6.1.5 Experimental Autoimmune Encephalopathy (EAE) Model

The EAE model is primarily a model for diseases causing autoimmune demyelination in the central nervous system, namely multiple sclerosis (MS). Autoimmunity to the myelin protein can be accomplished with a wide variety of proteins, such as myelin basic protein (MBP) which often shows a relapsing/remitting presentation [196, 197], and myelin oligodendrocyte glycoprotein (MOG) [198] which shows a chronic and progressive form of MS in rodents. Likewise, severity and time of RGC involvement varies depending on type of induction and type of animal treated for EAE. For example, RGC loss precedes optic nerve inflammation in rat chronic EAE models [37], whereas in mice optic neuritis comes first, and RGC loss follows afterward [199]. Typical induction time is 1–2 weeks. Current stem cell research regarding EAE models has been focused toward neuroprotection via NSCs [200] or MSCs [201], rather than direct replacement of RGCs or its precursors.

4.6.1.6 Models of Hereditary Diseases that Involve Optic Nerve Damage

Retinal degeneration from hereditary diseases such as Leber's hereditary optic neuropathy (LHON) and neurofibromatosis type I (NF1) does not have a definitive cure as of today. In these diseases, stem cell therapy has the promise of supplementing lost RGCs and/or protecting from further RGC damage. A number of animal models exist for LHON, including germline transfer of mutant human mitochondrial DNA [202], and rotenone-induced mitochondrial complex I inhibition [203]. Of note, the rotenone model causes immediate reduction in RGC layer thickness after 1 h post-injection, which is strikingly similar to the acute and abrupt onset of LHON. Recent work with retinal progenitor cells (RPCs) has shown that cultured cells are able to integrate into the RGC layer with positive activity evidenced by manganese-enhanced MRI [204].

Neurofibromatosis type I causes RGC loss mainly by the development of optic pathway gliomas (OPGs), whose mainstay treatment is currently limited to chemotherapy. Transgenic models of NF1 made via the Cre-lox system are able to target specific cell-lines, such as progential glial cells, to express mutant NF1. Currently the $Nf1^{flox/mut}$; GFAP-Cre mice [205] capture the pathogenesis of OPGs, including biallelic loss of NF1 gene in embryonic development, as well as microglial recruitment by CCL5/CXCR12. As with many transgenic models, creation of these mice requires careful breeding and can be difficult to use for larger studies (Table 4.3).

Table 4.3 Common animal models of RGC and optic nerve damage for stem cell research

Method	Commonly used species	Category	Characteristics	Previous stem cell research
Optic nerve crush/ axotomy	Rat [30] Monkey [162] Cat [206]	Surgical	1–2 week induction time, typically irreversible	RGC, [170], MSC [207, 208]
NMDA injection	Mouse [33]	Chemical	Damage more specific to RGCs	ESC [176, 177], Muller-derived RGC [178]
IOP-induced ischemia/ reperfusion model	Rat [36] Mouse [209] Rabbit [210]	Ischemic	Fast induction time, rapidly reversible	BMSC [188], RPC [189]
Venous occlusion	Rat [194] Cat [193]	Surgical or chemical	Similar technique used for arterial occlusion	N/A
Brn3b knockout via Cre-ER	Mouse [40]	Transgenic	Uses tamoxifen trigger; specific RGC ablation	RPC [15]
Experimental autoimmune encephalomyelitis	Rat [37] Mouse [199]	Chemical	Model of MS; Variable induction time and severity	MSC [201], NSC [200]
Rotenone-induced optic neuropathy	Mouse [203]	Chemical	Model of LHON; very fast and abrupt loss of RGC	RPC [204]
$Nf1^{flox/mut}$; GFAP-Cre	Mouse [205]	Transgenic	Model of OPG associated with NF1	N/A

4.6.2 Models of Glaucomatous Disease Conditions

Glaucoma is a heterogeneous disease entity with a multitude of causes and risk factors, sharing common characteristics of RGC loss and subsequent vision impairment. In terms of pathogenesis, glaucoma can be classified into open-angle and closed-angle forms. Animals with open-angle glaucoma have higher resistance to aqueous humor outflow, whereas animals with closed-angle glaucoma have a direct obstruction to outflow tracts via angle closure. The open- and closed-angle forms can again be subclassified as primary or secondary, depending on whether there is an identifiable cause for elevation of intraocular pressure (IOP). Rare congenital forms of glaucoma that fit into neither category exist as well. Animal models for these glaucoma subtypes have been discussed extensively in literature [211], and newer models are in development with recent advances in genetic engineering.

4.6.2.1 Animal Models of Primary Open-Angle Glaucoma (POAG) and Its Variants

Open-angle glaucoma is the most common subtype of glaucoma. It is characterized by a disruption of balance between aqueous production from the ciliary body and aqueous outflow via the trabecular meshwork and uveoscleral pathways. This mismatch may or may not lead to an elevation in IOP, as seen in the case with normal-tension glaucoma (NTG). Nonetheless, damage occurs to RGCs evidenced by pathologic changes.

4.6.2.1.1 Trabecular Meshwork (TM) and Episcleral Vein Occlusion Models

The trabecular meshwork accounts for most of aqueous humor outflow from the anterior chamber, draining into the Schlemn's canal and then episcleral veins. Dysfunction of this system is closely associated with the pathogenesis of POAG; however, depending on the degree of obstruction and time course, animal models created by occluding this system can be seen more of an angle closure-like model. (PACG is defined clinically by more than 270° of occlusion at the angle [212].)

TM is most commonly occluded with (1) laser photocoagulation [213], (2) injection of particles such as latex microbeads [31], or (3) injection with viscous materials such as methylcellulose [214] and hyaluronic acid [215] in the anterior chamber (intracameral). Laser photocoagulation has been initially introduced in monkeys [216], which has the benefit of human-like ocular anatomy, but more recently in rodents as well which showed stable IOP increase for 24 weeks [217]. With microbead models, the duration of IOP increase and extent of RGC damage is highly variable, ranging from a few weeks to a few months, and 10–70%, depending on the species and volume/size of microbeads [218]. One way to ensure a more prolonged response is by using multiple consecutive injections [219], or adding viscous materials. For example, methylcellulose can be added to further occlude the vein, and recent evidence with RGC-fluorescent (Thy1-CFP) reporter mice may suggest that addition of methylcellulose causes higher elevation in IOP and more severe RGC loss in mice [214].

The episcleral vein is typically occluded by either cauterization or saline injection [220]. Thermo- or electro-cauterization of the episcleral vein had been used in rodent [221, 222] and non-rodent species [223], and leads to a stable increase in IOP and subsequent RGC loss [224]. In fact, in pig models, episcleral vein cauterization is one of the most reliable methods of episcleral vein occlusion, with prolonged retinal damage [223]. On the other hand, early studies in rats have shown little difference in degree of RGC damage between episcleral cauterization or microbead occlusion of TM [225]. Although episcleral vein occlusion causes definitive glaucoma-like damage to the retina, interventions on the episcleral vein typically require more surgical precision than those on the trabecular meshwork, which can lead to high variability in symptoms.

4.6.2.1.2 Steroid-Induced IOP Elevation

Glucocorticoids are thought to increase IOP mainly via increased extracellular matrix deposition at the trabecular meshwork [226, 227]. Although best used as a model of steroid-induced glaucoma, it can be used as a substitute for subtypes of

POAG that are caused by various depositions at the Schlemm's canal. To induce ganglion cell depletion, steroid is preferably applied topical onto the animal: in mice, topical application of dexamethasone on the orbit leads to INL thinning [228], whereas systemic application does not lead to appreciable loss [229]. This model is easy to induce and to maintain; however, animals are at high risk of side effects, such as corneal damage and cataracts, with topical steroids.

4.6.2.1.3 Genetic Models of POAG in Rodents

Natural mutations with a POAG-like presentation have been discovered in monkeys [230] and dogs [231], but these models are rare and difficult to work with. Genetic models have been developed in rodents, with the knowledge of genes identified in these natural animal models [232] as well as in humans. One such example is the Tyr437His myocilin point mutation models. Myocilin (MYOC) is a gene that is implicated in 3–5% of POAG cases [233], and has been extensively studied since its discovery. There are multiple ways to introduce the mutant human gene into rodents, including pronuclear injection (*Tg-MYOC*^{Y437H}) [234, 235] and adenovirus-associated viral vectors [236, 237]. As discussed in Sect. 4.3, pronuclear injection of BAC DNA shows nonspecific integration of target gene, and extensive breeding/selecting is required to establish a model. On the other hand, injection of viral vectors is much easier to perform, but gene expression can be transient and often immunosuppression is needed to bolster mutant gene expression [238]. If viral vectors are used, intravitreal injection has been reported to yield better results due to a slow release of viral particles from the vitreous body [238, 239], and species differences among mice should be taken into account [236]. Other POAG genes transduced with this method include TGF-beta2 [240] and secreted frizzled-related protein-1 (sFRP-1) [241].

Unfortunately, many of viral vector models have not been able to demonstrate measurable RGC loss or damage to the superior colliculus [236], whereas models allowing transfer of mutant gene at germline stage such as pronuclear injection showed peripheral loss of RGCs but not robust IOP elevation [235]. Recently, Crispr/Cas9-mediated editing of MYOC was found to be able to reverse IOP changes and improve ERG results [242]. Because genetic models of POAG are faithful to the pathogenesis and progression of disease, successful cell therapy results in these models would be the most translatable to future clinical results. However, much work needs to be done in regards to cost and reliability of RGC damage until these models are used more often for cell therapy.

4.6.2.1.4 Genetic Models of Normal Tension Glaucoma (NTG) in Rodents

Normal tension glaucoma is a disease entity of ambiguous pathophysiology, associated with a number of risk factors such as systemic hypotension, migraine, Raynaud's phenomenon, etc. [243] Currently, transgenic mice that show RGC damage without IOP elevation are available. As discussed in Sect. 4.1, glutamate excitotoxicity in the retina is very specific to RGCs. Therefore, knockout models of glutamate transporters such as the GLAST mutant mice [244] have been revisited and were confirmed to show positive RGC and optic nerve damage [245]. It should be noted that the debilitating effects of GLAST knockout pervade over the entire CNS—therefore the shorter life span and neurological defects of these mice have to be considered in experimental design.

4.6.2.2 Animal Models of Primary Angle–Closure Glaucoma (PACG) and Its Variant

PACG in most cases is a chronic disease with gradual onset of visual symptoms, only distinguished from POAG by degree of angle closure. Only in select cases acute angle closure occurs, characterized by acute increase in IOP and conjunctival injection. Therefore, many of the occlusive models of the TM or episcleral vein listed in Sect. 4.1 can also represent damage from PACG, depending on the extent of outflow obstruction. Models listed below are those that primarily show only a PACG-like presentation.

4.6.2.2.1 Vav2/Vav3 Deficient Mice

The Vav proteins are a group of guanine exchange factors (GEFs) whose downstream effect includes alteration of cell behavior via actin regulation [246]. Although primarily used in immunological research, Fujikawa et al. have discovered that Vav2/vav3 knockout mice develop variable amounts of angle closure, IOP increase, and subsequent RGC loss [247]. The main strength of this model lies in the early onset of ocular symptoms (6 weeks for IOP increase), which is ideal for a mouse model. Furthermore, the Vav2/vav3 is a very well-characterized and common model, and no further surgical or chemical intervention is required in the orbit to develop symptoms. So far, this model has yet to be used for stem cell therapy research.

4.6.2.2.2 DBA/2J Mouse Line for Pigmentary Glaucoma

The DBA/2J mouse line is an inbred strain that has been discovered to develop pigmentary dispersion in the anterior chamber [44]. The human condition of pigmentary dispersion syndrome is a secondary POAG-like disease, but the DBA/2J mice show more of a PACG-like presentation with large accumulations of pigment-filled cells and synechia in the anterior chamber. The mice become symptomatic by 9 months of age, and RGC layer thinning and optic nerve damage is evident by 11–15 months. The DBA/2J mouse line is responsible for the discovery of major genes associated with pigmentary glaucoma such as *Tyrpl* [248] and *Gpnmb* [249]. On the other hand, stem cell therapy for this model is still at its early stage. Divya et al. [177] recently reported transplantation of ES-cell-derived neural progenitors (NP) in NMDA-ablated mice and DBA/2J mice, whereas NPs were able to differentiate into RGCs and integrate into host retina resulting in vision improvement in NMDA-ablated mice, no such phenomena was seen in DBA/2J mice. As discussed in Sect. 4.1, NMDA-mediated ablation causes a very specific damage to RGCs and no IOP change is present, whereas DBA/2J mice are primarily IOP increasing models and RGC loss is secondary. Based on these findings, Divya et al. suggest that it may be crucial to control the underlying glaucomatous process (including IOP changes) in order to make stem cell therapy more fruitful, which is especially relevant for the models of PACG-like presentation listed in this chapter (Table 4.4).

Table 4.4 Common animal models of glaucomatous disease for stem cell research

Method	Commonly used species	Category	Characteristics	Previous stem cell research
POAG phenotype				
Laser photo-coagulation of trabecular meshwork	Monkey [250, 251] Mouse [217] Rat [252] Pig [253]	Surgical	Easy to induce and manage in any animal, stable IOP increase	Muller-derived RGC [254] BMSC [255], RSC [256]
Episcleral vein cauterization	Mouse [221] Rat [222] Pig [223]	Surgical	Reliable method of inducing stable IOP increase, high degree of surgical finesse needed	MSC [257, 258]
Microbead/methylcellulose/hyaluronic acid injection	Mouse [31] Rat [31] Rabbit [219] Pig [223]	Surgical	Variable RGC damage	MSC [259]
Steroid-induced IOP	Rabbit [35] Mouse [229] Rat [260] Monkey [261] Cow [262] Sheep [263]	Chemical	Easy to induce and manage in almost any animal; various side effects	N/A
Viral vector-mediated mutant MYOC gene transfection	Mouse [237] Rat [240] Cat [264]	Transgenic	Inconsistent RGC damage	N/A
Glutamate transporter KO models	Mouse [245]	Transgenic	RGC damage without IOP increase; a model of NTG	N/A
PACG phenotype				
Vav2/vav3 deficient mice	Mouse [247]	Transgenic	Early onset	N/A
DBA/2J mice	Mouse [44]	Inbred mouse strain	Late onet; severe anterior chamber synechiae, unlike human counterpart	ES-derived neural progenitor [177]

References

1. Garg, A., et al. (2017). Stem cell therapies in retinal disorders. *Cell, 6*(1), E4.
2. Reichenbach, A., & Bringmann, A. (2013). New functions of Muller cells. *Glia, 61*(5), 651–678.
3. Burnight, E. R., et al. (2017). Using CRISPR-Cas9 to generate gene-corrected autologous iPSCs for the treatment of inherited retinal degeneration. *Molecular Therapy, 25*(9), 1999–2013.
4. Li, L. X., & Turner, J. E. (1988). Transplantation of retinal pigment epithelial cells to immature and adult rat hosts: Short-and long-term survival characteristics. *Experimental Eye Research, 47*(5), 771–785.
5. Vugler, A., et al. (2008). Elucidating the phenomenon of HESC-derived RPE: Anatomy of cell genesis, expansion and retinal transplantation. *Experimental Neurology, 214*(2), 347–361.
6. Kamao, H., et al. (2014). Characterization of human induced pluripotent stem cell-derived retinal pigment epithelium cell sheets aiming for clinical application. *Stem Cell Reports, 2*(2), 205–218.
7. Schwartz, S. D., et al. (2016). Subretinal transplantation of embryonic stem cell-derived retinal pigment epithelium for the treatment of macular degeneration: An assessment at 4 years. *Investigative Ophthalmology & Visual Science, 57*(5), ORSFc1–ORSFc9.
8. Berger, A. S., et al. (2003). Photoreceptor transplantation in retinitis pigmentosa: Short-term follow-up. *Ophthalmology, 110*(2), 383–391.
9. Eberle, D., et al. (2012). Outer segment formation of transplanted photoreceptor precursor cells. *PLoS One, 7*(9), e46305.
10. Warre-Cornish, K., et al. (2014). Migration, integration and maturation of photoreceptor precursors following transplantation in the mouse retina. *Stem Cells and Development, 23*(9), 941–954.
11. Pearson, R. A., et al. (2012). Restoration of vision after transplantation of photoreceptors. *Nature, 485*(7396), 99–103.
12. Santos-Ferreira, T., et al. (2015). Daylight vision repair by cell transplantation. *Stem Cells, 33*(1), 79–90.
13. Lakowski, J., et al. (2010). Cone and rod photoreceptor transplantation in models of the childhood retinopathy Leber congenital amaurosis using flow-sorted Crx-positive donor cells. *Human Molecular Genetics, 19*(23), 4545–4559.
14. Santos-Ferreira, T., et al. (2016). Stem cell-derived photoreceptor transplants differentially integrate into mouse models of cone-rod dystrophy. *Investigative Ophthalmology & Visual Science, 57*(7), 3509–3520.
15. Cho, J. H., Mao, C. A., & Klein, W. H. (2012). Adult mice transplanted with embryonic retinal progenitor cells: New approach for repairing damaged optic nerves. *Molecular Vision, 18*, 2658–2672.
16. Singhal, S., et al. (2012). Human Muller glia with stem cell characteristics differentiate into retinal ganglion cell (RGC) precursors in vitro and partially restore RGC function in vivo following transplantation. *Stem Cells Translational Medicine, 1*(3), 188–199.
17. Kador, K. E., et al. (2013). Tissue engineering the retinal ganglion cell nerve fiber layer. *Biomaterials, 34*(17), 4242–4250.
18. Kador, K. E., et al. (2014). Retinal ganglion cell polarization using immobilized guidance cues on a tissue-engineered scaffold. *Acta Biomaterialia, 10*(12), 4939–4946.
19. Kador, K. E., et al. (2016). Control of retinal ganglion cell positioning and neurite growth: Combining 3D printing with radial electrospun scaffolds. *Tissue Engineering. Part A, 22*(3–4), 286–294.
20. Venugopalan, P., et al. (2016). Transplanted neurons integrate into adult retinas and respond to light. *Nature Communications, 7*, 10472.

21. Stujenske, J. M., Dowling, J. E., & Emran, F. (2011). The bugeye mutant zebrafish exhibits visual deficits that arise with the onset of an enlarged eye phenotype. *Investigative Ophthalmology & Visual Science, 52*(7), 4200–4207.
22. Veth, K. N., et al. (2011). Mutations in zebrafish lrp2 result in adult-onset ocular pathogenesis that models myopia and other risk factors for glaucoma. *PLoS Genetics, 7*(2), e1001310.
23. Lauber, J. K. (1987). Light-induced avian glaucoma as an animal model for human primary glaucoma. *Journal of Ocular Pharmacology, 3*(1), 77–100.
24. Provis, J. M. (2001). Development of the primate retinal vasculature. *Progress in Retinal and Eye Research, 20*(6), 799–821.
25. Downs, J. C., & Girkin, C. A. (2017). Lamina cribrosa in glaucoma. *Current Opinion in Ophthalmology, 28*(2), 113–119.
26. De Schaepdrijver, L., Simoens, P., Lauwers, H., & De Geest, J. P. (1989). Retinal vascular patterns in domestic animals. *Research in Veterinary Science, 47*(1), 34–42.
27. May, C. A. (2008). Comparative anatomy of optic nerve head and inner retina in non-primate animal models used for glaucoma research. *The Open Ophthalmology Journal, 2*, 94–101.
28. Sugiyama, K., Gu, Z.-B., Kawase, C., Yamamoto, T., & Kitazawa, Y. (1999). Optic nerve and peripapillary choroidal microvasculature of the rat eye. *Investigative Ophthalmology & Visual Science, 40*, 3084–3090.
29. Ryan, S. J. (1979). The development of an experimental model of subretinal neovascularization in disciform macular degeneration. *Transactions of the American Ophthalmological Society, 77*, 707–745.
30. Berkelaar, M., Clarke, D. B., Wang, Y. C., Bray, G. M., & Aguayo, A. J. (1994). Axotomy results in delayed death and apoptosis of retinal ganglion cells in adult rats. *The Journal of Neuroscience, 14*(7), 4368–4374.
31. Sappington, R. M., et al. (2010). The microbead occlusion model: A paradigm for induced ocular hypertension in rats and mice. *Investigative Ophthalmology & Visual Science, 51*(1), 207–216.
32. Hariri, S., et al. (2013). Noninvasive imaging of the early effect of sodium iodate toxicity in a rat model of outer retina degeneration with spectral domain optical coherence tomography. *Journal of Biomedical Optics, 18*(2), 26017.
33. Li, Y., Schlamp, C. L., & Nickells, R. W. (1999). Experimental induction of retinal ganglion cell death in adult mice. *Investigative Ophthalmology & Visual Science, 40*(5), 1004–1008.
34. Shen, D., Wen, R., Tuo, J., Bojanowski, C. M., & Chan, C. C. (2006). Neovascularization induced by subretinal injection of matrigel in CCL2/MCP-1-deficient mice. *Ophthalmic Research, 38*(2), 71–73.
35. Lorenzetti, O. J. (1970). Effects of corticosteroids on ocular dynamics in rabbits. *The Journal of Pharmacology and Experimental Therapeutics, 175*(3), 763–772.
36. Hughes, W. F. (1991). Quantitation of ischemic damage in the rat retina. *Experimental Eye Research, 53*(5), 573–582.
37. Meyer, R., Weissert, R., Diem, R., Storch, M. K., de Graaf, K. L., Kramer, B., et al. (2001). Acute neuronal apoptosis in a rat model of multiple sclerosis. *The Journal of Neuroscience, 21*(16), 6214–6220.
38. Ufret-Vincenty, R. L., et al. (2010). Transgenic mice expressing variants of complement factor H develop AMD-like retinal findings. *Investigative Ophthalmology & Visual Science, 51*(11), 5878–5887.
39. Fernandez de Castro, J. P., et al. (2014). Cone photoreceptors develop normally in the absence of functional rod photoreceptors in a transgenic swine model of retinitis pigmentosa. *Investigative Ophthalmology & Visual Science, 55*(4), 2460–2468.
40. Cho, J. H., et al. (2009). Retinal ganglion cell death and optic nerve degeneration by genetic ablation in adult mice. *Experimental Eye Research, 88*(3), 542–552.
41. Wang, F., Rendel, K. G., Manning, W. C., Quiroz, D., Coyne, M., & Miller, S. S. (2003). AAV-mediated expression of vascular endothelial growth factor induces choroidal neovascularization in rat. *Investigative Ophthalmology & Visual Science, 44*(2), 781–790.

42. Bourne, M. C., Campbell, D. A., & Pyke, M. (1938). Cataract associated with an hereditary retinal lesion in rats. *The British Journal of Ophthalmology, 22*(10), 608–613.
43. Lolley, R. N., Rong, H., & Craft, C. M. (1994). Linkage of photoreceptor degeneration by apoptosis with inherited defect in phototransduction. *Investigative Ophthalmology & Visual Science, 35*(2), 358–362.
44. John, S. W., Smith, R. S., Savinova, O. V., Hawes, N. L., Chang, B., Turnbull, D., et al. (1998). Essential iris atrophy, pigment dispersion, and glaucoma in DBA/2J mice. *Investigative Ophthalmology & Visual Science, 39*, 951–962.
45. Liu, C. H., et al. (2017). Animal models of ocular angiogenesis: From development to pathologies. *The FASEB Journal, 31*(11), 4665–4681.
46. Baum, J. L., & Wise, G. N. (1965). Experimental subretinal neovascularization. *Transactions of the American Ophthalmological Society, 63*, 91–107.
47. Dobi, E., Puliafito, C. A., & Destro, M. (1989). A new model of experimental choroidal neovascularization in the rat. *Archives of Ophthalmology, 107*(2), 264–269.
48. Tobe, T., Ortega, S., Luna, J. D., Ozaki, H., Okamoto, N., Derevjanik, N. L., et al. (1998). Targeted disruption of the FGF2 gene does not prevent choroidal neovascularization in a murine model. *The American Journal of Pathology, 153*(5), 1641–1646.
49. Edelman, J. L., & Castro, M. R. (2000). Quantitative image analysis of laser-induced choroidal neovascularization in rat. *Experimental Eye Research, 71*(5), 523–533.
50. Shen, W. Y., Yu, M. J., Barry, C. J., Constable, I. J., & Rakoczy, P. E. (1998). Expression of cell adhesion molecules and vascular endothelial growth factor in experimental choroidal neovascularisation in the rat. *The British Journal of Ophthalmology, 82*(9), 1063–1071.
51. Takahashi, T., Nakamura, T., Hayashi, A., Kamei, M., Nakabayashi, M., Okada, A. A., et al. (2000). Inhibition of experimental choroidal neovascularization by overexpression of tissue inhibitor of metalloproteinases-3 in retinal pigment epithelium cells. *American Journal of Ophthalmology, 130*(6), 774–781.
52. Lukason, M., et al. (2011). Inhibition of choroidal neovascularization in a nonhuman primate model by intravitreal administration of an AAV2 vector expressing a novel anti-VEGF molecule. *Molecular Therapy, 19*(2), 260–265.
53. Wang, X., et al. (2017). TGF-beta participates choroid neovascularization through Smad2/3-VEGF/TNF-alpha signaling in mice with laser-induced wet age-related macular degeneration. *Scientific Reports, 7*(1), 9672.
54. Blaauwgeers, H. G. H., Holtkamp, G. M., Rutten, H., Witmer, A. N., Koolwijk, P., Partanen, T. A., et al. (1999). Polarized vascular endothelial growth factor secretion by human retinal pigment epithelium and localization of vascular endothelial growth factor receptors on the inner choriocapillaris. Evidence for a trophic paracrine relation. *The American Journal of Pathology, 155*(2), 421–428.
55. Li, F., et al. (2015). Subretinal transplantation of retinal pigment epithelium overexpressing fibulin-5 inhibits laser-induced choroidal neovascularization in rats. *Experimental Eye Research, 136*, 78–85.
56. Hou, H. Y., et al. (2010). A therapeutic strategy for choroidal neovascularization based on recruitment of mesenchymal stem cells to the sites of lesions. *Molecular Therapy, 18*(10), 1837–1845.
57. Tolentino, M. J., Miller, J. W., Gragoudas, E. S., Chatzistefanou, K., Ferrara, N., & Adamis, A. P. (1996). Vascular endothelial growth factor is sufficient to produce iris neovascularization and neovascularization and neovascular glaucoma in a nonhuman primate. *Archives of Ophthalmology, 114*(8), 964–970.
58. Zahn, G., et al. (2009). Preclinical evaluation of the novel small-molecule integrin alpha-5beta1 inhibitor JSM6427 in monkey and rabbit models of choroidal neovascularization. *Archives of Ophthalmology, 127*(10), 1329–1335.
59. Qiu, G., et al. (2006). A new model of experimental subretinal neovascularization in the rabbit. *Experimental Eye Research, 83*(1), 141–152.

60. Cao, J., Zhao, L., Li, Y., Liu, Y., Xiao, W., Song, Y., et al. (2010). A subretinal matrigel rat choroidal neovascularization (CNV) model and inhibition of CNV and associated inflammation and fibrosis by VEGF trap. *Investigative Ophthalmology & Visual Science, 51*(11), 6009–6017.
61. Kanemura, H., et al. (2014). Tumorigenicity studies of induced pluripotent stem cell (iPSC)-derived retinal pigment epithelium (RPE) for the treatment of age-related macular degeneration. *PLoS One, 9*(1), e85336.
62. Okamoto, N., Tobe, T., Hackett, S. F., Ozaki, H., Vinores, M. A., LaRochelle, W., et al. (1997). Transgenic mice with increased expression vascular endothelial growth factor in the retina: A new model of intraretinal and subretinal neovascularization. *The American Journal of Pathology, 151*(1), 281–291.
63. Ohno-Matsui, K., Hirose, A., Yamamoto, S., Saikia, J., Okamoto, N., Gehlbach, P., et al. (2002). Inducible expression of vascular endothelial growth factor in adult mice causes severe proliferative retinopathy and retinal detachment. *The American Journal of Pathology, 160*(2), 711–719.
64. Oshima, Y., et al. (2004). Increased expression of VEGF in retinal pigmented epithelial cells is not sufficient to cause choroidal neovascularization. *Journal of Cellular Physiology, 201*(3), 393–400.
65. Oshima, Y., et al. (2004). Angiopoietin-2 enhances retinal vessel sensitivity to vascular endothelial growth factor. *Journal of Cellular Physiology, 199*(3), 412–417.
66. Schmack, I., Berglin, L., Nie, X., Kang, S. J., Marcus, A. I., Yang, H., et al. (2009). Modulation of choroidal neovascularization by subretinal injection of retinal pigment epithelium and polystyrene microbeads. *Molecular Vision, 15*, 416–461.
67. Penn, J. S., Henry, M. M., & Tolman, B. L. (1994). Exposure to alternating hypoxia and hyperoxia causes severe proliferative retinopathy in the newborn rat. *Pediatric Research, 36*(6), 724–731.
68. Aiello, L. P., Pierce, E. A., Foley, E. D., Takagi, H., Chen, H., Riddle, L., et al. (1995). Suppression of retinal neovascularization in vivo by inhibition of vascular endothelial growth factor (VEGF) using soluble VEGF-receptor chimeric proteins. *Proceedings of the National Academy of Sciences of the United States of America, 92*(23), 10457–10461.
69. Taomoto, M., McLeod, D. S., Merges, C., & Lutty, G. A. (2000). Localization of adenosine A2a receptor in retinal development and oxygen-induced retinopathy. *Investigative Ophthalmology & Visual Science, 41*(1), 230–243.
70. Miller, J. W., Adamis, A. P., Shima, D. T., D'Amore, P. A., Moulton, R. S., O'Reilly, M. S., et al. (1994). Vascular endothelial growth factor: Vascular permeability factor is temporally and spatially correlated with ocular angiogenesis in a primate model. *The American Journal of Pathology, 145*(3), 574–584.
71. Cao, Z., et al. (2010). Hypoxia-induced retinopathy model in adult zebrafish. *Nature Protocols, 5*(12), 1903–1910.
72. Tso, M. O., Kurosawa, A., Benhamou, E., Bauman, A., Jeffrey, J., & Jonasson, O. (1988). Microangiopathic retinopathy in experimental diabetic monkeys. *Transactions of the American Ophthalmological Society, 86*, 389–421.
73. Weerasekera, L. Y., et al. (2015). Characterization of retinal vascular and neural damage in a novel model of diabetic retinopathy. *Investigative Ophthalmology & Visual Science, 56*(6), 3721–3730.
74. Barber, A. J., et al. (2005). The Ins2Akita mouse as a model of early retinal complications in diabetes. *Investigative Ophthalmology & Visual Science, 46*(6), 2210–2218.
75. Li, C. R., & Sun, S. G. (2010). VEGF expression and cell apoptosis in NOD mouse retina. *International Journal of Ophthalmology, 3*(3), 224–227.
76. Lee, S. E., et al. (2010). Ultrastructural features of retinal capillary basement membrane thickening in diabetic swine. *Ultrastructural Pathology, 34*(1), 35–41.
77. Intine, R. V., Olsen, A. S., & Sarras Jr., M. P. (2013). A zebrafish model of diabetes mellitus and metabolic memory. *Journal of Visualized Experiments, 72*, e50232.

78. Olivares, A. M., et al. (2017). Animal models of diabetic retinopathy. *Current Diabetes Reports, 17*(10), 93.
79. Kern, T. S., & Engerman, R. L. (1994). Comparison of retinal lesions in alloxan-diabetic rats and galactose-fed rats. *Current Eye Research, 13*(12), 863–867.
80. van Eeden, P. E., et al. (2006). Early vascular and neuronal changes in a VEGF transgenic mouse model of retinal neovascularization. *Investigative Ophthalmology & Visual Science, 47*(10), 4638–4645.
81. Danis, R. P., & Yang, Y. (1993). Microvascular retinopathy in the Zucker diabetic fatty rat. *Investigative Ophthalmology & Visual Science, 34*(7), 2367–2371.
82. Dawson, W. W., et al. (2008). Maculas, monkeys, models, AMD and aging. *Vision Research, 48*(3), 360–365.
83. Pauleikhoff, D., Zuels, S., Sheraidah, G. S., Marshall, J., Wessing, A., & Bird, A. C. (1992). Correlation between biochemical composition and fluorescein binding of deposits in Bruch's membrane. *Ophthalmology, 99*(10), 1548–1553.
84. Plump, A. S., Smith, J. D., Hayek, T., Aalto-Setala, K., Walsh, A., Verstuyft, J. G., et al. (1992). Severe hypercholesterolemia and atherosclerosis in apolipoprotein E-deficient mice created by homoloogous recombination in ES cells. *Cell, 71*(2), 343–353.
85. Klaver, C. C., et al. (1998). Genetic association of apolipoprotein E with age-related macular degeneration. *American Journal of Human Genetics, 63*(1), 200–206.
86. Dithmar, S., Curcio, C. A., Le NA, B. S., & Grossniklaus, H. E. (2000). Ultrastructural changes in Bruch's membrane of apolipoprotein E-deficient mice. *Investigative Ophthalmology & Visual Science, 41*(8), 2035–2042.
87. Malek, G., et al. (2005). Apolipoprotein E allele-dependent pathogenesis: A model for age-related retinal degeneration. *Proceedings of the National Academy of Sciences of the United States of America, 102*(33), 11900–11905.
88. Espinosa-Heidmann, D. G., Sall, J., Hernandez, E. P., & Cousins, S. W. (2004). Basal laminar deposit formation in APO B100 transgenic mice: Complex interactions between dietary fat, blue light, and vitamin E. *Investigative Ophthalmology & Visual Science, 45*(1), 260–266.
89. Sallo, F. B., et al. (2009). Bruch's membrane changes in transgenic mice overexpressing the human biglycan and apolipoprotein b-100 genes. *Experimental Eye Research, 89*(2), 178–186.
90. Heckenlively, J. R., Hawes, N. L., Friedlander, M., Nusinowitz, S., Hurd, R., Davisson, M., et al. (2003). Mouse model of subretinal neovascularization with choroidal anastomosis. *Retina, 23*(4), 518–522.
91. Chen, Y., et al. (2007). Very low density lipoprotein receptor, a negative regulator of the wnt signaling pathway and choroidal neovascularization. *The Journal of Biological Chemistry, 282*(47), 34420–34428.
92. Sun, Y., et al. (2017). Inflammatory signals from photoreceptor modulate pathological retinal angiogenesis via c-Fos. *The Journal of Experimental Medicine, 214*(6), 1753–1767.
93. Hageman, G. S., et al. (2005). A common haplotype in the complement regulatory gene factor H (HF1/CFH) predisposes individuals to age-related macular degeneration. *Proceedings of the National Academy of Sciences of the United States of America, 102*(20), 7227–7232.
94. Johnson, L. V., et al. (2001). Complement activation and inflammatory processes in drusen formation and age related macular degeneration. *Experimental Eye Research, 73*(6), 887–896.
95. Edwards, A. O., et al. (2005). Complement factor H polymorphism and age-related macular degeneration. *Science, 308*(5720), 421–424.
96. Coffey, P. J., Gias, C., McDermott, C. J., Lundh, P., Pickering, M. C., Sethi, C., et al. (2007). Complement factor H deficiency in aged mice causes retinal abnormalities and visual dysfunction. *Proceedings of the National Academy of Sciences of the United States of America, 104*(42), 16651–16656.

97. van der Schaft, T. L., Mooy, C. M., de Bruijn, W. C., & de Jong, P. T. (1993). Early stages of age-related macular degeneration: An immunofluorescence and electron microscopy study. *The British Journal of Ophthalmology, 77*(10), 657–661.
98. Killingsworth, M. C., Sarks, J. P., & Sarks, S. H. (1990). Macrophages related to Bruch's membrane in age-related macular degeneration. *Eye (London, England), 4*(Pt 4), 613–621.
99. Ambati, J., et al. (2003). An animal model of age-related macular degeneration in senescent Ccl-2- or Ccr-2-deficient mice. *Nature Medicine, 9*(11), 1390–1397.
100. Tuo, J., et al. (2004). The involvement of sequence variation and expression of CX3CR1 in the pathogenesis of age-related macular degeneration. *The FASEB Journal, 18*(11), 1297–1299.
101. Combadiere, C., et al. (2007). CX3CR1-dependent subretinal microglia cell accumulation is associated with cardinal features of age-related macular degeneration. *The Journal of Clinical Investigation, 117*(10), 2920–2928.
102. Tuo, J., Bojanowski, C. M., Zhou, M., Shen, D., Ross, R. J., Rosenberg, K. I., et al. (2007). Murine ccl2/cx3cr1 deficiency results in retinal lesions mimicking human age-related macular degeneration. *Investigative Ophthalmology & Visual Science, 48*(8), 3827–3836.
103. Pennesi, M. E., Neuringer, M., & Courtney, R. J. (2012). Animal models of age related macular degeneration. *Molecular Aspects of Medicine, 33*(4), 487–509.
104. Garcia-Layana, A., et al. (2017). Early and intermediate age-related macular degeneration: Update and clinical review. *Clinical Interventions in Aging, 12*, 1579–1587.
105. Imamura, Y., et al. (2006). Drusen, choroidal neovascularization, and retinal pigment epithelium dysfunction in SOD1-deficient mice: A model of age-related macular degeneration. *Proceedings of the National Academy of Sciences of the United States of America, 103*(30), 11282–11287.
106. Justilien, V., et al. (2007). SOD2 knockdown mouse model of early AMD. *Investigative Ophthalmology & Visual Science, 48*(10), 4407–4420.
107. Bonilha, V. L., et al. (2017). Absence of DJ-1 causes age-related retinal abnormalities in association with increased oxidative stress. *Free Radical Biology & Medicine, 104*, 226–237.
108. Bonilha, V. L., et al. (2015). Loss of DJ-1 elicits retinal abnormalities, visual dysfunction, and increased oxidative stress in mice. *Experimental Eye Research, 139*, 22–36.
109. Mao, H., Seo, S. J., Biswal, M. R., Li, H., Conners, M., Nandyala, A., et al. (2014). Mitochondrial oxidative stress in the retinal pigment epithelium leads to localized retinal degeneration. *Investigative Ophthalmology & Visual Science, 55*(7), 4613–4627.
110. Crabb, J. W., et al. (2002). Drusen proteome analysis: An approach to the etiology of age-related macular degeneration. *Proceedings of the National Academy of Sciences of the United States of America, 99*(23), 14682–14687.
111. Gu, J., Pauer, G. J. T., Yue, X., Narendra, U., Sturgill, G. M., Bena, J., et al. (2009). Assessing susceptibility to age-related macular degeneration with proteomic and genomic biomarkers. *Molecular & Cellular Proteomics, 8*(6), 1338–1349.
112. Hollyfield, J. G., et al. (2008). Oxidative damage-induced inflammation initiates age-related macular degeneration. *Nature Medicine, 14*(2), 194–198.
113. Kumar, S., & Zhuo, L. (2010). Longitudinal in vivo imaging of retinal gliosis in a diabetic mouse model. *Experimental Eye Research, 91*(4), 530–536.
114. Kim, J. M., et al. (2016). Perivascular progenitor cells derived from human embryonic stem cells exhibit functional characteristics of pericytes and improve the retinal vasculature in a rodent model of diabetic retinopathy. *Stem Cells Translational Medicine, 5*(9), 1268–1276.
115. Cerman, E., et al. (2016). Retinal electrophysiological effects of intravitreal bone marrow derived mesenchymal stem cells in streptozotocin induced diabetic rats. *PLoS One, 11*(6), e0156495.
116. Tso, M. O., Zhang, C., Abler, A. S., Chang, C. J., Wong, F., Chang, G. Q., et al. (1994). Apoptosis leads to photoreceptor degeneration in inherited retinal dystrophy of RCS rats. *Investigative Ophthalmology & Visual Science, 35*(6), 2693–2699.
117. Strauss, O., Stumpff, F., Mergler, S., Wienrich, M., & Widerholt, M. (1998). The Royal College of Surgeons rat: An animal model for inherited retinal degeneration with a still unknown genetic defect. *Acta Anatomica, 162*(2–3), 101–111.

118. D'Cruz, P. M., Yasumura, D., Weir, J., Matthes, M. T., Abderrahim, H., MM, L. V., et al. (2000). Mutation of the receptor tyrosine kinase gene Mertk in the retinal dystrophic RCS rat. *Human Molecular Genetics, 9*(4), 645–651.
119. LaVail, M. M. (2001). Legacy of the RCS rat: Impact of a seminal study on retinal cell biology and retinal degenerative diseases. *Progress in Brain Research, 131*, 617–627.
120. Kaitz, M. (1976). The effect of light on brightness perception in rats with retinal dystrophy. *Vision Research, 16*(2), 141–148.
121. Semo, M., et al. (2016). Efficacy and safety of human retinal progenitor cells. *Translational Vision Science & Technology, 5*(4), 6.
122. Qu, L., et al. (2017). Combined transplantation of human mesenchymal stem cells and human retinal progenitor cells into the subretinal space of RCS rats. *Scientific Reports, 7*(1), 199.
123. Pennesi, M. E., et al. (2008). The relationship of photoreceptor degeneration to retinal vascular development and loss in mutant rhodopsin transgenic and RCS rats. *Experimental Eye Research, 87*(6), 561–570.
124. Zambarakji, H. J., et al. (2006). High resolution imaging of fluorescein patterns in RCS rat retinae and their direct correlation with histology. *Experimental Eye Research, 82*(1), 164–171.
125. Ryals, R. C., et al. (2017). Long-term characterization of retinal degeneration in Royal College of Surgeons rats using spectral-domain optical coherence tomography. *Investigative Ophthalmology & Visual Science, 58*(3), 1378–1386.
126. Zhao, C., et al. (2017). Development of a refined protocol for trans-scleral subretinal transplantation of human retinal pigment epithelial cells into rat eyes. *Journal of Visualized Experiments, 126*, PMID:28829422.
127. McGill, T. J., et al. (2017). Long-term efficacy of GMP grade xeno-free hESC-derived RPE cells following transplantation. *Translational Vision Science & Technology, 6*(3), 17.
128. Davis, R. J., et al. (2017). The developmental stage of adult human stem cell-derived retinal pigment epithelium cells influences transplant efficacy for vision rescue. *Stem Cell Reports, 9*(1), 42–49.
129. Naskar, R., & Thanos, S. (2006). Retinal gene profiling in a hereditary rodent model of elevated intraocular pressure. *Molecular Vision, 12*, 1199–1210.
130. Chowers, G., et al. (2017). Course of sodium iodate-induced retinal degeneration in albino and pigmented mice. *Investigative Ophthalmology & Visual Science, 58*(4), 2239–2249.
131. Machalinska, A., et al. (2014). Dose-dependent retinal changes following sodium iodate administration: Application of spectral-domain optical coherence tomography for monitoring of retinal injury and endogenous regeneration. *Current Eye Research, 39*(10), 1033–1041.
132. Enzmann, V., et al. (2006). Behavioral and anatomical abnormalities in a sodium iodate-induced model of retinal pigment epithelium degeneration. *Experimental Eye Research, 82*(3), 441–448.
133. Wang, J., et al. (2014). Direct effect of sodium iodate on neurosensory retina. *Investigative Ophthalmology & Visual Science, 55*(3), 1941–1953.
134. Kadkhodaeian, H. A., et al. (2016). Histological and electrophysiological changes in the retinal pigment epithelium after injection of sodium iodate in the orbital venus plexus of pigmented rats. *Journal of Ophthalmic & Vision Research, 11*(1), 70–77.
135. Rosch, S., et al. (2014). Selective photoreceptor degeneration by intravitreal injection of N-methyl-N-nitrosourea. *Investigative Ophthalmology & Visual Science, 55*(3), 1711–1723.
136. Chen, Y. Y., Liu, S. L., Hu, D. P., Xing, Y. Q., & Shen, Y. (2014). N-methyl-N-nitrosourea-induced retinal degeneration in mice is independent of the p53 gene. *Experimental Eye Research, 121*, 102–113.
137. Yoshizawa, K., & Tsubura, A. (2005). Characteristics of N-methyl-N-nitrosourea-induced retinal degeneration in animals and application for the therapy of human retinitis pigmentosa. *Nippon Ganka Gakkai Zasshi, 109*(6), 327–337.
138. Ferrari, S., et al. (2011). Retinitis pigmentosa: Genes and disease mechanisms. *Current Genomics, 12*(4), 238–249.

139. McLaughlin, M. E., et al. (1993). Recessive mutations in the gene encoding the beta-subunit of rod phosphodiesterase in patients with retinitis pigmentosa. *Nature Genetics, 4*(2), 130–134.
140. Chang, B., et al. (2007). Two mouse retinal degenerations caused by missense mutations in the beta-subunit of rod cGMP phosphodiesterase gene. *Vision Research, 47*(5), 624–633.
141. Assawachananont, J., Mandai, M., Okamoto, S., Yamada, C., Eiraku, M., Yonemura, S., et al. (2014). Transplantation of embryonic and induced pluripotent stem cell-derived 3D retinal sheets into retinal degenerative mice. *Stem Cell Reports, 2*(5), 662–674.
142. Barnea-Cramer, A. O., Wang, W., Lu, S.-J., Singh, M. S., Luo, C., Huo, H., et al. (2016). Function of human pluripotent stem cell-derived photoreceptor progenitors in blind mice. *Scientific Reports, 6*, 29784.
143. Singh, M. S., et al. (2013). Reversal of end-stage retinal degeneration and restoration of visual function by photoreceptor transplantation. *Proceedings of the National Academy of Sciences of the United States of America, 110*(3), 1101–1106.
144. Leveillard, T., et al. (2004). Identification and characterization of rod-derived cone viability factor. *Nature Genetics, 36*(7), 755–759.
145. den Hollander, A. I., et al. (2004). CRB1 mutation spectrum in inherited retinal dystrophies. *Human Mutation, 24*(5), 355–369.
146. Yang, R. B., Robinson, S. W., Xiong, W. H., Yau, K. W., Birch, D. G., & Garbers, D. L. (1999). Disruption of a retinal guanylyl cyclase gene leads to cone-specific dystrophy and paradoxical rod behavior. *The Journal of Neuroscience, 19*(14), 5889–5897.
147. Daniels, D. M., Stoddart, C. W., Martin-Iverson, M. T., Lai, C. M., Redmond, T. M., & Rakoczy, P. E. (2003). Entrainment of circadian rhythm to a photoperiod reversal shows retinal dystrophy in RPE65(−/−) mice. *Physiology & Behavior, 79*(4–5), 701–711.
148. Charbel Issa, P., et al. (2013). Fundus autofluorescence in the Abca4(−/−) mouse model of Stargardt disease–correlation with accumulation of A2E, retinal function, and histology. *Investigative Ophthalmology & Visual Science, 54*(8), 5602–5612.
149. Sommer, J. R., et al. (2011). Production of ELOVL4 transgenic pigs: A large animal model for Stargardt-like macular degeneration. *The British Journal of Ophthalmology, 95*(12), 1749–1754.
150. Wiik, A. C., et al. (2008). A deletion in nephronophthisis 4 (NPHP4) is associated with recessive cone-rod dystrophy in standard wire-haired dachshund. *Genome Research, 18*(9), 1415–1421.
151. Goldstein, O., et al. (2013). IQCB1 and PDE6B mutations cause similar early onset retinal degenerations in two closely related terrier dog breeds. *Investigative Ophthalmology & Visual Science, 54*(10), 7005–7019.
152. Kostic, C., & Arsenijevic, Y. (2016). Animal modelling for inherited central vision loss. *The Journal of Pathology, 238*(2), 300–310.
153. Boye, S. L., et al. (2015). Gene therapy fully restores vision to the all-cone Nrl(−/−) Gucy2e(−/−) mouse model of leber congenital amaurosis-1. *Human Gene Therapy, 26*(9), 575–592.
154. Feathers, K. L., et al. (2008). Nrl-knockout mice deficient in Rpe65 fail to synthesize 11-cis retinal and cone outer segments. *Investigative Ophthalmology & Visual Science, 49*(3), 1126–1135.
155. Cho, B. J., et al. (2016). Monocular retinal degeneration induced by intravitreal injection of sodium iodate in rabbit eyes. *Japanese Journal of Ophthalmology, 60*(3), 226–237.
156. Park, U. C., et al. (2011). Subretinal transplantation of putative retinal pigment epithelial cells derived from human embryonic stem cells in rat retinal degeneration model. *Clinical and Experimental Reproductive Medicine, 38*(4), 216–221.
157. Rosch, S., et al. (2017). Photoreceptor degeneration by intravitreal injection of N-methyl-N-nitrosourea (MNU) in rabbits: A pilot study. *Graefe's Archive for Clinical and Experimental Ophthalmology, 255*(2), 317–331.
158. Allcutt, D., Berry, M., & Sievers, J. (1984). A qualitative comparison of the reactions of retinal ganglion cell axons to optic nerve crush in neonatal and adulte mice. *Brain Research, 318*(2), 231–240.

159. Levkovitch-Verbin, H. (2004). Animal models of optic nerve diseases. *Eye (London, England), 18*(11), 1066–1074.
160. Solomon, A. S., et al. (1996). Complete transection of rat optic nerve while sparing the meninges and the vasculature: An experimental model for optic nerve neuropathy and trauma. *Journal of Neuroscience Methods, 70*(1), 21–25.
161. Villegas-Perez, M. P., et al. (1993). Rapid and protracted phases of retinal ganglion cell loss follow axotomy in the optic nerve of adult rats. *Journal of Neurobiology, 24*(1), 23–36.
162. Agapova, O. A., et al. (2003). Differential expression of matrix metalloproteinases in monkey eyes with experimental glaucoma or optic nerve transection. *Brain Research, 967*(1–2), 132–143.
163. Koriyama, Y., Homma, K., & Kato, S. (2006). Activation of cell survival signals in the goldfish retinal ganglion cells after optic nerve injury. *Advances in Experimental Medicine and Biology, 572*, 333–337.
164. Ahmed, Z., et al. (2011). Ocular neuroprotection by siRNA targeting caspase-2. *Cell Death & Disease, 2*, e173.
165. Choudhury, S., et al. (2015). Caspase-7: A critical mediator of optic nerve injury-induced retinal ganglion cell death. *Molecular Neurodegeneration, 10*, 40.
166. Mansour-Robaey, S., Clarke, D. B., Wang, Y.-C., Bray, G. M., & Aguayo, A. J. (1994). Effects of ocular injury and administration of brain-derived neurotrophic factor on survival and regrowth of axotomized retinal ganglion cells. *Proceedings of the National Academy of Sciences of the United States of America, 91*(5), 1632–1636.
167. Ma, K., et al. (2009). Effect of brimonidine on retinal ganglion cell survival in an optic nerve crush model. *American Journal of Ophthalmology, 147*(2), 326–331.
168. Tsai, R. K., Chang, C. H., & Wang, H. Z. (2008). Neuroprotective effects of recombinant human granulocyte colony-stimulating factor (G-CSF) in neurodegeneration after optic nerve crush in rats. *Experimental Eye Research, 87*(3), 242–250.
169. Danesh-Meyer, H. V. (2011). Neuroprotection in glaucoma: Recent and future directions. *Current Opinion in Ophthalmology, 22*(2), 78–86.
170. de Lima, S. K. (2012). Full-length axon regeneration in the adult mouse optic nerve and partial reovery of simple visual behaviors. *Proceedings of the National Academy of Sciences of the United States of America, 109*(23), 9149–9154.
171. Fischer, D., et al. (2017). Optic nerve regeneration in mammals: Regenerated or spared axons? *Experimental Neurology, 296*, 83–88.
172. Izumi, Y., et al. (2002). Glutamate transporters and retinal excitotoxicity. *Glia, 39*(1), 58–68.
173. Lam, T. T., et al. (1997). Ameliorative effect of MK-801 on retinal ischemia. *Journal of Ocular Pharmacology and Therapeutics, 13*(2), 129–137.
174. Casson, R. J. (2006). Possible role of excitotoxicity in the pathogenesis of glaucoma. *Clinical & Experimental Ophthalmology, 34*(1), 54–63.
175. Kuehn, S., et al. (2017). Concentration-dependent inner retina layer damage and optic nerve degeneration in a NMDA model. *Journal of Molecular Neuroscience, 63*(3-4), 283–299.
176. Aoki, H., et al. (2008). Transplantation of cells from eye-like structures differentiated from embryonic stem cells in vitro and in vivo regeneration of retinal ganglion-like cells. *Graefe's Archive for Clinical and Experimental Ophthalmology, 246*(2), 255–265.
177. Divya, M. S., et al. (2017). Intraocular injection of ES cell-derived neural progenitors improve visual function in retinal ganglion cell-depleted mouse models. *Frontiers in Cellular Neuroscience, 11*, 295.
178. Becker, S., et al. (2016). Allogeneic transplantation of Muller-derived retinal ganglion cells improves retinal function in a feline model of ganglion cell depletion. *Stem Cells Translational Medicine, 5*(2), 192–205.
179. Park, Y. H., et al. (2015). AMPA receptor desensitization is the determinant of AMPA receptor mediated excitotoxicity in purified retinal ganglion cells. *Experimental Eye Research, 132*, 136–150.
180. Park, Y. H., Broyles, H. V., He, S., McGrady, N. R., Li, L., & Yorio, T. (2016). Involvement of AMPA receptor and its flip and flop isoforms in retinal ganglion cell death following oxygen/glucose deprivation. *Investigative Ophthalmology & Visual Science, 57*(2), 508–526.

181. Hayreh, S. S., & Weingeist, T. A. (1980). Experimental occlusion of the central artery of the retina. I. Ophthalmoscopic and fluorescein fundus angiographic studies. *The British Journal of Ophthalmology, 64*(12), 896–912.
182. Block, F., Grommes, C., Kosinski, C., Schmidt, W., & Schwarz, M. (1997). Retinal ischemia induced by the intraluminal suture method in rats. *Neuroscience Letters, 232*(1), 45–48.
183. Allen, R. S., et al. (2014). Severity of middle cerebral artery occlusion determines retinal deficits in rats. *Experimental Neurology, 254*, 206–215.
184. Lam, T. T., Abler, A. S., & Tso, M. O. (1999). Apoptosis and caspases after ischemia-reperfusion injury in rat retina. *Investigative Ophthalmology & Visual Science, 40*, 967–975.
185. Nakahara, T., et al. (2015). Structural and functional changes in retinal vasculature induced by retinal ischemia-reperfusion in rats. *Experimental Eye Research, 135*, 134–145.
186. Joachim, S. C., et al. (2017). Protective effects on the retina after ranibizumab treatment in an ischemia model. *PLoS One, 12*(8), e0182407.
187. Ko, M. L., et al. (2011). Simvastatin upregulates Bcl-2 expression and protects retinal neurons from early ischemia/reperfusion injury in the rat retina. *Experimental Eye Research, 93*(5), 580–585.
188. Mathew, B., et al. (2017). Bone-marrow mesenchymal stem-cell administration significantly improves outcome after retinal ischemia in rats. *Graefe's Archive for Clinical and Experimental Ophthalmology, 255*(8), 1581–1592.
189. Li, X., et al. (2014). Transplantation with retinal progenitor cells repairs visual function in rats with retinal ischemia-reperfusion injury. *Neuroscience Letters, 558*, 8–13.
190. Daugeliene, L., Niva, M., Hara, A., Matsuno, H., Yamamoto, T., Kitazawa, Y., et al. (2000). Transient ischemic injury in the rat retina caused by thrombotic occlusion-thrombolytic reperfusion. *Investigative Ophthalmology & Visual Science, 41*, 2743–2747.
191. Mosinger, J. L., Price, M. T., Bai, H. Y., Xiao, H., Wozniak, D. F., & Olney, J. W. (1991). Blockade of both NMDA and non-NMDA receptors is required for optimal protection against ischemic neuronal degeneration in the in vivo adult mammalian retina. *Experimental Neurology, 113*(1), 10–17.
192. Macrae, I. M. (2011). Preclinical stroke research- Advantages and disadvantages of the most common rodent models of focal ischaemia. *British Journal of Pharmacology, 164*(4), 1062–1078.
193. Hayashi, A., Imai, K., Kim, H. C., & de Juan, E. (1997). Activation of protein tyrosine phosphorylation after retinal branch vein occlusion in cats. *Investigative Ophthalmology & Visual Science, 38*, 372–380.
194. Zhang, Y., et al. (2008). Natural history and histology in a rat model of laser-induced photothrombotic retinal vein occlusion. *Current Eye Research, 33*(4), 365–376.
195. Ebneter, A., et al. (2015). Investigation of retinal morphology alterations using spectral domain optical coherence tomography in a mouse model of retinal branch and central retinal vein occlusion. *PLoS One, 10*(3), e0119046.
196. Fritz, R. B., Chou, C. H., & McFarlin, D. E. (1983). Relapsing murine experimental allergic encephalomyelitis induced by myelin basic protein. *Journal of Immunology, 130*(3), 1024–1026.
197. Panitch, H., & Ciccone, C. (1981). Induction of recurrent experimental allergic encephalomyelitis with myelin basic protein. *Annals of Neurology, 9*(5), 433–438.
198. Storch, M. K., Stefferl, A., Brehm, U., Weissert, R., Wallström, E., Kerschensteiner, M., et al. (1998). Autoimmunity to myelin oligodendrocyte glycoprotein in rats mimics the spectrum of multiple sclerosis pathology. *Brain Pathology, 8*(4), 681–694.
199. Quinn, T. A., Dutt, M., & Shindler, K. S. (2011). Optic neuritis and retinal ganglion cell loss in a chronic murine model of multiple sclerosis. *Front in Neurology, 2*, 50.
200. Pluchino, S., et al. (2009). Human neural stem cells ameliorate autoimmune encephalomyelitis in non-human primates. *Annals of Neurology, 66*(3), 343–354.
201. Lanza, C., et al. (2009). Neuroprotective mesenchymal stem cells are endowed with a potent antioxidant effect in vivo. *Journal of Neurochemistry, 110*(5), 1674–1684.

202. Yu, H., et al. (2015). Consequences of zygote injection and germline transfer of mutant human mitochondrial DNA in mice. *Proceedings of the National Academy of Sciences of the United States of America, 112*(42), E5689–E5698.
203. Zhang, X. J., Jones, D., & Gonzalez-Lima, F. (2002). Mouse model of optic neuropathy caused by mitochondrial complex I dysfunction. *Neuroscience Letters, 326*(2), 97–100.
204. Mansergh, F. C., et al. (2014). Cell therapy using retinal progenitor cells shows therapeutic effect in a chemically-induced rotenone mouse model of Leber hereditary optic neuropathy. *European Journal of Human Genetics, 22*(11), 1314–1320.
205. Toonen, J. A., Ma, Y., & Gutmann, D. H. (2017). Defining the temporal course of murine neurofibromatosis-1 optic gliomagenesis reveals a therapeutic window to attenuate retinal dysfunction. *Neuro-Oncology, 19*(6), 808–819.
206. Weber, A. J., et al. (2010). Combined application of BDNF to the eye and brain enhances ganglion cell survival and function in the cat after optic nerve injury. *Investigative Ophthalmology & Visual Science, 51*(1), 327–334.
207. Mesentier-Louro, L. A., et al. (2014). Distribution of mesenchymal stem cells and effects on neuronal survival and axon regeneration after optic nerve crush and cell therapy. *PLoS One, 9*(10), e110722.
208. Zwart, I., Hill, A. J., Al-Allaf, F., Shah, M., Girdlestone, J., Sanusi, A. B., et al. (2009). Umbilical cord blood mesenchymal stromal cells are neuroprotective and promote regeneration in a rat optic tract model. *Experimental Neurology, 216*(2), 439–448.
209. Harada, C., Harada, T., Slusher, B. S., Yoshida, K., Matsuda, H., & Wada, K. (2000). N-acetylated-alpha-linked-acidic dipeptidase inhibitor has a neuroprotective effect on mouse retinal ganglion cells after pressure-induced ischemia. *Neuroscience Letters, 292*(2), 134–136.
210. Gohdo, T., Ueda, H., Ohno, S., Iijima, H., & Tsukahara, S. (2001). Heat shock protein 70 expression increased in rabbit müller cells in the ischemia-reperfusion model. *Opthalmic Research, 33*(5), 298–302.
211. Bouhenni, R. A., et al. (2012). Animal models of glaucoma. *Journal of Biomedicine & Biotechnology, 2012*, 692609.
212. Weinreb, R. N., Aung, T., & Medeiros, F. A. (2014). The pathophysiology and treatment of glaucoma: A review. *JAMA, 311*(18), 1901–1911.
213. Levkovitch-Verbin, H., Quigley, H. A., Martin, K. R., Valenta, D., Baumrind, L. A., & Pease, M. E. (2002). Translimbal laser photocoagulation to the trabecular meshwork as a model of glaucoma in rats. *Investigative Ophthalmology & Visual Science, 43*, 402–410.
214. Liu, H., & Ding, C. (2017). Establishment of an experimental glaucoma animal model: A comparison of microbead injection with or without hydroxypropyl methylcellulose. *Experimental and Therapeutic Medicine, 14*(3), 1953–1960.
215. Moreno, M. C., et al. (2005). A new experimental model of glaucoma in rats through intracameral injections of hyaluronic acid. *Experimental Eye Research, 81*(1), 71–80.
216. Gaasterland, D., & Kupfer, C. (1974). Experimental glaucoma in the rhesus monkey. *Investigative Ophthalmology, 13*(6), 455–457.
217. Yun, H., et al. (2014). A laser-induced mouse model with long-term intraocular pressure elevation. *PLoS One, 9*(9), e107446.
218. Morgan, J. E., & Tribble, J. R. (2015). Microbead models in glaucoma. *Experimental Eye Research, 141*, 9–14.
219. Zhao, J., et al. (2016). Optic neuropathy and increased retinal glial fibrillary acidic protein due to microbead-induced ocular hypertension in the rabbit. *International Journal of Ophthalmology, 9*(12), 1732–1739.
220. Morrison, J. C., et al. (1997). A rat model of chronic pressure-induced optic nerve damage. *Experimental Eye Research, 64*(1), 85–96.
221. Ruiz-Ederra, J., & Verkman, A. S. (2006). Mouse model of sustained elevation in intraocular pressure produced by episcleral vein occlusion. *Experimental Eye Research, 82*(5), 879–884.

222. Danias, J., Shen, F., Kavalarakis, M., Chen, B., Goldblum, D., Lee, K., et al. (2006). Characterization of retinal damage in the episcleral vein cauterization rat glaucoma model. *Experimental Eye Research, 82*(2), 219–228.
223. Ruiz-Ederra, J., et al. (2005). The pig eye as a novel model of glaucoma. *Experimental Eye Research, 81*(5), 561–569.
224. Anders, F., et al. (2017). Proteomic profiling reveals crucial retinal protein alterations in the early phase of an experimental glaucoma model. *Graefe's Archive for Clinical and Experimental Ophthalmology, 255*(7), 1395–1407.
225. Urcola, J. H., Hernandez, M., & Vecino, E. (2006). Three experimental glaucoma models in rats: Comparison of the effects of intraocular pressure elevation on retinal ganglion cell size and death. *Experimental Eye Research, 83*(2), 429–437.
226. Johnson, D., Gottanka, J., Flügel, C., Hoffmann, F., Futa, R., & Lütjen-Drecoll, E. (1997). Ultrastructural changes in the trabecular meshwork of human eyes treated with corticosteroids. *Archives of Ophthalmology, 115*(3), 375–383.
227. Tawara, A., et al. (2008). Immunohistochemical evaluation of the extracellular matrix in trabecular meshwork in steroid-induced glaucoma. *Graefe's Archive for Clinical and Experimental Ophthalmology, 246*(7), 1021–1028.
228. Zode, G. S., et al. (2014). Ocular-specific ER stress reduction rescues glaucoma in murine glucocorticoid-induced glaucoma. *The Journal of Clinical Investigation, 124*(5), 1956–1965.
229. Whitlock, N. A., et al. (2010). Increased intraocular pressure in mice treated with dexamethasone. *Investigative Ophthalmology & Visual Science, 51*(12), 6496–6503.
230. Dawson, W. W., Brooks, D. E., Hope, G. M., Samuelson, D. A., Sherwood, M. B., Engel, H. M., et al. (1993). Primary open angle glaucomas in the rhesus monkey. *The British Journal of Ophthalmology, 77*(5), 302–310.
231. Gelatt, K. N., Gum, G. G., Gwin, R. M., Bromberg, N. M., Merideth, R. E., & Samuelson, D. A. (1981). Primary open angle glaucoma: Inherited primary open angle glaucoma in the beagle. *The American Journal of Pathology, 102*(2), 292–295.
232. Kuchtey, J., Olson, L. M., Rinkoski, T., Mackay, E. O., Iverson, T. M., Gelatt, K. N., et al. (2011). Mapping of the disease locus and identification of ADAMTS10 as a candidate gene in a canine model of primary open angle glaucoma. *PLoS Genetics, 7*(2), e1001306.
233. Stone, E. M., Fingert, J. H., Alward, W. L., Nguyen, T. D., Polansky, J. R., Sunden, S. L., et al. (1997). Identification of a gene that causes primary open angle glaucoma. *Science, 275*(5300), 668–670.
234. Zode, G. S., et al. (2011). Reduction of ER stress via a chemical chaperone prevents disease phenotypes in a mouse model of primary open angle glaucoma. *The Journal of Clinical Investigation, 121*(9), 3542–3553.
235. Senatorov, V., et al. (2006). Expression of mutated mouse myocilin induces open-angle glaucoma in transgenic mice. *The Journal of Neuroscience, 26*(46), 11903–11914.
236. McDowell, C. M., et al. (2012). Mutant human myocilin induces strain specific differences in ocular hypertension and optic nerve damage in mice. *Experimental Eye Research, 100*, 65–72.
237. Shepard, A. R., et al. (2007). Glaucoma-causing myocilin mutants require the Peroxisomal targeting signal-1 receptor (PTS1R) to elevate intraocular pressure. *Human Molecular Genetics, 16*(6), 609–617.
238. Millar, J. C., Pang, I. H., Wang, W. H., Wang, Y., & Clark, A. F. (2008). Effect of immunomodulation with anti-CD40L antibody on adenoviral-mediated transgene expression in mouse anterior segment. *Molecular Vision, 14*, 10–19.
239. Pang, I. H., Millar, J. C., & Clark, A. F. (2015). Elevation of intraocular pressure in rodents using viral vectors targeting the trabecular meshwork. *Experimental Eye Research, 141*, 33–41.
240. Shepard, A. R., et al. (2010). Adenoviral gene transfer of active human transforming growth factor-{beta}2 elevates intraocular pressure and reduces outflow facility in rodent eyes. *Investigative Ophthalmology & Visual Science, 51*(4), 2067–2076.

241. Wang, W. H., et al. (2008). Increased expression of the WNT antagonist sFRP-1 in glaucoma elevates intraocular pressure. *The Journal of Clinical Investigation, 118*(3), 1056–1064.
242. Jain, A., et al. (2017). CRISPR-Cas9-based treatment of myocilin-associated glaucoma. *Proceedings of the National Academy of Sciences of the United States of America, 114*(42), 11199–11204.
243. Mallick, J., et al. (2016). Update on normal tension glaucoma. *J. Ophthalmic Vis. Res., 11*(2), 204–208.
244. Watase, K., Hashimoto, K., Kano, M., Yamada, K., Watanabe, M., Inoue, Y., et al. (1998). Motor discoordination and increased susceptibility to cerebellar injury in GLAST mutant mice. *The European Journal of Neuroscience, 10*(3), 976–988.
245. Harada, T., et al. (2007). The potential role of glutamate transporters in the pathogenesis of normal tension glaucoma. *The Journal of Clinical Investigation, 117*(7), 1763–1770.
246. Bustelo, X. R. (2014). Vav family exchange factors: An integrated regulatory and functional view. *Small GTPases, 5*(2), 9.
247. Fujikawa, K., et al. (2010). VAV2 and VAV3 as candidate disease genes for spontaneous glaucoma in mice and humans. *PLoS One, 5*(2), e9050.
248. Chang, B., et al. (1999). Interacting loci cause severe iris atrophy and glaucoma in DBA/2J mice. *Nature Genetics, 21*(4), 405–409.
249. Anderson, M. G., et al. (2002). Mutations in genes encoding melanosomal proteins cause pigmentary glaucoma in DBA/2J mice. *Nature Genetics, 30*(1), 81–85.
250. Wang, R. F., Schumer, R. A., Serle, J. B., & Podos, S. M. (1998). A comparison of argon laser and diode laser photocoagulation of the trabecular meshwork to produce trabecular meshwork to produce the glaucoma monkey model. *Journal of Glaucoma, 7*(1), 45–49.
251. Zhang, Y., et al. (2009). Morphological and hydrodynamic correlates in monkey eyes with laser induced glaucoma. *Experimental Eye Research, 89*(5), 748–756.
252. Ueda, J., Sawaguchi, S., Hanyu, T., Yaoeda, K., Fukuchi, T., Abe, H., et al. (1998). Experimental glaucoma model in the rat induced by laser trabecular photocoagulation after an intracameral injection of India ink. *Japanese Journal of Ophthalmology, 42*(5), 337–344.
253. Rosolen, S. G., Rigaudière, F., & Le Gargasson, J. F. (2003). A new model of induced ocular hyperpressure using the minipig. *Journal Français d'Ophtalmologie, 26*(3), 259–267.
254. Song, W. T., Zhang, X. Y., & Xia, X. B. (2015). Atoh7 promotes the differentiation of Muller cells-derived retinal stem cells into retinal ganglion cells in a rat model of glaucoma. *Experimental Biology and Medicine (Maywood, N.J.), 240*(5), 682–690.
255. Hu, Y., et al. (2013). Bone marrow mesenchymal stem cells protect against retinal ganglion cell loss in aged rats with glaucoma. *Clinical Interventions in Aging, 8*, 1467–1470.
256. Zhou, X., Xia, X. B., & Xiong, S. Q. (2013). Neuro-protection of retinal stem cells transplantation combined with copolymer-1 immunization in a rat model of glaucoma. *Molecular and Cellular Neurosciences, 54*, 1–8.
257. Harper, M. M., et al. (2011). Transplantation of BDNF-secreting mesenchymal stem cells provides neuroprotection in chronically hypertensive rat eyes. *Investigative Ophthalmology & Visual Science, 52*(7), 4506–4515.
258. Roubeix, C., et al. (2015). Intraocular pressure reduction and neuroprotection conferred by bone marrow-derived mesenchymal stem cells in an animal model of glaucoma. *Stem Cell Research & Therapy, 6*, 177.
259. Emre, E., et al. (2015). Neuroprotective effects of intravitreally transplanted adipose tissue and bone marrow-derived mesenchymal stem cells in an experimental ocular hypertension model. *Cytotherapy, 17*(5), 543–559.
260. Sawaguchi, K., et al. (2005). Myocilin gene expression in the trabecular meshwork of rats in a steroid-induced ocular hypertension model. *Ophthalmic Research, 37*(5), 235–242.
261. Fingert, J. H., Clark, A. F., Criag, J. E., Alward, W. L. M., Snibson, G. R., McLaughlin, M., et al. (2001). Evaluation of the myocilin (MYOC) glaucoma gene in monkey and human steroid-induced ocular hypertension. *Investigative Ophthalmology & Visual Science, 42*, 145–152.

262. Tektas, O. Y., Hammer, C. M., Danias, J., Candia, O., Gerometta, R., Podos, S. M., et al. (2010). Morphologic changes in the outflow pathways of bovine eyes treated with corticosteroids. *Investigative Ophthalmology & Visual Science, 51*(8), 4060–4066.
263. Candia, O. A., et al. (2010). Suppression of corticosteroid-induced ocular hypertension in sheep by anecortave. *Archives of Ophthalmology, 128*(3), 338–343.
264. Zhang, Z., et al. (2014). Outflow tract ablation using a conditionally cytotoxic feline immunodeficiency viral vector. *Investigative Ophthalmology & Visual Science, 55*(2), 935–940.

Chapter 5
Limbal Stem Cells and the Treatment of Limbal Stem Cell Deficiency

Bruce R. Ksander, Markus H. Frank, and Natasha Y. Frank

Abstract Many organs (skin, stomach, intestines, colon, and eye) possess an epithelial layer that is short-lived and rapidly lost, requiring a source of adult stem cells that produces a continual supply of epithelial cells to replenish the lost cells. Maintaining these rapidly self-renewing epithelial surfaces during normal homeostasis is therefore dependent upon the health of the adult stem cell population. One important challenge in regenerative medicine is replacing these adult stem cells when they are eliminated following an injury or disease. The eye contains two highly specialized stratified squamous epithelia, the conjunctival epithelium and the corneal epithelium, which are separated by the limbal epithelium (Fig. 5.1). A healthy corneal epithelium is essential for maintaining a clear cornea and normal vision. The limbus contains a small subpopulation of rare *LSC* (*L*imbal *S*tem *C*ells) that continually repopulates the corneal epithelium. Patients with a *LSCD* (*L*imbal *S*tem *C*ell *D*eficiency) are unable to regenerate the corneal epithelium, resulting in migration of the conjunctival epithelium over the corneal stroma, called "conjunctivalization," that triggers neovascularization, chronic inflammation, and corneal opacity. A *complete* LSCD results in the total loss of the corneal epithelium and blindness due to an irreversibly opaque cornea. The extent of LSC loss can range

B. R. Ksander (✉)
Schepens Eye Research Institute, Massachusetts Eye and Ear Infirmary, Boston, MA, USA
e-mail: bruce_ksander@meei.harvard.edu; bruce.ksander@schepens.harvard.edu

M. H. Frank
Transplantation Research Center, Boston Children's Hospital, Boston, MA, USA

Department of Dermatology, Brigham and Women's Hospital, Boston, MA, USA
e-mail: markus.frank@childrens.harvard.edu

N. Y. Frank
Department of Medicine, VA Boston Healthcare System, Harvard Medical School, Boston, MA, USA

Division of Genetics, Brigham and Women's Hospital, Boston, MA, USA
e-mail: nfrank@partners.org

B. G. Ballios, M. J. Young (eds.), *Regenerative Medicine and Stem Cell Therapy for the Eye*, Fundamental Biomedical Technologies,
https://doi.org/10.1007/978-3-319-98080-5_5

from partial to complete and can be either unilateral or bilateral with a corresponding range in the loss of vision. LSCD can be caused by a variety of injuries or diseases: chemical or thermal burns [1], Stevens-Johnson syndrome [2, 3], aniridia [3], contact lens-induced keratopathy [4], multiple surgeries [5], cryotherapy of the limbus [5], chronic peripheral corneal inflammation [6], and lysosomal storage disease. However, corneal burns are by far the most frequent cause of a LSCD [5].

Keywords Limbal stem cells · Limbal stem cell deficiency · Cultured limbal epithelial transplantation · Allogeneic transplantation · Transplant immunology

5.1 Introduction

Many organs (skin, stomach, intestines, colon, and eye) possess an epithelial layer that is short-lived and rapidly lost, requiring a source of adult stem cells that produces a continual supply of epithelial cells to replenish the lost cells. Maintaining these rapidly self-renewing epithelial surfaces during normal homeostasis is therefore dependent upon the health of the adult stem cell population. One important challenge in regenerative medicine is replacing these adult stem cells when they are eliminated following an injury or disease. The eye contains two highly specialized stratified squamous epithelia, the conjunctival epithelium and the corneal epithelium, which are separated by the limbal epithelium (Fig. 5.1). A healthy corneal epithelium is essential for maintaining a clear

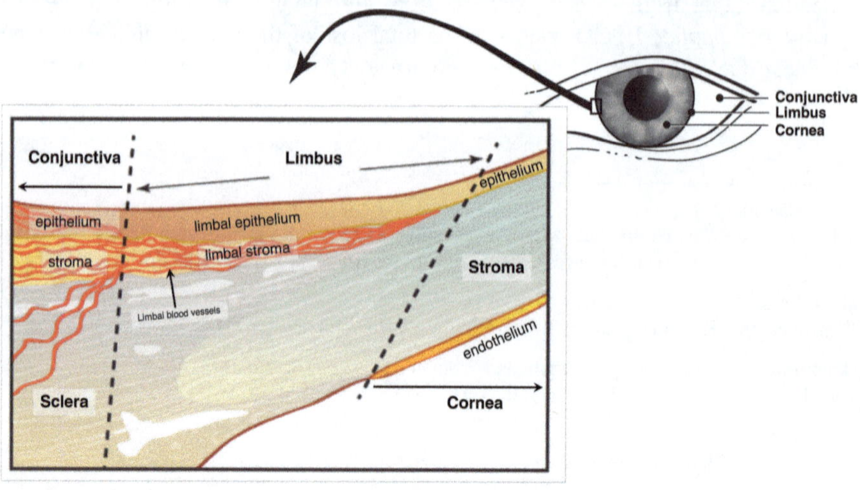

Fig. 5.1 *Anatomy of the limbus*. The limbus separates the conjunctiva from the cornea. The limbal blood vessels in the stroma can be used to identify where the corneal epithelium ends, and the limbal epithelium begins

cornea and normal vision. The limbus contains a small subpopulation of rare *LSC* (*L*imbal *S*tem *C*ells) that continually repopulates the corneal epithelium. Patients with a *LSCD* (*L*imbal *S*tem *C*ell *D*eficiency) are unable to regenerate the corneal epithelium, resulting in migration of the conjunctival epithelium over the corneal stroma, called "conjunctivalization," that triggers neovascularization, chronic inflammation, and corneal opacity. A *complete* LSCD results in the total loss of the corneal epithelium and blindness due to an irreversibly opaque cornea. The extent of LSC loss can range from partial to complete and can be either unilateral or bilateral with a corresponding range in the loss of vision. LSCD can be caused by a variety of injuries or diseases: chemical or thermal burns [1], Stevens-Johnson syndrome [2, 3], aniridia [3], contact lens-induced keratopathy [4], multiple surgeries [5], cryotherapy of the limbus [5], chronic peripheral corneal inflammation [6], and lysosomal storage disease. However, corneal burns are by far the most frequent cause of a LSCD [5].

There has been remarkable clinical success in regenerating the corneal epithelium in patients with a *unilateral* LSCD, which dates back to the first successful report 28 years ago by Kenyon and Tseng [7] and continues to the present with the recent announcement of Holoclar® as the first advanced therapy medicinal product containing stem cells approved by the EMA (European Medical Agency) for treating unilateral LSCD due to ocular burns. Importantly, Holoclar® is the first such approval for *any* cell therapy treating an ocular or *non*-ocular disease [8]. Although, currently there is considerable interest in the development of novel stem cell therapies to treat retinal disease, these exciting retinal studies are in the beginning stages and have yet to provide any significant clinical success. For this reason, it could be argued that the long-standing achievements in corneal regeneration are underappreciated by both the field of regenerative medicine in general and even within the ophthalmologic community. Notwithstanding this clinical success, there is still room for improvement of the clinical LSCD treatments and still much to be learned about the molecular mechanisms by which LSC maintain and restore the ocular surface. In many respects, the knowledge of the molecular mechanisms that control adult stem cells that maintain the hair follicle, intestines, and colon are considerably more advanced even though there has yet to be any clinical success in regenerating these tissues [9]. The advances in these other tissues are mainly due to the relatively recent lineage tracing studies used to identify and track the stem cells in these target tissues which are only beginning to be conducted in the limbus [10–13]. In summary, while successful long-term regeneration of the corneal epithelium has been achieved in some unilateral LSCD patients, there are still many unmet clinical challenges, such as the treatment of complete *bilateral* LSCD and patients with a LSCD due to chronic ocular surface inflammation. Recently, there have been a number of research advances in identifying LSC and reprograming cells which may lead to new clinical approaches that will in turn lead to advances in our ability to regenerate the ocular surface in patients that are afflicted with the most severe and difficult to treat forms of ocular surface diseases.

5.2 Historical Perspective

5.2.1 Tissue Transplants

In 1983, Thoft and Friend proposed the X, Y, Z hypothesis, which was one of the first proposals that identified the limbus as the source of corneal epithelial stem cells [14]. However, the first experimental evidence that stem cells were contained within the limbus was reported in 1989 by Cotsarelis and Lavker who observed that label-retaining cells, a marker of quiescent stem cells, were present within the limbus but not present within the central cornea of mice [15]. These label-retaining cells were located within a subpopulation of basal limbal epithelial cells, providing the first evidence of the location of limbal stem cells within this tissue. This initial experimental evidence was followed within the same year by the first report of a clinical success in restoring the corneal epithelium in unilateral LSCD burn patients by Kenyon and Tseng [7], who based their surgical approach, in part, on the then recent discoveries of Cotsarelis and Lavker (K.K. personal communication). They removed a small biopsy of normal limbus from the unaffected eye of a patient with unilateral LSCD caused by an alkali burn and transplanted this tissue directly onto the limbal area of the effected eye, resulting in the long-term sustained regeneration of the corneal epithelium and improved vision in the affected eye. The sustained regeneration and continued maintenance of a clear cornea in some of these early patients treated by Kenyon and Tseng is demonstrated by fact that the patient's corneas still remain clear now, almost three decades after their surgery (K.K. personal communication). Subsequently, this procedure was used by a variety of ophthalmologists and, although the procedure was successful in some patients, it became apparent that the procedure had several drawbacks. We now know that LSC are not evenly distributed around the limbus [16], which caused the biopsies to not always contain sufficient numbers of LSC and, if larger limbal biopsies were removed, there was an increased risk of triggering a partial LSCD and reduced vision in the one remaining normal eye of the patient. Therefore, the low frequency and uneven distribution of LSC made the success of this approach difficult to predict. One solution to this problem was to develop a method that would increase the number of limbal stem cells obtained from the biopsy tissue, which lead to the next series of studies in the 1990s on culturing and expanding limbal epithelial cells in vitro.

5.2.2 In Vitro Expanded Cell Transplants

In an effort to increase the success of autologous limbal transplants by increasing the number of LSC obtained from the original small biopsy tissue, Pellegrini and co-workers worked to develop techniques to expand LSC by culturing biopsy-derived limbal epithelial cells in vitro [17]. They reported that the limbus, but not the central cornea, contained cells with the proliferative capacity to form holoclone colonies in vitro, suggesting these colonies contained cells with extensive self-renewing capacity, a benchmark for identifying stem cells. Additional studies indicated that a subpopulation of basal limbal epithelial cells expressed the ΔNp63α transcription factor and that holoclone cultures contained a subpopulation of ΔNp63α$^+$ cells [18, 19], leading to the proposal by Pellegrini et al. that ΔNp63α was a marker for limbal stem cells. p63 is a multi-isoform family of transcription factors that are required for epidermal development and also required for maintaining the adult epidermis. The majority of the described activity of p63 is contained within the ΔNp63 isoforms α, β, and γ [20, 21]. Therefore, it is logical that limbal stem cells would express ΔNp63α, but experimental proof for this was ultimately provided by Paolo Rama and co-workers who correlated limbal cell transplant success and corneal regeneration with the frequency of ΔNp63α positive cells within the donor graft, as described next.

The in vitro expansion of limbal stem cells led to the first report in 1997 of treating unilateral LSCD patients, not with a limbal biopsy tissue transplant, but with a form of autologous cell therapy in which the biopsy tissue was digested, the epithelial cells recovered and expanded within in vitro cultures, and then the expanded cells were transplanted back to the patient on a fibrin gel or amnionic membrane. During the last two decades, this approach has been used in experimental clinical studies at a number of different institutions with considerable success [3, 22–24]. However, the most extensive study was reported in 2010 by Paolo Rama et al. who described the results of 112 burn patients with a unilateral LSCD [25]. Seventy-seven percent of the patients displayed regeneration of the corneal epithelium with a median follow-up time of approximately 2 years and a maximum follow-up time of 10 years. Successful regeneration of the cornea led to loss of corneal neovascularization and opacity and a significant increase in the patient's visual acuity. Importantly, the success of these autologous transplants coincided with the frequency of p63 positive cells within companion samples of the in vitro expanded cultures. If the cultures contained <3% p63 positive cells, the transplants failed to regenerate the cornea. Cultures that contained >3% p63 positive cells successfully regenerated the corneal epithelium in 77% of the patients. It is interesting to note that the *maximum* in vitro expansion of p63 positive cells in these cultures was limited to only 10% with a median frequency of 5.5% positive cells, indicating that, although the overall number of LSC was increased in these cultures, the ability to expand LSC was not unlimited and restricted to a relatively small percentage of the overall cells in the donor graft. It is tempting to speculate that increasing the number of LSC within the donor graft so that it is over this 10% limit would result in even more effective regeneration of the ocular surface. Another important and surprising aspect of this study was that transplant success and regeneration of the cornea *did not coincide* with the frequency of holoclones, or clonogenic cells, within the cultures. This is important because a change in the holoclone frequency is used by many laboratories to indicate they have achieved an effect on the frequency of functional LSC, even though this transplantation study indicates there is no direct correlation between holoclone frequency and the regenerative capacity of the cultured LSC. This is clearly a topic that needs to be investigated and studied further.

While this in vitro culture technique increased the frequency of LSC in the autologous transplants and also increased the clinical success of this procedure as compared with the whole-tissue limbal biopsy transplant used originally by Kenyon and Tseng, this approach also had several potential drawbacks. Currently, there is no agreement on the optimal culture conditions needed to expand functionally active LSC, making it difficult to compare the clinical data obtained from different institutions [26]. Moreover, the original limbal biopsies recovered from the patients are heterogeneous and contain many different cell subpopulations, in addition to LSC. When these heterogeneous cells are cultured in vitro, there is variable expansion of the different cell types. In addition, since p63 is an intracellular transcription factor that can only be identified by permeabilizing and killing the cell, the frequency of p63 positive cells with the cultures can only be determined in companion cultures and not within the cells that are transplanted to the recipient eye. Therefore, it is unclear: (1) how many LSC are present originally prior to expansion, (2) what degree of stem cell expansion is achieved, (3) what other populations of cells are present and what effect these cells might have, if any, on corneal regeneration, and (4) whether the in vitro expanded LSC are functionally the same as the original stem cells? Therefore, the variability in the success of this approach reported in the literature between different institutions is likely due to the variations between techniques used to harvest, digest, and culture limbal epithelial cells. In spite of these potential issues, it is important to remember this is the first example of a cell therapy using laboratory cultured and expanded stem cells to successfully produce the long-term regeneration of the corneal epithelium in patients with a unilateral LSCD, resulting in restoration of vision.

5.3 Current Status and Future Treatment of LSCD Patients

5.3.1 Regeneration of the Corneal Epithelium in Unilateral *LSCD Patients*

5.3.1.1 Tissue Transplants

While significant advances have been made in the treatment of LSCD patients, treatment of *unilateral* LSCD patients is still more successful than treatment of patients with bilateral disease and this success is also limited mainly to burn patients. Autologous limbal biopsy tissue transplants from the unaffected eye are still the standard of care in the United States, since no in vitro expanded cell therapy has been approved by the FDA. Conjunctival limbal grafts and keratolimbal grafts are used with success rates of approximately 73% [27].

5.3.1.2 Simple Limbal Epithelial Transplant, SLET

One of the newest advances in treating unilateral burn LSCD patients has been the development of a procedure called *SLET* (*S*imple *L*imbal *E*pithelial *T*ransplant) by Sayon Basu and colleagues [28] which was developed in India as an alternative to the in vitro expansion of LSC that requires extensive GMP laboratory facilities in order to culture the cells under the new guidelines for cell therapies (European Medicines Agency, Doc. Ref. EMEA/CHMP/410869/2006) SLET is a modification of the autologous limbal biopsy transplant procedure originally developed by Kenyon and Tseng in which the small biopsy from the patient's normal contralateral eye is dissected into small pieces that are then glued onto the central corneal stroma of the affected eye after the pannus is removed, instead of placing the whole tissue over the limbal area as was done previously. Another modification used by the SLET technique appears to be that a deeper biopsy is taken so that it includes more of the limbal stroma. The effects of these modifications are significant with success rates reported to be at 80% which are as good as those obtained with in vitro expanded cell therapies. The improved success of the SLET technique may be related to the transplantation of more limbal stroma, resulting in establishment of multiple small stem cell niches around the separate dissected biopsy transplants placed in the central cornea. This idea is supported by the observation of "swirls" of epithelium emanating from the pieces of transplanted tissue. The mechanism of regeneration via this technique is incompletely understood and whether this is a temporary niche that is maintained until the limbus is re-established or is a permanent new niche is unknown. It seems unlikely that a permanent stem cell niche could be established in the central cornea, since a normal cornea is avascular and most stem cell niches are found within close proximity to a rich vascular supply [29]. If more widespread use of the SLET technique confirms the high success rate of this technique over the whole-tissue transplants (conjunctival limbal grafts and kerato-limbal grafts), then it will be important to determine the scientific basis for this increased regenerative capacity, which may provide important insights into how the donor stem cells are engrafted into the recipient graft bed and how the supporting stem cell niche is re-established.

5.3.1.3 In Vitro Expanded Cultured Cells, Holoclar®

As mentioned in the beginning of this article, in 2015 Holoclar® was the first advanced therapy medicinal product containing stem cells approved by the EMA (European Medical Agency) for treating unilateral limbal stem cell deficiency due to ocular burns and this was the first approval granted for *any* type of cell therapy by the EMA since the new guidelines for cell therapies were implemented. Holoclar® is not available in the United States and has not yet been approved by the FDA, so there is currently no in vitro expanded cell therapy available for patients in the

USA. However, there is a new clinical trial that will compare autologous whole tissue transplants with in vitro cultured limbal cells using a modification of the approach used by Holoclar® to treat unilateral LSCD patients (ClinicalTrials.gov; identifier: NCT02592330). In addition, Sophie Deng and co-workers have reported the development of a feeder-cell-free and xenobiotic-free culture system to expand limbal epithelial cells in vitro, as well as, a novel Wnt activator with the potential to stimulate LSC [30–32]. These new developments may lead to improved methods to expand LSC in vitro for autologous transplants in unilateral LSCD burn patients.

5.3.1.4 In Vitro Expanded Oral Mucosal Epithelium

Transplants containing in vitro cultured autologous oral mucosal epithelial cells were originally developed and used primarily in patients with bilateral LSCD that have no source of autologous LSC. However, the rationale for using this approach in *unilateral* LSCD patients is that it eliminates the risk associated with removing normal limbus from the only remaining normal eye of the patient [6]. While this approach has been used with some level of success, one of the main complications is corneal neovascularization.

5.3.2 Regeneration of the Corneal Epithelium in Bilateral *LSCD Patients Using Allogeneic Tissue*

Patients with bilateral LSCD have no source of autologous LSC, since the reservoir of stem cells within the limbus of both eyes has been depleted, and therefore investigators have attempted to treat these patients with transplants using *allogeneic* limbal tissue or in vitro cultured limbal epithelial cells that were obtained from cadaveric donor corneas. In general, allogeneic limbal transplants in bilateral LSCD patients have a worse outcome as compared with the success observed with autologous limbal transplants in unilateral LSCD patients. This is most likely of result of allograft rejection due to the activation of donor alloantigen-specific immunity.

5.3.2.1 In Vitro Expanded Cells, Holoclar®

Paolo Rama and colleagues have attempted to use allogeneic in vitro expanded limbal cells (Holoclar®) to treat bilateral LSCD patients without success. However, they have reported some success with a limited number of bilateral LSCD patients in which they recovered a few remaining *autologous* LSC from a limbal biopsy recovered from one of the damaged eyes. These autologous cells were then cultured and expanded in vitro and transplanted back to the patient [25]. It is unclear how successful this approach will be in treating bilateral LSCD patients, since it will depend upon the frequency of bilateral LSCD patients that have residual LSC that can be recovered and successfully cultured in vitro. This approach will also depend upon whether these residual LSC retain their regenerative capacity. However, these results indicate the interesting possibility that nonfunctional LSC that have lost their ability to maintain the corneal epithelium may still be found within the damaged limbal tissue of patients with the diagnosis of a "complete" bilateral LSCD and that these LSC can be "resuscitated" into fully functional stem cells with regenerative capacity. This also highlights the current problems with diagnosing patients with a LSCD. By definition, a patient with a complete LSCD has lost all LSC. However, this diagnosis is subjective and based completely upon the ophthalmologist's visual evaluation of the ocular surface and whether conjunctivalization has occurred. The presence of conjunctival epithelium covering the cornea identifies a loss of LSC function but does not *prove* a complete loss of all stem cells, which may still be present with either reduced or no functional activity. This also highlights the need to develop methods to identify LSC in situ which would greatly enhance the ability to diagnose partial and complete LSCD patients and also identify potential autologous LSC that are present but nonfunctional that could be recovered and revived in vitro. A more complete review of the problems associated with diagnosis of LSCD has been provided by Sophie Deng et al. [33, 34], who has also found evidence of LSC "crypts" that remain protected in patients diagnosed with a complete LSCD and may therefore be another valuable source of residual LSC in patients diagnosed with a complete LSCD [35].

5.3.2.2 Tissue Transplants

5.3.2.2.1 SLET: Simple Limbal Epithelial Transplants

Allogeneic SLET is not primarily used as a long-term treatment for patients with bilateral LSCD because of the need for long-term immunosuppressive therapy and the potential for allograft rejection but has been used as a short-term treatment of bilateral ocular burn patients to stabilize the ocular surface while a more permanent second treatment is arranged [28]. This second treatment is either an in vitro expanded autologous oral mucosal epithelial cell transplant or Boston KPro. The fact that allogeneic SLET can stabilize the ocular surface of these patients suggests the potential pro-wound-healing properties of limbal epithelial cells even though this tissue contains numerous highly immunogenic cells capable of inducing alloimmunity. While the mechanism of this protective/stabilizing effect of allogeneic SLET in bilateral burn patients is unknown, it is tempting to speculate this could be due to the immunosuppressive properties of LSC.

5.3.2.2.2 Keratolimbal Allografts (KLAL)

Ed Holland is a leader in the field of allogeneic tissue transplantation for patients with LSCD and has treated patients with keratolimbal allografts (KLAL) and conjunctival limbal allografts (CLAL) in which the limbal tissue is attached to the corneoscleral carrier tissue [27, 36]. Since allogeneic tissue from cadaveric donors is used, it is possible to transplant large grafts covering the entire limbus, providing the opportunity to transplant more LSC as compared with the relatively small number of LSC transplanted onto unilateral LSCD patients that receive a small autologous biopsy graft. Although the larger allografts deliver a considerably higher number of potential LSC, they also provide a potent stimulus of anti-donor alloimmunity that can result in allograft rejection and destruction of all donor tissue. For this reason, preservation of the donor tissue and prevention of allograft immune-mediated rejection requires long-term systemic immunosuppression similar to the treatment used to prevent kidney allograft rejection. Using this approach, Ed Holland and colleagues have reported a 73% success rate in treating LSCD patients with KLAL (94 eyes at 4.7 years) [27, 36]. While other investigators, such as Baradaran-Rafii et al. have reported 73% of patients with a stable ocular surface after repeated KLAL surgeries due to a dry ocular surface [37]. Han et al. reported a stable corneal epithelium in 33% of 46 KLAL surgeries on 24 eyes at 2 years [38]. While this treatment approach can allow long-term allograft survival in bilateral LSCD patients, there is a significant risk of toxicity from the systemic immune suppression regimen.

5.4 Activation of Anti-donor Alloimmunity to Limbal Allografts

As will be discussed in more detail in the subsequent section, the central cornea is an immune privileged tissue that does not trigger destructive allograft immunity and therefore corneal allograft recipients do not need prolonged immunosuppressive treatment. By contrast, the limbus and conjunctiva are *not immune privileged* tissues and contain the normal array of immune cells found within the skin: macrophages, dendritic cells, and Langerhans cells, which all express MHC class I and class II that can trigger donor alloantigen-specific immunity via either the Direct and/or Indirect Pathways (Fig. 5.2). The *Direct Pathway* is activated when donor MHC-peptide complexes are recognized directly by recipient T cells [39]. Typically, the MHC-peptide complexes are presented on the surface of the donor cells, which in the case of limbal transplants would be when "passenger" donor dendritic cells migrate out of the limbal graft and travel to secondary lymphoid organs, where recipient allospecific precursor CD4+ T cells and CD8+ T cells recognize the foreign MHC-peptide complexes and are activated and clonally expanded. It is well known that within the general population there is a high-precursor frequency of T cells with direct MHC allospecificity.

The *Indirect Pathway* is activated when donor MHC proteins are broken down into peptides and presented via recipient antigen presenting cells [39]. This occurs when recipient dendritic cells migrate into the donor limbal transplant, resulting in phagocytosis of dead donor cells and reprocessing of the foreign donor class I and/or class II proteins into peptide fragments that are presented on the cell surface of the

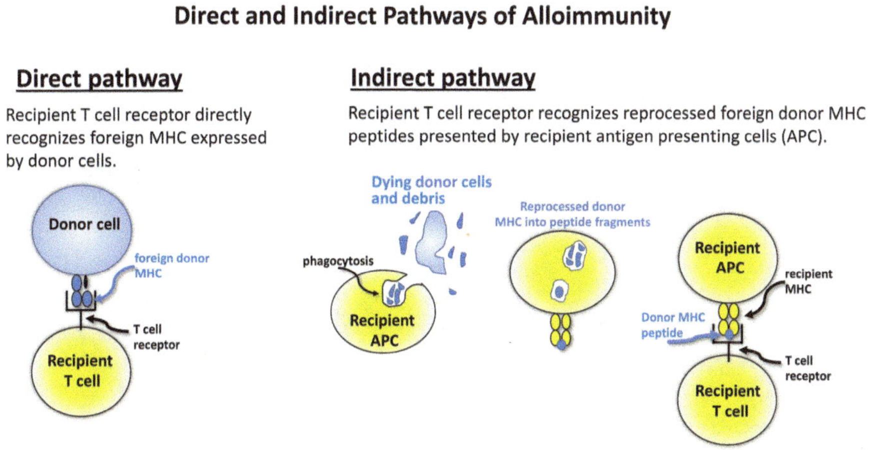

Fig. 5.2 Direct and indirect pathways of alloimmunity. Major Histocompatibility Complex (MHC) incompatible donor transplants are recognized as foreign by the recipient's immune system, which responds through two pathways. In the Direct Pathway, recipient T cell receptors directly recognize the foreign donor MHC. By contrast, the Indirect Pathway is activated when recipient antigen presenting cells (APC) migrate into the donor graft and phagocytize dead donor cells and/or debris, which is then reprocessed into small peptide fragments that are ultimately presented via the recipient's MHC to T cells

recipient macrophages via self MHC. These recipient macrophages cells carry the alloantigen from the graft site to the draining secondary lymphoid organs where they activate self-MHC restricted CD4+ and CD8+ T cells that are specific for the alloantigenic peptides. Therefore, two populations of effector T cells can be activated in the lymphoid organs against the donor tissue, T cells that can recognize either (1) foreign donor MHC-peptide complexes directly, and/or (2) foreign donor MHC peptides presented indirectly in the context of self MHC. Once these effector T cells are activated and expanded within the lymph nodes, they migrate systemically throughout the host via the blood vasculature in search of the targeted foreign cells.

The targets of the *direct allospecific T cells* would be donor class I and/or class II positive cells, which would include all limbal epithelial cells (class I positive) and passenger immune cells (class I and II positive) within the transplant. In addition, direct allospecific T cells would also target the fully differentiated mature class I positive corneal epithelial cells that are produced by donor LSC and migrate from the limbal transplant onto the recipient corneal surface (Fig. 5.3). Activated *indirect allospecific T cells* have a more restricted group of target cells and would only recognize *recipient* macrophages within the donor graft that express reprocessed donor allogeneic peptides. Importantly, stem cells, in general, *do not express* either MHC class I or class II and would therefore not be recognized by recipient alloreactive T cells activated via either the Direct or Indirect Pathways. Moreover, they would also not be capable of inducing alloimmunity, unless it was via expression of minor histocompatibility alloantigens [40]. Therefore, the only possible way that immune destruction of donor LSC within the limbal transplant is possible would be via the

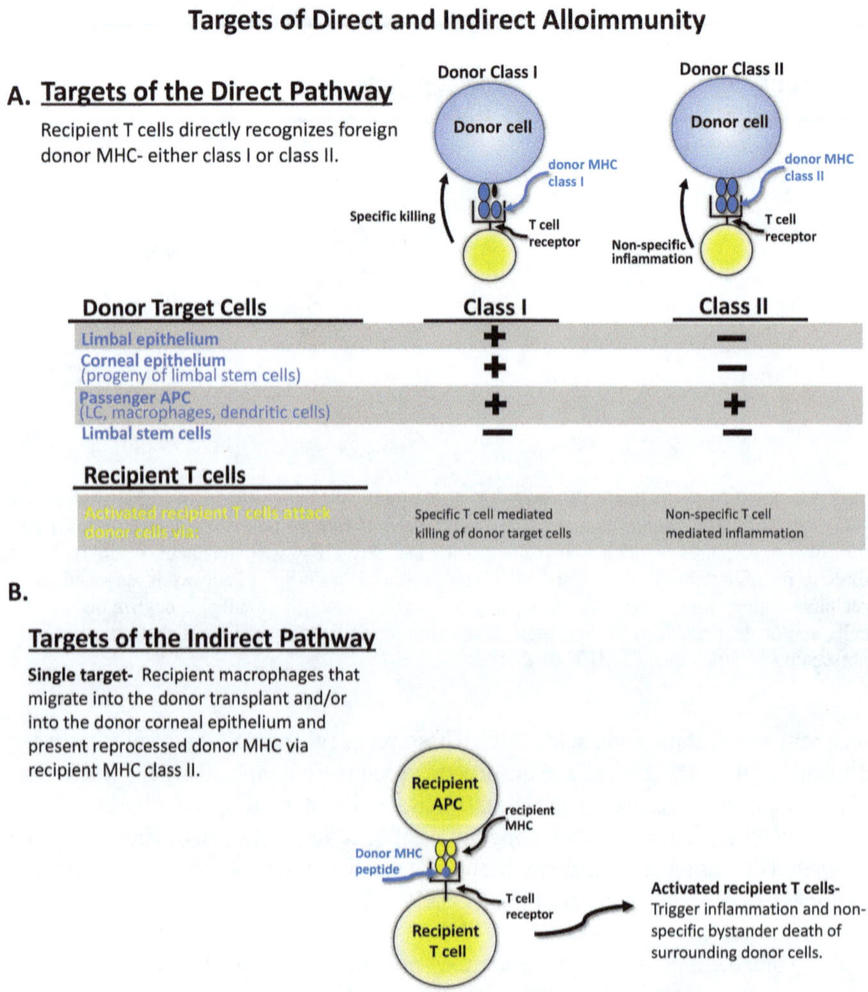

Fig. 5.3 Targets of direct and indirect alloimmunity. Once the recipient's immune system has activated T cells via either the Direct and/or Indirect pathways, the activated T cells recirculate through the blood vasculature in search of the foreign donor cells. T cells activated via the Direct Pathway can recognize a variety of donor target cells (**a**), whereas T cells activated via the Indirect Pathway have a very limited range of target cells and only respond to recipient APC that have migrated into the donor transplant and are presenting reprocessed donor MHC peptides (**b**)

"collateral damage" effect when inflammatory CD4+ allospecific T cells release cytokines that nonspecifically damage and/or kill nearby cells, such as LSC. As discussed in the following section, this has important implications for the transplantation of purified allogeneic LSC.

Undisputed circumstantial evidence that rejection of allogeneic limbal transplants is due to anti-donor immunity is provided by the requirement of systemic immunosuppression and the loss of donor grafts when this suppressive therapy is either too short or stopped prematurely [37, 41]. However, conclusive proof of allograft rejection, such as studies that detect expanded frequencies of donor-specific alloreactive T cells in the peripheral blood of these patients has yet to be conducted. A study that identified the donor cells that trigger anti-graft alloimmunity and the contribution of effector T cells activated via the Direct versus Indirect pathways would provide important information that could lead to more targeted forms of immune suppression that would result in a higher frequency of allograft acceptance and lower toxicity from the immunosuppressive treatment. This could reduce the difficulty in managing the current long-term systemic immunosuppressive therapy and potential toxicity associated with KLAL.

There are at least two potential solutions to the problem of allograft rejection and treatment of bilateral LSCD patients: (1) overcoming the allogeneic immune barriers that prevent successful engraftment of allogeneic donor limbal stem cells, or (2) use an alternative source of *autologous* adult stem cells that are either reprogramed or redirected into becoming limbal stem cells that successfully regenerate and maintain the corneal epithelium.

5.5 Overcoming Allogeneic Immune Barriers to LSC Allografts

5.5.1 Immune Privilege

While the central cornea is an immune privileged tissue that escapes from allograft immunity, the limbus is *not immune privileged* and can induce allograft immunity. Therefore, an important question is whether it is possible to extend immune privilege to allow the engraftment of allogeneic limbal stem cells? This may be possible, if the limbal stem cells display immunosuppressive/immune privilege properties and these cells can be separated from the cells within the limbus that are potent inducers of alloimmunity; the donor passenger macrophages, dendritic cells, and Langerhans cells. In other words, although the limbal tissue in aggregate does not display immune privilege, it may contain within this tissue a few rare cells, such as LSC, that possess immune privilege. Therefore, if the allogeneic donor immune privileged LSC are purified and separated from the surrounding donor immunogenic cells, then these purified donor LSC allografts would engraft and escape immune destruction. This possibility will be discussed below, but first, the basis for immune privilege in the central cornea will be briefly summarized.

Peter Medawar first proposed the concept of "immune privilege" over 50 years ago which was based upon the idea that certain anatomical sites allowed transplanted allografts to survive for an extended period of time, as compared with the survival of the same allograft transplanted into a non-privileged site [42–44]. The extent of immune privilege was determined by how long the allograft survived, which varied from "limited" immune privilege where allograft survival was only extended for a brief time, to "absolute" immune privilege where the allograft was completely accepted and survived indefinitely within the host. In general, the level of immune privilege is *inversely related* to the immunogenicity of the allograft. Highly immunogenic allografts display more limited survival within privileged sites, as compared with weakly immunogenic allografts that survive longer within the same privileged site. Historically, the most well-known and studied immune privileged anatomical sites are the eye, brain, testis, and in the maternal/fetal interface [45–47].

The best known and most dramatic clinical example of immune privilege is the long-term survival of allogeneic full-thickness corneal transplants. However, even with these transplants the extent of immune privilege varies and depends upon whether the recipient is "low-risk" or "high-risk." Low-risk recipients that lack corneal inflammation and neovascularization in the graft bed display high levels of immune privilege and accept 90–95% of corneal allografts with only transient steroid treatment. By contrast, high-risk recipients with an inflamed and/or vascularized graft bed, reject approximately 70–75% of corneal allografts within the first year, in spite of local or systemic corticosteroid treatment (reviewed in [48]). Thus, the important factor that determines allograft survival in the cornea is the status of the recipient graft bed.

Extensive experimental studies revealed that a variety of overlapping mechanisms contribute to immune privilege that allows the survival of corneal allografts in low-risk recipients [43, 44, 49, 50]. Adaptive immunity is inhibited by activation of regulatory T cells (TREG) and innate immunity is inhibited by local factors within the aqueous humor and tear film. In addition, the delivery of corneal alloantigens to the local lymph nodes and spleen is diminished because the central cornea lacks: (1) vascular and lymphatic drainage, and (2) mature antigen presenting cells. Finally, corneal epithelial cells display low levels of MHC class I and no class II, reducing their immunogenic potential. Together, these mechanisms contribute to the long-term survival of corneal allografts in low-risk recipients.

LSCD patients have lost the immune-privileged environment of the central cornea. The limbus forms the barrier between the immune-privileged cornea and the surrounding non-privileged conjunctiva, which is highly vascularized and contains numerous resident pro-inflammatory cells. When the limbus is damaged or destroyed, the LSC are lost and, equally important, the barrier function of the limbal tissue is also lost. Without the limbal barrier, the conjunctival epithelium migrates over the cornea, resulting in conjunctivalization and a highly vascularized and inflamed cornea that no longer possesses immune privilege. Restoration of a normal clear avascular corneal epithelium in treated LSCD patients requires that the barrier between the cornea and conjunctiva is reestablished, as well as, the immune privileged environment of the central cornea.

5.5.2 Stem Cells and Immune Privilege

Since adult stem cells are quiescent long-lived cells that are essential to maintaining normal cell homeostasis and wound repair in a variety of organs, there is a strong rationale supporting the idea that adult stem cells should be protected by immune privilege so they cannot be inadvertently destroyed by nonspecific local inflammation. In support of this, there is considerable evidence indicating that adult stem cells and their supporting niche are immune privileged. Mesenchymal stem cells are well known to suppress inflammation via a variety of mechanisms and in some instances are reported to induce tolerance to alloantigens [51, 52]. However, results that are highly relevant to LSC transplantation were recently reported indicating that bone marrow (BM) is an immune privileged tissue. These studies were originally based upon the finding that BM contains a higher frequency of CD4+ CD25+ FoxP3+ TREG cells, as compared with other lymphoid organs [53]. Based on this observation, Fujisaki et al. [54] tested whether hematopoietic stem cell/progenitor cells (HSPC) established an immune privilege niche within the bone marrow, in which allogeneic donor HSPC could survive. They demonstrated that allogeneic HSPC injected into the BM of recipient mice survived equally as well as syngeneic HSPC without the benefit of any immunosuppressive therapy. The immune privilege environment within the BM was maintained by TREG cells that colocalize with the donor HSPC within the bone marrow niche. The TREG cells created a localized immune privileged "zone" where the allogeneic HSPC survived and, if the TREG cells were depleted, immune privilege was lost, resulting in rejection of the donor HSPC. Further studies indicated the unique niche-associated TREG cells blocked allograft rejection via adenosine [55]. Therefore, this is an example of how a localized "immune privileged zone" can be established within a tissue that protects a valuable adult stem cell population. Is it possible that a similar mechanism is present within the limbus that protects the equally valuable LSC, in which a localized immune privileged zone is established that protects the LSC and their surrounding niche? This possibility can now be tested directly, since LSC can be purified and separated from the surrounding immunogenic cells using antibodies specific for ABCB5.

5.5.3 ABCB5+ Limbal Stem Cells

Our studies demonstrated that the ABCB5 gene, a relatively new member of the ATP-binding cassette (ABC) superfamily of active transporters, is expressed by LSC in both mouse and human limbus [56]. The normal function of ABCB5+ LSC is required for proper corneal development and wound repair via its essential roles in LSC maintenance and survival, as demonstrated by the fact that knockout mice that lack ABCB5 do not develop a fully differentiated mature corneal epithelium and are unable to properly repair epithelial wounds. Importantly, ABCB5 is a *cell surface protein* and specific monoclonal antibodies are capable of isolating pure ABCB5-positive cells from the limbus. Transplantation of the purified ABCB5^{positive} (but not ABCB5^{negative}) limbal epithelial cells onto the corneal stroma of mice with an induced LSCD resulted in long-term regeneration of the corneal epithelium (Fig. 5.4). Therefore, not only do these data demonstrate the regenerative capacity of limbal epithelial cells is contained within the cells that express ABCB5, but they also demonstrate that anti-ABCB5 antibodies can be used to purify ABCB5+ LSC

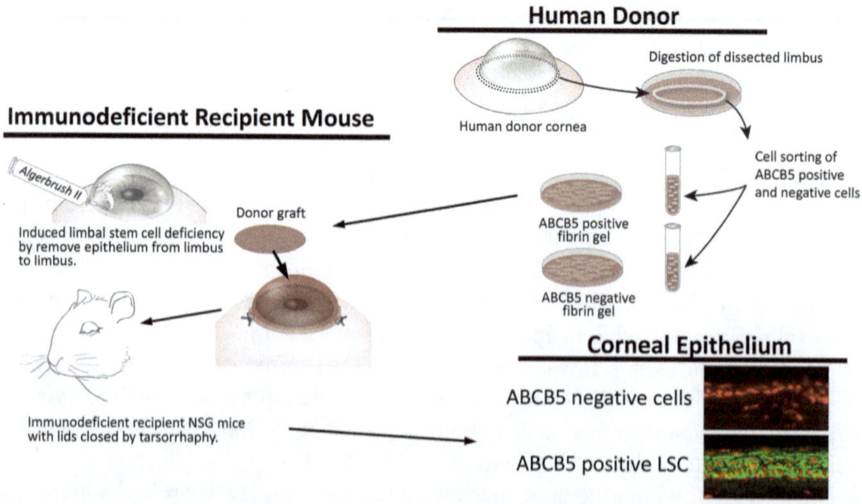

Fig. 5.4 Purification and transplantation of ABCB5 positive limbal stem cells. Evidence of the regenerative capacity of purified ABCB5 positive limbal stem cells was provided by xenogeneic transplantation experiments in which human donor grafts containing either purified ABCB5 positive or ABCB5 negative cells were transplanted onto immunodeficient mice with an induced limbal stem cell deficiency. Mice receiving grafts containing ABCB5 positive LSC, but not ABCB5 negative cells, were capable of regenerating Krt12 positive corneal epithelial cells (green staining). Dapi stained nuclei (red)

and separate LSC from the other immunogenic cells contained within the limbus. Using this approach in the murine model system, it is now possible to determine whether purified allogeneic ABCB5+ LSC can establish immune privilege and long-term engraftment in recipients without inducing destructive alloimmunity and without receiving any immunosuppressive therapy.

The possibility that the limbus contains immunosuppressive/immunoregulatory ABCB5+ LSC is supported by the observation that mouse and human *dermal skin* contains a small subpopulation of previously unrecognized ABCB5+ immunoregulatory cells that are CD45 negative and therefore distinct from the known immunoregulatory hematopoietic-linage cells within the skin [57]. These ABCB5+ cells express the programmed cell death protein 1 receptor (PD-1) which is an immune checkpoint molecule that is primarily expressed on T cells and, when PD-1 on T cells is triggered by its ligand PD-L1, inhibits immunity by suppressing inflammatory T cells and promotes immune tolerance by enhancing the survival of immunosuppressive regulatory T cells [58]. Unexpectedly, PD-1 expressed on purified ABCB5+ skin cells also triggered potent inhibition of primary immune responses in vivo, inhibiting both mitogen-driven and alloantigen-driven T cell activity. As a result of this inhibitory activity, allogeneic donor ABCB5+ skin cells were successfully engrafted into fully MHC-mismatched recipient mice in the absence of any immunosuppressive therapy, demonstrating the immune privilege capacity of ABCB5+ skin cells. This data supports the possibility that ABCB5+ LSC may also display immunosuppressive/immune privilege capabilities and, in following with this prediction, we recently detected the PD-1 receptor on ABCB5+ LSC.

However, even if allogeneic ABCB5+ LSC are capable of establishing a local immune privileged niche within the limbus, an important remaining issue is whether this immune privilege would also be extended outside the limbus and protect the progeny of LSC, the differentiated mature corneal epithelial cells that migrate onto the corneal stroma. There are at least two possible reasons why mature corneal epithelial cells may also be protected and experience immune privilege. First, if ABCB5+ LSC are similar to ABCB5+ skin stem cells, they may be capable of inducing systemic immune tolerance that protects allogeneic cells throughout the host. Second, allogeneic corneal epithelial cells may be protected by their own immune privilege properties which are an established component of the overall immune privilege environment within the central cornea. These two possibilities are discussed below.

First, the possibility that ABCB5+ LSC can induce immune tolerance is based on the recent discovery that treatment of recipient mice with an intravenous injection of purified allogeneic ABCB5+ *skin* cells induced immune tolerance that prolonged the survival of a subsequent heterotopic cardiac allograft [57]. Immune tolerance was mediated by CD4+ CD25+ Foxp3+ TREG cells that were activated via the PD-1 receptor on ABCB5+ skin cells, which was demonstrated by experiments in which recipient mice were treated with allogeneic ABCB5+ skin cells that were harvested from donor PD-1 knockout mice. These donor PD-1 receptor *negative* ABCB5 skin cells were unable to induce: TREG cell activity, allogeneic T cell tolerance, and prolonged allograft survival. Thus, PD-1 on allogeneic PD-1+ ABCB5+ skin cells was capable of inducing systemic tolerance that provided immune privilege to fully differentiated mature allogeneic cardiac myocytes that were transplanted into the recipient at a distant anatomical site. Therefore, it is possible that allogeneic PD-1+ ABCB5+ LSC may also trigger systemic tolerance via a similar mechanism that protects the allogeneic donor corneal epithelial cells. Important to the current discussion of the immune privilege within the limbus and cornea, we detected PD-1 on ABCB5+ LSC and previous studies from another laboratory indicated that PDL-1 was also expressed on corneal epithelial cells [59]. Evidence the PD-1/PDL-1 pathway was required for corneal immune privilege was provided by experiments from Shen et al. [60] in which treatment of recipient mice with anti-PDL-1 antibodies that blocked activation of PD-1 resulted in termination of immune privilege and accelerated corneal allograft rejection. Thus, the PD-1/PDL-1 pathway is involved in maintaining corneal immune privilege and the possibility exists that PD-1+ ABCB5+ LSC contribute to the overall immune privilege environment in the central cornea.

A second possible reason that allogeneic corneal epithelial cells may be protected by immune privilege is based upon data indicating that corneal epithelial cells themselves contribute to establishing and maintaining the immune privilege environment of the central cornea. This was demonstrated in experiments that were designed to determine the relative allogenicity of the donor corneal epithelium versus the donor corneal stroma [61–63]. The survival of two types of donor corneal allografts was compared using a mouse model system: (1) full-thickness allografts (epithelium, stroma, and endothelium), and (2) epithelial-cell-deficient allografts (containing only stroma and endothelium). If the corneal epithelium did not contribute to immune privilege and/or was immunogenic, then you would expect the epithelium-deficient transplants to survive better than transplants with an intact epithelium. However, the exact opposite result was observed. Epithelium-deficient allografts triggered *accelerated rejection*, as compared with epithelium-intact full-thickness allografts, indicating that the corneal epithelium contributed to immune privilege and allograft survival. The corneal epithelium has been shown to contribute to immune privilege via multiple mechanisms: (1) expressing sFlt (a soluble form of the VEGF receptor 1) that is anti-angiogenic and prevents neovascularization of the cornea, an important property of immune privilege [64], (2) expressing Fas Ligand that maintains immune privilege by depletion of infiltrating Fas + immune cells [65, 66], (3) upregulating PD-L1 on corneal epithelial cells exposed to inflammatory cytokines, which inhibits or induces apoptosis of PD-1+ infiltrating T cells [60], and (4) secretion of anti-inflammatory cytokines [50]. In summary, there are several possible scenarios in which allogeneic purified ABCB5+ LSC may establish immune privilege within the limbal stem cell niche that also extends protection to the progeny of these LSC, the allogeneic corneal epithelial cells that migrate onto the corneal surface.

5.6 Alternative Sources of Non-limbus-Derived Autologous Stem Cells

5.6.1 Reprograming Autologous Stem Cells

The generation and use of iPS cells in regeneration of the retina is discussed at length in other chapters of this book. However, it is important to briefly discuss here the unique problems associated with using iPS cells to treat patients with bilateral LSCD and to highlight how this approach is different from the use of iPS cells to treat retinal diseases. An important advance in regenerative medicine was the discovery of the "Yamanaka factors," a group of four genes (Oct4, Sox2, cMyc, and Klf4) that, when transfected into adult somatic cells, triggered dedifferentiation of the cells into an "induced" pluripotent stem (iPS) cell with the potential to produce differentiated cells from all three developmental lineages (ectoderm, mesoderm, endoderm) [67]. Subsequently, the in vitro culture conditions and differentiation protocols were developed for driving the differentiation of iPS cells into specific cell types, which included a variety of ocular cells, such as RGCs [68–71], photoreceptors [72, 73], retinal pigment epithelial cells (RPE) [74–77], and more recently corneal epithelial cells [78]. Moreover, the large in vitro proliferative capacity of iPS cells meant that an essentially unlimited source of autologous iPS cells could be produced for any specific donor. In other words, adult cells such as skin fibroblasts could be harvested, transfected, and expanded to produce large numbers of iPS cells which could then be differentiated in vitro to produce any mature cell type. It was initially believed this approach would solve the problems associated with rejection of transplanted allogeneic donor cells, since patient-specific *autologous* iPS-derived cells could be produced. However, it remains unclear if the iPS-derived autologous cells are truly non-immunogenic. Researchers reported that mouse iPS-derived cardiomyocytes were immunogenic and eliminated when transplanted into syngeneic recipients [79], while other researchers reported mouse iPS-derived vascular endothelial cells were not immunogenic [80, 81], suggesting the immunogenicity of iPS-derived differentiated cells may depend upon the type of cell produced. Data supporting this for *human* iPS-derived cells was provided by Zhao et al. [82] who demonstrated that iPS-derived cardiomyocytes, but not iPS-derived retinal pigment epithelial cells (RPE), were immunogenic in mice engrafted with an autologous human immune system. The gene expression profiles of iPS-derived cardiomyocytes and normal human cardiomyocytes indicated the immunogenicity of the iPS-derived cells was due to a higher expression of specific antigens. Interestingly, this study also confirmed that the subretinal space of the eye has the advantage of being an immune privilege site that permits the survival of iPS-derived cells, even the iPS-derived cardiomyocytes that were rejected when transplanted outside the eye. Therefore, the immunogenicity of iPS-derived differentiated cells and whether they can be transplanted into autologous recipients without immune suppressive therapy depends upon the type of cell produced and where the cells are transplanted.

The best clinical example of using iPS-derived cells to replace damaged tissue is the transplantation of iPS-produced RPE cells into the subretinal space of patients with age-related macular degeneration (AMD), a topic discussed at length in other chapters of this book. The initial clinical trials using this approach have highlighted the safety of this technique, as well as, some of the difficulties that include the extensive genetic testing required to ensure that transfection of cells does not alter normal gene function, and the long culture period required to produce enough fully differentiated RPE cells for safety studies, genetic analysis, and the transplant for the patient. However, it is important to note that these studies use iPS-derived differentiated cells to replace damaged tissue in the recipient that *does not turnover*, which is not the case for replacing the damage cornea in bilateral LSCD patients. The RPE cell layer in the retina is believed to survive for the life of the organism and has little or no cell turnover. Therefore, the transplanted mature RPE cells essentially survive indefinitely. Importantly, the current rules for using iPS cell-derived cell transplants, such as RPE cells, require that *all undifferentiated cells are removed* from the transplanted cells. In other words, all stem cells must be removed and only fully differentiate RPE cells can be grafted into the patient. This is due to the danger of tumor formation from undifferentiated pluripotent iPS cells that have a high capacity to trigger growth of teratomas.

Some investigators have proposed to use iPS-derived cells to treat bilateral LSCD patients. Transplanting iPS-derived differentiated mature corneal epithelial cells would be futile, since these cells turnover rapidly and the transplanted cells would quickly sluff off and not be replaced. What is needed for this approach to be successful is to produce something novel that has yet to be reported, turning multipotent iPS cells into *unipotent adult stem cells*, such as LSC, that: (1) remain in an undifferentiated state, (2) are long-lived cells, (3) can be induced into asymmetrical proliferation that produces mature corneal epithelial cells without depleting the undifferentiated stem cell population, and (4) most importantly, completely lack any tumorigenic potential. Several articles in the literature report using iPS cells to produce mature corneal epithelial cells but provide no evidence they have produced LSC with the properties described above. They fail to show the cells possess long-term regenerative potential which is used to define functional adult stem cells. However, the study by Hayashi et al. [83] demonstrated the production of iPS-derived corneal "organoid-like" structures that possessed LSC-like cells capable of generating mature corneal epithelial cells in vitro. Moreover, the potential regenerative capacity of these LSC-like cells was demonstrated in short-term transplantation studies, suggesting future studies may demonstrate the long-term corneal regenerative capacity of these cells and their significant clinical potential. However, it is important to note that, if this approach is successful and adult stem cells are produced from iPS cells, then new regulatory guidelines will have to be developed before they can be translated into the clinic, since there are currently no guidelines on how to produce iPS-derived stem cells that remain undifferentiated. The current guidelines require that transplants contain only mature fully differentiated cells and no undifferentiated stem cells.

A second possible route to reprograming autologous stem cells was provided by Hong Ouyang et al. [84] who reported a method to reprogram skin stem cells into LSC-like cells with the capacity to produce and maintain the corneal epithelium. By comparing differences between corneal epithelial cells ($K3^+$ $K12^+$ $K1^-$ $K10^-$) and keratinizing skin epithelial cells ($K3^-$ $K12^-$ $K1^+$ $K10^+$) they determined that WNT7A/FZD5 and PAX6 were critical genes that controlled the differentiation of LSC into corneal epithelial cells. In the absence of WNT7A and PAX6, LSC differentiation pathway led to skin-like $K3^-$ $K12^-$ $K1^+$ $K10^+$ corneal epithelial cells. Furthermore, expression of either PAX6a or PAX6b in skin stem cells changed the cell fate of progeny cells into corneal epithelial-like cells that expressed K3 and K12. Transplantation of PAX6+ skin epithelial cells onto the cornea of rabbits with an induced LSCD resulted in short-term restoration and maintenance of a clear corneal epithelium, indicating this may be a therapeutic approach for patients with bilateral LSCD. Therefore, there are at least two potential routes to reprograming autologous stem cells into LSC, either via reprograming iPS cells or transfecting skin stem cells with PAX6. Further studies of the feasibility and safety will be needed to determine whether either or both of these approaches will be viable clinical options. The most important safety issue arises from the genetic manipulation—either insertion of the four Yamanaka genes into somatic cells or insertion of the PAX6 gene into skin stem cells. While the initial safety data of iPS-derived RPE cells is encouraging, the feasibility of this approach for individual patient-specific autologous RPE transplants remains a challenge due to the time and expense required to develop safe clinical grade RPE cells.

5.6.2 Non-reprogramed Autologous Stem Cells

Because of the potential issues associated with genetic manipulation, researchers have also attempted to use non-reprogramed autologous stem cells from peripheral blood mesenchymal stem cells or oral mucosal epithelial stem cells to treat patients with bilateral LSCD [6, 85]. The rationale is that these adult stem cells may have some degree of plasticity so that when they are transplanted into the ocular environment the cell fate of progeny cells will become more like corneal epithelial cells than skin or oral mucosal epithelium. There is considerable clinical data from patients receiving autologous mucosal epithelial transplants that indicate some level of success using this approach, however, corneal neovascularization appears to be an important complication.

The recent discovery of ABCB5+ dermal skin cells also raises the possibility that these cells are a potential alternative source of autologous adult stem cells for patients with bilateral LSCD [57]. Since ABCB5 is expressed on a restricted number of adult stem cells in the limbus, skin, hair follicle, intestines, and stomach, it is possible that these ABCB5+ cells display shared cell fate determining pathways that might lead to increased plasticity when these cells are transplanted into the ocular environment, resulting in the reprograming of these ABCB5+ skin stem cells into stem cells that produce progeny that are similar to corneal epithelial cells. However, whether these ABCB5+ skin stem cells are capable of restoring and maintaining the corneal epithelium has yet to be determined.

5.7 Summary and Conclusions

A shining example of a successful sight-saving stem cell treatment is the long-term regeneration and maintenance of a clear avascular cornea in burn patients with a unilateral LSCD. From the first successful limbal tissue transplants in 1989 by Ken Kenyon and Shaffer Tseng to Holoclar®, the first cell therapy approved by the EMA in 2014, these clinical studies have been leading the field of regenerative medicine. However, significant barriers still remain to expand this success to patients with the more common bilateral LSCD where all autologous LSC are lost and allogeneic LSC transplants require long-term immunosuppressive therapy. In addition, there are also significant barriers to using this approach to treat patients with LSCD caused by chronic inflammatory disease and severe autoimmune ocular surface diseases such as Steven-Johnson's syndrome. New research advances may provide solutions to overcoming these barriers via the production of new sources of autologous LSC through iPS cells or reprograming skin stem cells that would restore patient-specific LSC transplants that are not subject to immune-mediated rejection. In addition, it may also be possible to overcome the allogeneic immune barriers to using third-party purified LSC that display immunosuppressive properties. The clinical translation of these more recent research studies should provide additional hope for developing new treatments that will regenerate the cornea in patients with even the most severe ocular surface diseases.

References

1. Shanbhag, S. S., Saeed, H. N., Paschalis, E. I., & Chodosh, J. (2017). Keratolimbal allograft for limbal stem cell deficiency after severe corneal chemical injury: A systematic review. *The British Journal of Ophthalmology, pii*, bjophthalmol–2017–311249. https://doi.org/10.1136/bjophthalmol-2017-311249
2. Kim, Y. H., Kim, D. H., Shin, E. J., Lee, H. J., Wee, W. R., Jeon, S., et al. (2016). Comparative analysis of substrate-free cultured oral mucosal epithelial cell sheets from cells of subjects with and without Stevens- Johnson syndrome for use in ocular surface reconstruction. *PLoS One, 11*, e0147548. https://doi.org/10.1371/journal.pone.0147548
3. Shortt, A. J., Bunce, C., Levis, H. J., Blows, P., Dore, C. J., Vernon, A., et al. (2014). Three-year outcomes of cultured limbal epithelial allografts in aniridia and Stevens-Johnson syndrome evaluated using the clinical outcome assessment in surgical trials assessment tool. *Stem Cells Translational Medicine, 3*, 265–275. https://doi.org/10.5966/sctm.2013-0025
4. Rossen, J., Amram, A., Milani, B., Park, D., Harthan, J., Joslin, C., et al. (2016). Contact lens-induced limbal stem cell deficiency. *The Ocular Surface, 14*, 419–434. https://doi.org/10.1016/j.jtos.2016.06.003
5. Vazirani, J., Nair, D., Shanbhag, S., Wurity, S., Ranjan, A., & Sangwan, V. (2018). Limbal stem cell deficiency-demography and underlying causes. *American Journal of Ophthalmology, 188*, 99–103. https://doi.org/10.1016/j.ajo.2018.01.020
6. Sotozono, C., Inatomi, T., Nakamura, T., Koizumi, N., Yokoi, N., Ueta, M., et al. (2014). Cultivated oral mucosal epithelial transplantation for persistent epithelial defect in severe ocular surface diseases with acute inflammatory activity. *Acta Ophthalmologica, 92*, e447–e453. https://doi.org/10.1111/aos.12397
7. Kenyon, K. R., & Tseng, S. C. (1989). Limbal autograft transplantation for ocular surface disorders. *Ophthalmology*. https://doi.org/10.1016/S0161-6420(89)32833-8
8. Pellegrini, G., Ardigò, D., Milazzo, G., Iotti, G., Guatelli, P., Pelosi, D., et al. (2018). Navigating market authorization: The path holoclar took to become the first stem cell product approved in the European Union. *Stem Cells Translational Medicine, 7*, 146–154. https://doi.org/10.1002/sctm.17-0003
9. Barker, N., Bartfeld, S., & Clevers, H. (2010). Tissue-resident adult stem cell populations of rapidly self-renewing organs. *Cell Stem Cell, 7*, 15–15. https://doi.org/10.1016/j.stem.2010.11.016
10. Richardson, A., Lobo, E. P., Delic, N. C., Myerscough, M. R., Lyons, J. G., Wakefield, D., et al. (2017). Keratin-14-positive precursor cells spawn a population of migratory corneal epithelia that maintain tissue mass throughout life. *Stem Cell Reports, 9*, 1081–1096. https://doi.org/10.1016/j.stemcr.2017.08.015
11. Kasetti, R. B., Gaddipati, S., Tian, S., Xue, L., Kao, W. W., Lu, Q., et al. (2016). Study of corneal epithelial progenitor origin and the Yap1 requirement using keratin 12 lineage tracing transgenic mice. *Scientific Reports, 6*, 35202. https://doi.org/10.1038/srep35202
12. Dora, N. J., Hill, R. E., Collinson, J. M., & West, J. D. (2015). Lineage tracing in the adult mouse corneal epithelium supports the limbal epithelial stem cell hypothesis with intermittent periods of stem cell quiescence. *Stem Cell Research, 15*(3), 665–677. https://doi.org/10.1016/j.scr.2015.10.016
13. Gonzalez, G., Sasamoto, Y., Ksander, B. R., Frank, M. H., & Frank, N. Y. (2018). Limbal stem cells: Identity, developmental origin, and therapeutic potential. *Wiley Interdisciplinary Reviews: Developmental Biology, 7*, e303. https://doi.org/10.1002/wdev.303
14. Thoft, R. A. R., & Friend, J. J. (1983). The X, Y, Z hypothesis of corneal epithelial maintenance. *Investigative Ophthalmology & Visual Science, 24*, 1442–1443.
15. Cotsarelis, G., Cheng, S.-Z., Dong, G., Sun, T. T., & Lavker, R. M. (1989). Existence of slow-cycling limbal epithelial basal cells that can be preferentially stimulated to proliferate: Implications on epithelial stem cells. *Cell, 57*, 201–209. https://doi.org/10.1016/0092-8674(89)90958-6

16. Pellegrini, G., Golisano, O., Paterna, P., Lambiase, A., Bonini, S., Rama, P., et al. (1999). Location and clonal analysis of stem cells and their differentiated progeny in the human ocular surface. *The Journal of Cell Biology, 145*, 769–782. https://doi.org/10.1083/jcb.145.4.769
17. Pellegrini, G., Traverso, C. E., Franzi, A. T., Zingirian, M., Cancedda, R., & De Luca, M. (1997). Long-term restoration of damaged corneal surfaces with autologous cultivated corneal epithelium. *Lancet, 349*, 990–993. https://doi.org/10.1016/S0140-6736(96)11188-0
18. Pellegrini, G., Dellambra, E., Golisano, O., Martinelli, E., Fantozzi, I., Bondanza, S., et al. (2001). p63 identifies keratinocyte stem cells. *Proceedings of the National Academy of Sciences of the United States of America, 98*, 3156–3161. https://doi.org/10.1073/pnas.061032098
19. Di Iorio, E., Barbaro, V., Ruzza, A., Ponzin, D., Pellegrini, G., & De Luca, M. (2005). Isoforms of DeltaNp63 and the migration of ocular limbal cells in human corneal regeneration. *Proceedings of the National Academy of Sciences of the United States of America, 102*, 9523–9528. https://doi.org/10.1073/pnas.0503437102
20. Melino, G., Memmi, E. M., Pelicci, P. G., & Bernassola, F. (2015). Maintaining epithelial stemness with p63. *Science Signaling, 8*, re9–re9. https://doi.org/10.1126/scisignal.aaa1033
21. Truong, A. B., Kretz, M., Ridky, T. W., Kimmel, R., & Khavari, P. A. (2006). p63 regulates proliferation and differentiation of developmentally mature keratinocytes. *Genes & Development, 20*, 3185–3197. https://doi.org/10.1101/gad.1463206
22. Liang, L., Sheha, H., Li, J., & Tseng, S. C. G. (2009). Limbal stem cell transplantation: New progresses and challenges. *Eye (London, England), 23*, 1946–1953. https://doi.org/10.1038/eye.2008.379
23. Pellegrini, G., Rama, P., Matuska, S., Lambiase, A., Bonini, S., Pocobelli, A., et al. (2013). Biological parameters determining the clinical outcome of autologous cultures of limbal stem cells. *Regenerative Medicine, 8*, 553–567. https://doi.org/10.2217/rme.13.43
24. Shortt, A. J., Tuft, S. J., & Daniels, J. T. (2010). Ex vivo cultured limbal epithelial transplantation. A clinical perspective. *The Ocular Surface, 8*, 80–90.
25. Rama, P., Matuska, S., Paganoni, G., Spinelli, A., De Luca, M., & Pellegrini, G. (2010). Limbal stem-cell therapy and long-term corneal regeneration. *The New England Journal of Medicine, 363*, 147–155. https://doi.org/10.1056/NEJMoa0905955
26. Szabó, D. J., Noer, A., Nagymihály, R., Josifovska, N., Andjelic, S., Vereb, Z., et al. (2015). Long-term cultures of human cornea limbal explants form 3D structures ex vivo–implications for tissue engineering and clinical applications. *PLoS One, 10*, e0143053. https://doi.org/10.1371/journal.pone.0143053
27. Cheung, A. Y., & Holland, E. J. (2017). Keratolimbal allograft. *Current Opinion in Ophthalmology, 28*, 377–381. https://doi.org/10.1097/ICU.0000000000000374
28. Basu, S., Sureka, S. P., Shanbhag, S. S., Kethiri, A. R., Singh, V., & Sangwan, V. S. (2016). Simple limbal epithelial transplantation: Long-term clinical outcomes in 125 cases of unilateral chronic ocular surface burns. *Ophthalmology, 123*, 1000–1010. https://doi.org/10.1016/j.ophtha.2015.12.042
29. Sasine, J. P., Yeo, K. T., & Chute, J. P. (2017). Concise review: Paracrine functions of vascular niche cells in regulating hematopoietic stem cell fate. *Stem Cells Translational Medicine, 6*, 482–489. https://doi.org/10.5966/sctm.2016-0254
30. González, S., Chen, L., & Deng, S. X. (2017). Comparative study of xenobiotic-free media for the cultivation of human limbal epithelial stem/progenitor cells. *Tissue Engineering. Part C, Methods, 23*, 219–227. https://doi.org/10.1089/ten.tec.2016.0388
31. Mei, H., Nakatsu, M. N., Baclagon, E. R., & Deng, S. X. (2014). Frizzled 7 maintains the undifferentiated state of human limbal stem/progenitor cells. *Stem Cells, 32*, 938–945. https://doi.org/10.1002/stem.1582
32. Nakatsu, M. N., Ding, Z., Ng, M. Y., Truong, T. T., Yu, F., & Deng, S. X. (2011). Wnt/β-catenin signaling regulates proliferation of human cornea epithelial stem/progenitor cells. *Investigative Ophthalmology & Visual Science, 52*, 4734–4741. https://doi.org/10.1167/iovs.10-6486
33. Chan, E., Le, Q., Codriansky, A., Hong, J., Xu, J., & Deng, S. X. (2016). Existence of normal limbal epithelium in eyes with clinical signs of total limbal stem cell deficiency. *Cornea, 35*, 1483–1487. https://doi.org/10.1097/ICO.0000000000000914

34. Le, Q., Xu, J., & Deng, S. X. (2018). The diagnosis of limbal stem cell deficiency. *The Ocular Surface, 16*, 58–69. https://doi.org/10.1016/j.jtos.2017.11.002
35. Zarei-Ghanavati, S., Ramirez-Miranda, A., & Deng, S. X. (2011). Limbal lacuna: A novel limbal structure detected by in vivo laser scanning confocal microscopy. *Ophthalmic Surgery, Lasers & Imaging, 42*, e129–e131. https://doi.org/10.3928/15428877-20111201-07
36. Holland, E. J., Mogilishetty, G., Skeens, H. M., Hair, D. B., Neff, K. D., Biber, J. M., et al. (2012). Systemic immunosuppression in ocular surface stem cell transplantation: Results of a 10-year experience. *Cornea, 31*, 655–661. https://doi.org/10.1097/ICO.0b013e31823f8b0c
37. Eslani, M., Haq, Z., Movahedan, A., Moss, A., Baradaran-Rafii, A., Mogilishetty, G., et al. (2017). Late acute rejection after allograft limbal stem cell transplantation: Evidence for long-term donor survival. *Cornea, 36*, 26–31. https://doi.org/10.1097/ICO.0000000000000970
38. Han, E. S., Wee, W. R., Lee, J. H., & Kim, M. K. (2011). Long-term outcome and prognostic factor analysis for keratolimbal allografts. *Graefe's Archive for Clinical and Experimental Ophthalmology, 249*, 1697–1704. https://doi.org/10.1007/s00417-011-1760-3
39. Lin, C. M., & Gill, R. G. (2016). Direct and indirect allograft recognition: Pathways dictating graft rejection mechanisms. *Current Opinion in Organ Transplantation, 21*, 40–44. https://doi.org/10.1097/MOT.0000000000000263
40. Almoguera, B., Shaked, A., & Keating, B. J. (2014). Transplantation genetics: Current status and prospects. *American Journal of Transplantation, 14*, 764–778. https://doi.org/10.1111/ajt.12653
41. Djalilian, A. R., Mahesh, S. P., Koch, C. A., Nussenblatt, R. B., Shen, D., Zhuang, Z., et al. (2005). Survival of donor epithelial cells after limbal stem cell transplantation. *Investigative Ophthalmology & Visual Science, 46*, 803–807. https://doi.org/10.1167/iovs.04-0575
42. Medawar, P. B. (1948). Immunity to homologous grafted skin; the fate of skin homografts transplanted to the brain, to subcutaneous tissue, and to the anterior chamber of the eye. *British Journal of Experimental Pathology, 29*, 58–69.
43. Streilein, J. W. (2003). Ocular immune privilege: Therapeutic opportunities from an experiment of nature. *Nature Reviews. Immunology, 3*, 879–889. https://doi.org/10.1038/nri1224
44. Niederkorn, J. Y. (2006). See no evil, hear no evil, do no evil: The lessons of immune privilege. *Nature Immunology, 7*, 354–359. https://doi.org/10.1038/ni1328
45. Louveau, A., Harris, T. H., & Kipnis, J. (2015). Revisiting the mechanisms of CNS immune privilege. *Trends in Immunology, 36*, 569–577. https://doi.org/10.1016/j.it.2015.08.006
46. Engelhardt, B., Vajkoczy, P., & Weller, R. O. (2017). The movers and shapers in immune privilege of the CNS. *Nature Immunology, 18*, 123–131. https://doi.org/10.1038/ni.3666
47. Spadoni, I., Fornasa, G., & Rescigno, M. (2017). Organ-specific protection mediated by cooperation between vascular and epithelial barriers. *Nature Reviews. Immunology, 17*, 761–773. https://doi.org/10.1038/nri.2017.100
48. Tan, D. T. H., Dart, J. K. G., Holland, E. J., & Kinoshita, S. (2012). Corneal transplantation. *Lancet, 379*, 1749–1761. https://doi.org/10.1016/S0140-6736(12)60437-1
49. Amouzegar, A., Chauhan, S. K., & Dana, R. (2016). Alloimmunity and tolerance in corneal transplantation. *Journal of Immunology, 196*, 3983–3991. https://doi.org/10.4049/jimmunol.1600251
50. Niederkorn, J. Y. (2013). Corneal transplantation and immune privilege. *International Reviews of Immunology, 32*, 57–67. https://doi.org/10.3109/08830185.2012.737877
51. Casiraghi, F., Perico, N., & Remuzzi, G. (2017). Mesenchymal stromal cells for tolerance induction in organ transplantation. *Human Immunology*. https://doi.org/10.1016/j.humimm.2017.12.008
52. Hua, F., Chen, Y., Yang, Z., Teng, X., Huang, H., & Shen, Z. (2018). Protective action of bone marrow mesenchymal stem cells in immune tolerance of allogeneic heart transplantation by regulating CD45RB+ dendritic cells. *Clinical Transplantation, 6*, e13231. https://doi.org/10.1111/ctr.13231
53. Zou, L., Barnett, B., Safah, H., Larussa, V. F., Evdemon-Hogan, M., Mottram, P., et al. (2004). Bone marrow is a reservoir for CD4 +CD25 +regulatory T cells that traffic through CXCL12/

CXCR4 signals. *Cancer Research, 64*, 8451–8455. https://doi.org/10.1158/0008-5472.CAN-04-1987
54. Fujisaki, J., Wu, J., Carlson, A. L., Silberstein, L., Putheti, P., Larocca, R., et al. (2011). In vivo imaging of Treg cells providing 888 immune privilege to the haematopoietic stem-cell niche. *Nature, 474*, 216–219. https://doi.org/10.1038/nature10160
55. Hirata, Y., Furuhashi, K., Ishii, H., Li, H. W., Pinho, S., Ding, L., et al. (2018). CD150highbone marrow tregs maintain hematopoietic stem cell quiescence and immune privilege via adenosine. *Cell Stem Cell, 22*, 445–453.e5. https://doi.org/10.1016/j.stem.2018.01.017
56. Ksander, B. R., Kolovou, P. E., Wilson, B. J., Saab, K. R., Guo, Q., Ma, J., et al. (2014). ABCB5 is a limbal stem cell gene required for corneal development and repair. *Nature, 511*, 353–357. https://doi.org/10.1038/nature13426
57. Schatton, T., Yang, J., Kleffel, S., Uehara, M., Barthel, S. R., Schlapbach, C., et al. (2015). ABCB5 identifies immunoregulatory dermal cells. *Cell Reports, 12*, 1564–1574. https://doi.org/10.1016/j.celrep.2015.08.010
58. Sharpe, A. H., & Pauken, K. E. (2018). The diverse functions of the PD1 inhibitory pathway. *Nature Reviews. Immunology, 18*, 153–167. https://doi.org/10.1038/nri.2017.108
59. Hori, J., Wang, M., Miyashita, M., Tanemoto, K., Takahashi, H., Takemori, T., et al. (2006). B7-H1-induced apoptosis as a mechanism of immune privilege of corneal allografts. *Journal of Immunology, 177*, 5928–5935.
60. Shen, L., Jin, Y., Freeman, G. J., Sharpe, A. H., & Dana, M. R. (2007). The function of donor versus recipient programmed death-ligand 1 in corneal allograft survival. *Journal of Immunology, 179*, 3672–3679.
61. Hori, J., & Streilein, J. W. (2001). Dynamics of donor cell persistence and recipient cell replacement in orthotopic corneal allografts in mice. *Investigative Ophthalmology & Visual Science, 42*, 1820–1828.
62. Hori, J., & Streilein, J. W. (2003). Survival in high-risk eyes of epithelium-deprived orthotopic corneal allografts reconstituted in vitro with syngeneic epithelium. *Investigative Ophthalmology & Visual Science, 44*, 658–664.
63. Hori, J. (2008). Mechanisms of immune privilege in the anterior segment of the eye: What we learn from corneal transplantation. *Journal of Ocular Biology, Diseases, and Informatics, 1*, 94–100. https://doi.org/10.1007/s12177-008-9010-6
64. Ambati, B. K., Nozaki, M., Singh, N., Takeda, A., Jani, P. D., Suthar, T., et al. (2006). Corneal avascularity is due to soluble VEGF receptor-1. *Nature, 443*, 993–997. https://doi.org/10.1038/nature05249
65. Griffith, T. S., Brunner, T., Fletcher, S. M., Green, D. R., & Ferguson, T. A. (1995). Fas ligand-induced apoptosis as a mechanism of immune privilege. *Science, 270*, 1189–1192.
66. Stuart, P. M., Griffith, T. S., Usui, N., Pepose, J., Yu, X., & Ferguson, T. A. (1997). CD95 ligand (FasL)-induced apoptosis is necessary for corneal allograft survival. *The Journal of Clinical Investigation, 99*, 396–402. https://doi.org/10.1172/JCI119173
67. Takahashi, K., & Yamanaka, S. (2016). A decade of transcription factor-mediated reprogramming to pluripotency. *Nature Reviews. Molecular Cell Biology, 17*, 183–193. https://doi.org/10.1038/nrm.2016.8
68. Huang, L., Chen, M., Zhang, W., Sun, X., Liu, B., & Ge, J. (2018). Retinoid acid and taurine promote NeuroD1-induced differentiation of induced pluripotent stem cells into retinal ganglion cells. *Molecular and Cellular Biochemistry, 438*, 67–76. https://doi.org/10.1007/s11010-017-3114-x
69. Teotia, P., Van Hook, M. J., Wichman, C. S., Allingham, R. R., Hauser, M. A., & Ahmad, I. (2017). Modeling glaucoma: Retinal ganglion cells generated from induced pluripotent stem cells of patients with SIX6 risk allele show developmental abnormalities. *Stem Cells, 35*, 2239–2252. https://doi.org/10.1002/stem.2675
70. Kobayashi, W., Onishi, A., Tu, H. Y., Takihara, Y., Matsumura, M., Tsujimoto, K., et al. (2018). Culture systems of dissociated mouse and human pluripotent stem cell-derived retinal ganglion

cells purified by two-step immunopanning. *Investigative Ophthalmology & Visual Science, 59*, 776–787. https://doi.org/10.1167/iovs.17-22406
71. Yokoi, T., Tanaka, T., Matsuzaka, E., Tamalu, F., Watanabe, S. I., Nishina, S., et al. (2017). Effects of neuroactive agents on axonal growth and pathfinding of retinal ganglion cells generated from human stem cells. *Scientific Reports, 7*, 16757. https://doi.org/10.1038/s41598-017-16727-1
72. Ramsden, C. M., Powner, M. B., Carr, A.-J. F., Smart, M. J., da Cruz, L., & Coffey, P. J. (2014). Neural retinal regeneration with pluripotent stem cells. *Developments in Ophthalmology, 53*, 97–110. https://doi.org/10.1159/000357363
73. Boucherie, C., Sowden, J. C., & Ali, R. R. (2011). Induced pluripotent stem cell technology for generating photoreceptors. *Regenerative Medicine, 6*, 469–479. https://doi.org/10.2217/rme.11.37
74. Kamao, H., Mandai, M., Okamoto, S., Sakai, N., Suga, A., Sugita, S., et al. (2014). Characterization of human induced pluripotent stem cell-derived retinal pigment epithelium cell sheets aiming for clinical application. *Stem Cell Reports, 2*, 205–218. https://doi.org/10.1016/j.stemcr.2013.12.007
75. Leach, L. L., & Clegg, D. O. (2015). Concise review: Making stem cells retinal: Methods for deriving retinal pigment epithelium and implications for patients with ocular disease. *Stem Cells, 33*, 2363–2373. https://doi.org/10.1002/stem.2010
76. Croze, R. H., & Clegg, D. O. (2014). Differentiation of pluripotent stem cells into retinal pigmented epithelium. *Developments in Ophthalmology, 53*, 81–96. https://doi.org/10.1159/000357361
77. Westenskow, P. D., Kurihara, T., & Friedlander, M. (2014). Utilizing stem cell-derived RPE cells as a therapeutic intervention for age-related macular degeneration. *Advances in Experimental Medicine and Biology, 801*, 323–329. https://doi.org/10.1007/978-1-4614-3209-8_41
78. Kamarudin, T. A., Bojic, S., Collin, J., Yu, M., Alharthi, S., Buck, H., et al. (2018). Differences in the activity of endogenous bone morphogenetic protein signaling impact on the ability of induced pluripotent stem cells to differentiate to corneal epithelial-like cells. *Stem Cells, 36*, 337–348. https://doi.org/10.1002/stem.2750
79. Zhao, T., Zhang, Z.-N., Rong, Z., & Xu, Y. (2011). Immunogenicity of induced pluripotent stem cells. *Nature, 474*, 212–215. https://doi.org/10.1038/nature10135
80. de Almeida, P. E., Meyer, E. H., Kooreman, N. G., Diecke, S., Dey, D., Sanchez-Freire, V., et al. (2014). Transplanted terminally differentiated induced pluripotent stem cells are accepted by immune mechanisms similar to self-tolerance. *Nature Communications, 5*, 3903. https://doi.org/10.1038/ncomms4903
81. Araki, R., Uda, M., Hoki, Y., Sunayama, M., Nakamura, M., Ando, S., et al. (2013). Negligible immunogenicity of terminally differentiated cells derived from induced pluripotent or embryonic stem cells. *Nature, 494*, 100–104. https://doi.org/10.1038/nature11807
82. Zhao, T., Zhang, Z.-N., Westenskow, P. D., Todorova, D., Hu, Z., Lin, T., et al. (2015). Humanized mice reveal differential immunogenicity of cells derived from autologous induced pluripotent stem cells. *Cell Stem Cell, 17*, 353–359. https://doi.org/10.1016/j.stem.2015.07.021
83. Hayashi, R., Ishikawa, Y., Sasamoto, Y., Katori, R., Nomura, N., Ichikawa, T., et al. (2016). Co-ordinated ocular development from human iPS cells and recovery of corneal function. *Nature, 531*, 376–380. https://doi.org/10.1038/nature17000
84. Ouyang, H., Xue, Y., Lin, Y., Zhang, X., Xi, L., Patel, S., et al. (2014). WNT7A and PAX6 define corneal epithelium homeostasis and pathogenesis. *Nature, 511*, 358–361. https://doi.org/10.1038/nature13465
85. Galindo, S., Herreras, J. M., Lopez-Paniagua, M., Rey, E., de la Mata, A., Plata-Cordero, M., et al. (2017). Therapeutic effect of human adipose tissue-derived mesenchymal stem cells in experimental corneal failure due to limbal stem cell niche damage. *Stem Cells, 35*, 2160–2174. https://doi.org/10.1002/stem.2672

Chapter 6
Stem Cell Therapy and Regenerative Medicine in the Cornea

Christopher D. McTiernan, Isabelle Brunette, and May Griffith

Abstract Currently, full-thickness transplantation with human donor corneas is the most widely accepted treatment for corneal blindness. However, due to a severe shortage of human donor corneas as well as problems associated with the storage, screening, and immune response to allogeneic tissues, there has been a push to develop alternative therapies and materials for corneal tissue repair. Here, we review a range of stem cell-based therapies, prosthetics, and extracellular matrix-derived scaffolds, which have been utilized or are being developed for corneal regeneration in vitro, animal models, and human clinical trials.

Keywords Cornea · Regenerative medicine · Stem cell · Biomaterial · Implant

6.1 Introduction

The cornea, which is the transparent front covering of the human eye, plays a vital role in vision. Not only does it serve as the lens that transmits and focuses light into the eye to generate vision, it also acts as a mechanical barrier to protect the inner contents of the eye [1]. The human cornea measures 11.5–12.5 mm horizontally, it is thinnest in its center (approximately 550 μm), thickest at its periphery (approximately 750 μm) [2], and is comprised of three distinct cellular layers, an outermost epithelium, a middle stroma, and an innermost endothelial layer [3].

The epithelium is the primary protective barrier of the eye against foreign materials such as water, dust, and micro-organisms. This layer is composed of stratified non-keratinized epithelial cells. The protective function of the epithelium also involves the secretion of anti-inflammatory and anti-microbial peptides [4]. As surface cells are shed, new layers of epithelium are generated from the corneal stem

C. D. McTiernan · I. Brunette · M. Griffith (✉)
Maisonneuve-Rosemont Hospital Research Centre, Montreal, QC, Canada

Department of Ophthalmology, University of Montreal, Montreal, QC, Canada

B. G. Ballios, M. J. Young (eds.), *Regenerative Medicine and Stem Cell Therapy for the Eye*, Fundamental Biomedical Technologies,
https://doi.org/10.1007/978-3-319-98080-5_6

cells located at the basal layer of the corneal-scleral junction, referred to as the limbus [5].

The stroma is a hydrophilic structure, consisting of a mostly collagenous extracellular matrix (ECM) containing an interpenetrating network of fibroblast-like cells called keratocytes. The shape, elasticity, and strength exhibited by the cornea is derived from the collagenous ECM from which this layer is shaped [6]. The transparency of this avascular layer is dictated by the unique spacing and arrangement afforded by the lamellar type organization of collagenous fibrils. The keratocytes are responsible for synthesizing and organizing this optically transparent collagenous matrix of the stroma. Keratocytes are quite sensitive, undergoing apoptosis at the site of injury.

Finally, the innermost endothelial monolayer is responsible for osmoregulation of the cornea. It does this by pumping out the excess liquid that leaks from the anterior chamber of the eye into the stromal layer of the cornea [7]. This regulation is important as excessive stromal swelling results in corneal opacification and vision loss. The endothelial cells do not normally regenerate in vivo and any damage to this layer results in irreversible endothelial cell loss.

Due to its external location, the cornea is prone to both injury and infection that can cause irreversible loss of transparency and lead to vision loss or blindness. In fact, corneal diseases are one of the main causes of vision loss worldwide. There are an estimated five million people suffering from bilateral blindness and another 23 million unilaterally blind from corneal disease or damage. The current widely accepted treatment for corneal blindness is transplantation with donor tissue, but there is a drastic shortage of human donor corneas [8]. For example, it is estimated that approximately 12.7 million patients are waiting for a corneal transplantation worldwide, with a high percentage of these cases located primarily in low to middle income countries [9]. The need for donated tissue outpaces the rate of tissue donation in almost all regions of the world, which highlights the need for new therapeutic alternatives. Stem cell therapy, regenerative medicine, and advanced biomaterials technology hold the promise of a major evolution in the concept of corneal healing and regeneration, as they offer an entirely new spectrum of opportunities for solving current limitations with corneal transplantation.

6.2 Traditional Treatment

The current gold standard in treating diseased corneas in the clinical setting is transplantation, where the pathological tissue is excised and replaced by a donated human cornea. While the procedure does have an overall high success rate of approximately 85–90%, corneal transplantation is not devoid of side effects [10]. A full-thickness corneal transplantation, for instance, usually involves a prolonged rehabilitation period, with wound healing being delayed by the 360° denervation of the graft and by the suboptimal apposition of the wound. The sutures, which typically induce severe surface distortion, are rarely removed before 1 year, further delaying prescription of glasses and functional rehabilitation of the patient. Sutures also increase the risk for infection, inflammation, and vascularization, all of which are associated with an increased risk for immune rejection and irreversible graft failure. And finally, the new grafts also suffer from gradual decreases in endothelial cell population, with a large percentage of cells lost during the first 10 years [11].

Success rate significantly decreases in high-risk patients [10]. These patients presenting with inflammation or vascularization from severe pathologies such as chemical burns, autoimmune diseases, or rejected grafts are at a much higher risk of experiencing complications and graft failure. Additional surgeries are often needed and the overall prognosis decreases with each subsequent graft.

Lamellar transplantation instead of full-thickness replacement is gaining in popularity [12]. The idea here is to replace only the damaged and/or diseased tissue, leaving the healthy layers untouched. There are two types of lamellar transplantation, anterior (aimed at the replacement of the stroma) and posterior (essentially aimed at the replacement of the corneal endothelium).

Anterior lamellar transplantation offers the significant advantage of preserving the patient's endothelium, thus eliminating the risk for corneal endothelial immune rejection, minimizing postoperative endothelial cell loss, and avoiding entering the eye at the time of surgery, which in turn minimizes the risk for postoperative intraocular infection. Anterior lamellar transplantation, however, shares several disadvantages with full-thickness corneal transplantation, namely, the need for sutures leading to corneal astigmatism, delayed and incomplete functional rehabilitation, infection, inflammation, vascularization, and stromal rejection.

As for posterior lamellar transplantation, the idea is to replace the endothelium while keeping the majority or all the rest of the cornea intact. The advantage of this technique is that it does not involve the corneal surface. The need for sutures is thus eliminated, induced astigmatism is minimal, healing time is reduced, and the resulting visual acuity is significantly enhanced.

However, all of the above techniques require donor corneas, which implies expensive eyebanking procedures for tissue quality assessment and screening for diseases potentially transmitted by donors such as syphilis, HIV, or Creutzfeldt-Jacob disease. Transplantation of an allogeneic human cornea also implies the risk of immune rejection [13], which requires long-term immunosuppressing therapy and regular ophthalmology follow-ups for the patient. And finally, as discussed above, wait time for a donor cornea constitutes a growing concern [14].

6.3 Stem Cell Graft

The stem cells of the corneal epithelium are located within limbus [5]. Normal turnover of the epithelium, as well as healing from injury, occurs through the regenerative proliferation of the corneal limbal stem cells. These cells repopulate the non-keratinized corneal epithelium in a centripetal fashion and further serve to halt the encroachment of conjunctival epithelium which would obscure vision should it cover the cornea. There are many causes of ocular limbal stem cell deficiency, including chemical burns, radiation therapy, use of contact lenses, infections, neoplasia, Stevens-Johnson syndrome, mucous membrane pemphigoid, chronic limbitis, chronic bullous keratopathy, diabetic keratopathy, aniridia, and epidermal dysplasia [15]. Limbal stem cell deficiency results in a nonhealing corneal epithelium with subsequent conjunctivalization, vascularization, opacification, and vision loss. It must be noted that the prognosis of traditional corneal transplantation in patients with limbal stem cell deficiency is very poor, which explains why penetrating keratoplasty is avoided in these patients until preliminary rehabilitation of the corneal surface has been achieved. Depending on the degree of damage to the corneal and limbal tissues, in some cases it is sufficient to simply remove some of the invading conjunctival epithelium to allow limbal stem cells from adjacent regions to migrate to the site of damage [16]. Coupling the removal of conjunctival tissue with the laying down of amniotic membrane also increases healing [17]. For more severe cases of limbal stem cell deficiency, a number of viable clinical solutions have also been developed, involving techniques of autologous or allogeneic conjunctival and conjunctival-limbal grafts [18–22].

6.3.1 Conjunctival and Conjunctival-Limbal Autografts

The use of autologous conjunctival tissue as a source of stem cells in patients suffering from unilateral eye injuries was first suggested in the 1960s and then followed up by Richard Thoft in the late 1970s [23]. The major drawback of conjunctival transplantation is the absence of corneal epithelial stem cells. As it became clearer that the limbal tissues played a key role in the regeneration of the corneal surface, Kenyon and Tseng adapted the conjunctival autografts to include limbal tissue in 1989 [24]. Since this time, conjunctival limbal autografts have become one of the standard treatments for unilateral eye injuries, with success rates reported to reach 94% [24–26]. Typically, a large graft (approximately 7 mm large) comprising both epithelial and limbal tissues is harvested and transplanted into the diseased area. The key to the technique is harvesting enough limbal tissue in the graft to promote regeneration in the damaged eye without destabilizing surface homeostasis in the healthy eye. Autografts offer the benefit of not requiring immunosuppressive therapy since the stem cells and tissue are derived from the patient.

6.3.2 Conjunctival–Limbal Allografts

Bilateral limbal stem cell deficiency is more challenging to address because of the lack of available autologous stem cells and tissue. Two types of transplantation are used in these patients, based on the origin of the stem cells and limbal tissue transplanted (Holland and Schwartz classification) [27]: the cadaveric ketarolimbal allograft (KLAL) and the living-related conjunctival-limbal allograft (IrCLAL) [28, 29] The Cincinnati procedure was developed to treat bilateral limbal stem cell deficiency with combined living-related CLAL and cadaveric KLAL. The modified Cincinnati procedure for unilateral limbal stem cell deficiency uses a conjunctival-limbal autograft and KLAL. The goal of adding conjunctival-limbal autograft to KLAL was to introduce more conjunctival stem cells than KLAL alone which is important in patients with severe inflammation [30, 31].

Of the two methods cadaveric KLAL has the advantage that a complete limbus can be transplanted, allowing for a higher concentration of progenitor stem cells to be introduced into diseased eye. However, there are major drawbacks to the KLAL technique: In many cases it can be difficult or impossible to find immune histocompatible tissue for transplantation. It is important that these tissues be histocompatible as the limbus is a highly vascularized site, thus increasing the risk of graft rejection. These grafts mandate patients undergo prolonged systemic immunosuppressive therapy in the postoperative period and often indefinitely. Other challenges to the use of cadaveric tissue include quality of preservation method and time between death and transplantation. In many cases by the time the tissue has been implanted and allowed to vascularize, a large percentage of the transplanted stem cells have died. While the amount of tissue that can be transplanted in the case of IrCAL is limited, just as in the case of conjunctival-limbal autografts, it offers many advantages over cadaveric tissues. Due to the higher likelihood of histocompatibility, the prolonged need for high dosages of systemic immunosuppressive agents is reduced. In addition, a higher proportion of healthy stem cells is theoretically transplanted as the time between excision, transplantation, and revascularization is typically shorter allowing for greater survival rates [32].

6.3.3 Simple Limbal Epithelial Transplantation

Recently, it has been found that only small amounts of tissue are needed to restore the ocular surface in a procedure called simple limbal epithelial transplantation (SLET). In SLET, a 2 × 2 mm donor graft is taken and sectioned into small pieces which are then placed over a prepared corneal bed using human amniotic membrane and fibrin glue, which precludes the need for sutures. This acts as an in-vivo incubator for the stem cells to grow. Mid-term durability of this procedure shows success as high as 84% in a number of studies with only minor variations in technique [33–35]. Perez's group has also demonstrated that one can further improve this technique by sandwiching the sectioned limbal tissue between two sheets of amniotic membrane to allow more protection for the explant [36].

6.4 Cell Culture Techniques

6.4.1 *Ex Vivo Cultivated Limbal Epithelial Transplantation*

Cultured epithelial stem cell therapies have been available since the 1970s [37], but it was not until 1997 that Pellegrini and co-workers first applied the technique to corneal regeneration [38]. Cultured limbal epithelial transplantation (CLET) is now among the most common procedures for the treatment of unilateral limbal stem cell deficiency. The use of stem cells from allogeneic sources, such as living-related or deceased donors, has not obtained as much success, mainly because of the long-term immunosuppressing drugs required after surgery to prevent rejection of the implanted cells [39].

Small amounts of limbal tissue are taken from the contralateral healthy eye. Cells can either be isolated by enzymatic digestion and expanded in culture prior to seeding on the carrier (suspension technique) or they can be expanded directly on the carrier (explant techniques). The graft is then transplanted onto the prepared corneal bed of the diseased eye. Several carriers have been proposed, including fibrin, human amniotic membranes, layers of cultured 3T3 fibroblasts, silk fibroin, collagen membranes, and electrospun biopolymer mats (e.g., poly D,L-lactide-co-glycolide, PLGA) [38, 40]. Cryopreservation of some of the expanded cells for potential future use (in case regrafting is needed) is an interesting advantage of CLET, while high cost of production under good manufacturing practices remains the main drawback. In comparison to the CLAL technique, the CLET technique tends to result in less inflammation and quicker epithelialization. It also requires a significantly smaller amount of tissue (2 × 2 mm) and as such, is less likely to disrupt the homeostasis in the healthy donor eye [18]. Long-term success of this method ranges from 71% to 85% in a number of studies which employ differing substrates such as fibrin/3T3, and denuded or intact hAM [39, 41–44].

6.4.2 *Cultivated Oral Mucosal Epithelial Transplantation*

A variety of stem cell therapies have been developed based on the trans-differentiation of progenitor cells derived from oral mucosa or other mucous membranes, or from the conversion of fibroblasts or mesenchymal stromal cells into autologous corneal lineages through induced pluripotency [45]. Cultivated oral mucosal epithelium transplantation (COMET) is a viable option for patients suffering from bilateral limbal stem cell deficiency. The first successful cases were described in 2004 by Nishida and co-workers. Four patients with total limbal stem cell deficiency caused by Stevens-Johnson Syndrome or autoimmune blistering were grafted with cell sheets of autologous oral mucosal epithelial cells cultured on temperature responsive cell-culture surfaces with 3T3 feeder cells. Complete reepithelialization of the

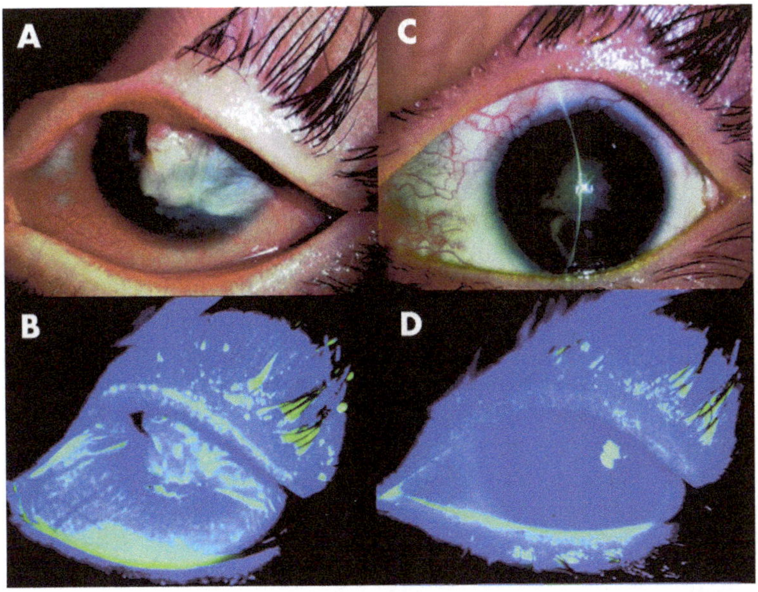

Fig. 6.1 Transplantation of autologous oral epithelial cells cultivated on amniotic membrane to a patient with SJS. Representative slit lamp photographs taken before transplantation without (**a**) and with fluorescein (**b**). The photographs in (**c**, **d**) were taken at the last follow-up visit without (**c**) and with fluorescein (**d**). Before transplantation, the eye manifested inflammatory subconjunctival fibrosis with neovascularization, conjuntivalization, and severe symblepharon. At the last follow-up visit, the corneal surface was stable without defects. Reproduced with permission from Nakamura et al. [47]

corneal surfaces occurred within one week in all four eyes and all corneas remained transparent during the 14 month study period [46].

In 2008, Nakamura and co-workers set out to improve the technique by culturing the oral mucosa on human amniotic membrane, which could then be used as carrier to stabilize the ocular surface [47]. Rapid reepithelialization was observed in the six human eyes suffering from severe limbal stem cell deficiency due to either chemical burns or Stevens-Johnson Syndrome. The corneal surface was still stable 14 months after transplantation with only mild peripheral neovascularization, likely due to the angiogenic potential of oral mucosa cells (see Fig. 6.1).

Overall, COMET has shown success rates as high as 79% [46, 48–50] and it is now used for a variety of conditions such as temperature-related burns and aniridia. It has been suggested that the minor neovascularization could be treated through the use of anti-angiogenic factors or drugs [51, 52]. Long-term stability of these grafts might be limited by the inability of the transplanted cells to undergo full transdifferentiation from oral into the limbal phenotype [53].

6.4.3 Other Sources of Therapeutic Stem Cells

The success of COMET has inspired the development of new strategies that utilize autologous mesenchymal stem cells from a variety of different adult tissues, such as stromal cells derived from adipose or bone marrow [54, 55]. While these autologous sources have great promise as translatable technologies, there are very few examples of these cell types being employed in the clinical setting [55]. Other interesting sources of pluripotent stems cells include hair follicles and dental pulp from deciduous teeth [56, 57].

Induced pluripotent stem cells can also be isolated and differentiated into limbal stem cells [58, 59]. Again the autologous nature of the cells reduces the risk of immune rejection. Typically, the cultured cells are induced into pluripotency and reprogrammed to a variety of different cell fates through the introduction of transcription factors. Unlike embryonic stem cells, which can be difficult to isolate, induced pluripotent stem cells are typically derived from differentiated fibroblasts and keratinocytes obtained through minimally invasive techniques and then reprogramed into pluripotency.

A final approach involves the direct reprogramming of cells into a somatic state of a different cell lineage without going through the classical pluripotent state [60]. This is accomplished through the induction of a short partial pluripotency state, which is brought on by the transient overexpression of transcriptional factors [61]. A combination of environmental factors such as growth factors, cytokines, and other inducing agents can then be introduced to direct reprogramming to the desired cell type. Avoiding the potentially carcinogenic pluripotency stage is a major advantage. Furthermore, the resulting limbal stem cells are much easier to produce and characterize in comparison to a number of pluripotent stem cell lines [61]. Considering that repair of the cornea does not require a large number of cells or cell types, it would appear as though the limited expansion ability of reprogrammed cells would not make this technique prohibitive.

6.4.4 Stromal and Endothelial Stem Cells

Up to this point, we have focused primarily on limbal stem cells, which are responsible for the regeneration of the corneal epithelium. The other two cellular layers of the cornea, the stroma and endothelium, do not contain well-defined sources of stem cells for regeneration as does the epithelium. Stromal keratocytes are responsible for generating and maintaining the extracellular matrix of the stroma and thus its transparency [62]. The problem with keratocytes is their tendency under in vitro conditions to undergo phenotypic differentiation into fibroblasts, which are not capable of constructing the organized ECM framework required for stromal tissue [63, 64]. In addition, when expanded in vivo, keratocytes eventually undergo morphological changes and lack expression of certain genes, which ultimately results in a hazy stroma [65].

Du and coworkers have demonstrated the presence of a population of cells in the human corneal stroma expressing stem cell markers and exhibiting multipotent differentiation potential [66]. They also found that when cultured as free floating cells (no carrier or rigid matrix) in serum-free media, these human corneal stromal stem cells can differentiate into a cell type that resembles human keratocytes in both gene expression and ability to secrete a highly organized ECM network with orthogonally arranged collagen fibrils [67]. Interestingly, when injected into mouse corneal wounds, these human corneal stromal stem cells prevented formation of scar tissue and induced regeneration of ablated stroma with lamellar structure and collagen organization similar to that of native tissue [68].

The endothelial layer of the cornea is comprised of a singular layer of corneal endothelial cells (CECs) that do not regenerate [69]. As lost CECs are not replaced, the neighboring cells migrate and increase their size to cover the deficit and maintain the monolayer. The normal slow attrition of CECs with age typically does not interfere with the endothelial pump function. However, any excessive cell loss due to a trauma or a disease and resulting in endothelial cell counts below an approximate threshold of 500 cells/mm^2 is susceptible to jeopardize endothelial cell function and lead to corneal edema, decreased vision, and eventually, painful blindness. The only current treatment for these eyes is transplantation of a healthy endothelium from a human donor. While younger donors are preferred, as they typically will contain a larger population of CECs, selecting only corneas from younger donors is not affordable, especially in the context of donor shortage currently experienced worldwide. This, once more, reiterates the need for alternatives to donor corneas.

The rapidly evolving surgical techniques currently used for corneal endothelial transplantation (see above) now allow restitution of an anatomy and a functionality very close to normal [70]. Allogeneic endothelial grafts, however, remain at risk of detachment (during the early postoperative period), immune rejection, and endothelial failure. And in all cases, they necessitate a human donor cornea.

Tissue engineering now allows for the in vitro production of a corneal endothelium that remains functional during and one week after transplantation in the living animal [71]. Several challenges are still to be met prior to the transfer of this technology to the clinic, including long-term follow-up of the transplanted animal and optimization of the carrier on which the CECs are cultured and transplanted. Limited proliferation ability, cellular senescence, and fibroblastic transformation during culture [72] are also problems associated with the cultivation of CECs. A detailed review of reported methods being developed for the preparation, delivery, and transplantation of a tissue-engineered corneal endothelium, as well as the optimization steps that are still required, is available in Proulx et al. [73].

Cultured CECs can also be seeded in the anterior chamber of the eye without a carrier [74]. The first clinical results in 11 patients were recently reported by Kinoshita's group in Japan. The main challenge with this technique is preventing the mesenchymal transition of these cells [75]. Rho-associated kinase inhibitor eye drops have been proposed as an adjuvant modality to modulate endothelial cell adhesion and migration in corneal endothelial diseases and/or surgery [74, 76, 77]. Safety and efficacy, however, remain to be confirmed.

Another possible source of CECs involves differentiation from human Embryonic Stem Cells (hESCs). Both McCabe and co-workers as well as Song and co-workers have developed independent methods of deriving CECs from the neural crest phase of hESCs [78, 79]. Additionally, Zhang and co-workers have demonstrated that CECs can be derived from the periocular mesenchymal phase of hESCs. In all cases the hESC-derived CECs show morphology and endothelial cell marker expression similar to those of adult human CECs. In this study, the resulting CEC-like cells, transplanted into rabbits using a DSEK technique, allowed gradual restoration of transparency [80].

6.5 Keratoprostheses

6.5.1 Clinical Use of Keratoprostheses

In the clinical setting, the first generation of artificial corneas to be developed were keratoprostheses (KPros) [81]. They were developed as an alternative to donor tissue, which as we have already mentioned can be in severe shortage in many countries, and as an alternative for patients in which the use of an allogeneic graft is of too high risk.

While there are some differences between the available KPros, for the most part these devices are based on a core and skirt model. The central core is composed of an optically transparent material such as poly methyl methacrylate (PMMA), to allow transmission of light, while the skirt, typically porous or perforated, allows for the in-growth of cells and anchoring of the prosthetic to the corneal surface.

The two most popular keratoprostheses in the clinical setting are the Boston KPro [82] and the osteodonto-keratoprostheses (OOKP) [83]. Interestingly, both of these prosthetics contain a biological interface. In the case of the Boston KPro a corneal rim is employed, whereas in the OOKP oral mucosal tissue is employed. While the OOKP and BostonKPro are the more popular devices, there are a number of variants that have been tested clinically, including devices from the Filatov institute [84]. One of the more recent devices consists of a fused core and skirt, which is comprised entirely of poly hydroxyethyl methacrylate (pHEMA) [85, 86].

At present, keratoprostheses are mostly indicated for patients with multiple failed corneal grafts, stem cell deficiencies, aniridic patients and patients with cicatrizing disease Steven-Johnson syndrome and ocular cicatricial pemphigoid [87, 88]. While KPros are in general safe and efficient, they are typically employed in end-stage eyes due to the irreversible nature of the surgery, the inability of the implant to promote regeneration, and the risks of complications relating to glaucoma, infection, necrosis, and extrusion. Patients must be treated with sustained antibiotics and immunosuppressive drugs for the life of the device. While the devices have achieved much success, there continues to be significant progress in their design, for example, through the addition of holes in the backplate, replacement of the threads with a lock ring (click-on-model), and the use of titanium back-plates in the Boston KPro [89, 90].

6.5.2 Regenerative Keratoprostheses

Recently, there has been a push to develop Kpros that promote or allow for regeneration of the corneal surface, the idea being that reepithelialization of the implanted device would allow for a better coverage by the tear film, which will prevent infection, commonly incurred in classical Kpros, as well as extrusion of the implant itself. Typically, the regenerative properties of these next generation KPros are brought about through either lithographic patterning or surface modification chemistries. For example, it has been demonstrated that when the PMMA surface of the Boston KPro is modified through covalent attachment of ECM-derived components such as fibronectin, epidermal growth factors, collagen, laminin, or other cell adhesion promoting peptides (e.g., RGD, YIGSR, and IKVAV), the growth of epithelial cells over the KPro is enhanced [91–96]. Interestingly, the ECM components need to be covalently attached to the surface of the KPro as non-covalent attachment of either laminin or fibronectin was shown incapable of promoting the regeneration of host tissues. Furthermore, it has been demonstrated that using flexible polyethylene glycol (PEG) linkers between the surface of the device and the ECM components can significantly enhance epithelial cell attachment and growth in comparison to direct attachment in vitro [97, 98]. The enhancement observed when the factors are tethered via a linker may be due to a decrease in steric constraints, which allows the tethered motifs to orient themselves in such a way that produces an optimal micro-environment for the attachment and proliferation of epithelial cells. It has also been demonstrated that collagen-based peptides, such as Gly-Pro-NLeu as well as a combination of growth factors tethered to poly 2-hydroxyethyl methacrylate (PHEMA), can enhance both epithelial growth and attachment in vitro [99].

A series of lithographically patterned KPros have been developed by Myung and co-workers. Through patterning of the KPro surface they have been able to enhance epithelial attachment to the device. In the first generation of these KPros, photo-lithographic techniques were utilized to pattern the surface of the PEG/PAA central core, which was then coupled to a poly hydroxyethyl acrylate (PHEA) micro-perforated skirt [100]. From here they then coupled type I collagen to the hydrogel and found that the resulting material was capable of promoting epithelial growth both in vitro and in vivo in rabbits [101]. The newest iteration of the device consists of a single-piece KPro. Through photolithographic polymerization techniques they were able to generate a core-skirt form of cross-linked PEG. This PEG-derived core-skirt mold was then utilized as template for the polymerization and cross-linking of acrylic acid [102]. It has also been demonstrated that polyelectrolytes such as chitosan and heparin can be layered onto the skirt of KPros to create a layer that is ideal for the adsorption of small peptides, which can improve the ingrowth of epithelial cells [102].

6.6 ECM-Derived Materials

6.6.1 Decellularized Implants

The use of decellularized corneas in regenerative methods relies on the ability of the various components of the extracellular matrix (ECM) to promote regeneration. In the decellularization process, the cells of the donor tissue are eliminated through a variety of different techniques [103]. In general most procedures utilize a combination of detergents; however, there exist protocols that utilize enzymes, chelating agents, acid/base treatments, alcohols, as well as physical manipulations such as hydrostatic pressure to kill the cell populations [103]. While the implanted ECM can itself induce the regeneration of tissue from endogenous cell populations, it is also possible to seed the decellularized ECM with autologous cells prior to transplantation. In any case, the use of decellularized corneas in regenerative medicine remains limited by the availability of donor tissue, as well as by stringent screening protocols that are in place to prevent the transmission of disease through allogeneic tissue transplantation. It should also be mentioned that due to the varying efficacies of the decellularization protocols, it is possible in some instances that certain populations of cells remain trapped with the ECM or that key components of the ECM be removed.

In a short-term noncontrolled retrospective clinical study, Daoud and co-workers described the use of gamma-irradiated decellularized corneas (no donor keratocytes or endothelial cells) as patch graft (79%), in anterior lamellar keratoplasty (11%) or in keratoprostheses (10%) for cases where a viable endothelium was not necessary. Reported advantages included sterilization of donor corneas and longer shelf life, which can be especially helpful in emergency situations or in remote areas [104].

More recently, Zhang and co-workers reported on 47 patients who received a xenogeneic decellularized porcine anterior lamellar stromal graft to treat a fungal infection. In 41 of the 47 patients the implant became transparent, with no immunogenic response or recurrence of fungal infection observed after 6 months [105].

Another decellularized xeno-derived corneal implant that has found use in the clinical setting is fish scale. Being comprised of a high percentage of connective tissue proteins and collagens, fish scales offer an interesting alternative as ECM scaffolds in corneal repair [106–108]. Currently, decellularized fish scale implants marketed as ologen™ BioCornea are being employed to temporarily seal corneal perforations and in trabeculectomy procedures for patients suffering from glaucoma [109].

6.6.2 Cell-Derived Self-Assembling ECM Implants

ECM corneal constructs can also be derived through the self-assembly of proteins secreted by corneal or skin fibroblasts. Through the culture of fibroblast in the presence of ascorbic acid or ascorbic acid derivatives, one can induce the fibroblasts to

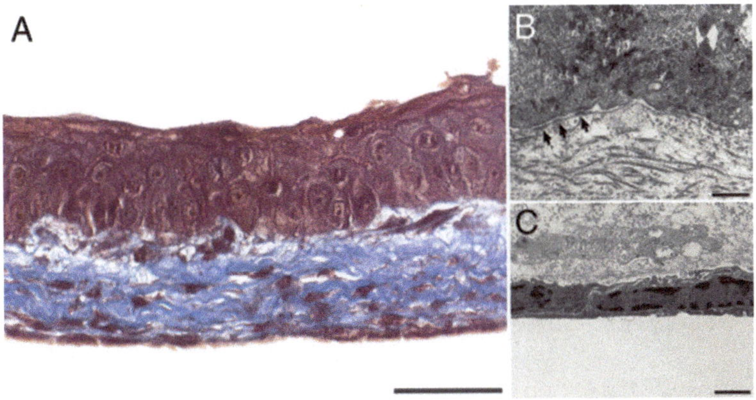

Fig. 6.2 Tissue-engineered human cornea. (**a**) Histology (Masson's Trichrome staining) of the tissue-engineered cornea, showing a well-differentiated epithelium on top, and a monolayer of endothelial cells underneath, both adhered to the self-assembled stromal matrix. (**b**) Transmission electron microscopy of the epithelial basal membrane showing many hemidesmosomes (arrows). (**c**) Transmission electron microscopy of the corneal endothelium, showing a monolayer of flattened cells. The bar in (**a**) equals 50 μm, in (**b**) equals 1 μm, and in (**c**) equals 0.5 μm. Reproduced with permission from Proulx et al. [111]

secrete ECM macromolecules [110]. After weeks in culture the resulting ECM tissue layers can be stacked together to form a lamellar stromal-like tissue. Epithelial and endothelial cells can then be seeded onto the artificial stroma and allowed to expand in culture to give rise to both endothelium and epithelium layers [110, 111] (see Fig. 6.2).

It has also been demonstrated that human umbilical cord mesenchymal stem cells can be induced to secrete stroma-like ECM components in the presence of ascorbic acid. The addition of transforming growth factor-beta (TGF-β) to the culture media increased ECM secretion resulting in thicker sheets of tissue. Interestingly, the cord stem cells used to generate the stroma-like material, differentiated themselves into cells resembling stromal fibroblasts [112]. The influence of macromolecular crowding has also been employed to induce and enhance the secretion of ECM components by human corneal fibroblasts [113]. The addition of carrageenan, a negatively charged galactose derivative, to the culture media, induced a 12-fold increase in ECM deposition without changes in the cellular morphology phenotype [114].

These types of self-assembled ECM-based corneal constructs have now been tested in animal models. For example, self-assembled corneal-stromal constructs derived from both feline and human donor corneal cells have been implanted into the eyes of cats through an intra-stromal pocket [115]. The corneas grafted with the allogeneic and xenogeneic tissue-engineered stromas were transparent and innervated within 1 month of implantation, without vascularization [115].

Considering that there is a population of progenitor cells in the stromal layer, it seems plausible that stromal ECM and corneal constructs could be prepared in vitro from autologous cells and then implanted into the patient eye. However, one needs to take into consideration the time-scale of self-assembly (several weeks), the need for the patient to undergo surgery for tissue isolation, and the high costs associated with manufacturing the constructs under Good Manufacturing Practice (GMP) guidelines for Advanced Medical Therapy Products.

6.6.3 Collagen Derived and Other Cell-Free Implants

As collagen is the main structural component of the corneal ECM, the wide interest in its use for the design of artificial corneas and as scaffolds for the promotion of tissue regeneration is understandable. Wray and Orwin first demonstrated that type I collagen could be electrospun and crosslinked with gluteraldehyde to produce highly aligned fibers that give rise to a lamellae type structure similar to that found in the stroma of native corneas [116]. They further showed that corneal fibroblasts seeded on the top of this material tended to elongate along the long axis of the fibers and respond to changes in both the structure and organization of the matrix. It has also been demonstrated that hydrogels prepared with a relatively high concentration (18% w/w) of recombinant human collagen (RHC) type III, self-assemble into highly aligned fibers [117]. Hybrid corneal implants consisting of a blend of chitosan and collagen have been manufactured and tested in animal models [118]. The chitosan filler is believed to promote regeneration and possesses anti-microbial properties [119]. Fibrin-agarose scaffolds seeded with stromal fibroblasts and epithelial cells have also shown success in animal models [120–122].

The first collagen-based artificial cornea, which was designed to both regenerate and reinnervate corneal tissue, was evaluated in ten patients presenting with significant vision impairment due to keratoconus and scarring. These biomimetic corneas were prepared through the crosslinking of RHC type III with 1-ethyl-3-(3-dimethylaminopropyl) carbodiimide (EDC) and N-hydroxysuccinimide (NHS) [123]. Once implanted into the pathologic eyes, the cell-free construct served as a template for the regeneration of the damaged corneal epithelium, stroma, and nerves. At the 4 year postoperative follow-up, it was found that the artificial corneas were capable of promoting the proliferation and differentiation of endogenous corneal cell populations. As these cells migrated into the artificial scaffold, they generated a highly integrated neo-cornea that resembled the normal and healthy tissue [124].

Preclinical studies in the rabbit model have shown that the RHC type III implants were not as effective in corneas that were highly inflamed or vascularized due to alkali chemical burns [125]. Interestingly, the addition of a synthetic phospholipid, 2-methacryloyloxyethyl phosphorylcholine (MPC), was shown to decrease inflammation and inhibit corneal vascularization when crosslinked into the matrix through a PEG diacrylate (PEGDA) polymer using ammonium persulfate and tetramethylethylenediamine [126]. A full-thickness graft of the RHC type III-MPC corneal construct was capable of promoting the regeneration of corneal tissue and nerves in a guinea pig study [127]. It was also capable of inhibiting the vascularization of rabbit corneas damaged by alkali burns [125].

RHC type III-MPC implants have now been grafted in patients with severe pathologies, to relieve the symptoms of pain and discomfort resulting from corneal ulceration after severe infection, chemical burns, failed grafts and neurotrophic keratitis. Within 1–2 weeks after implantation, the discomfort had subsided. The epithelium and stroma of all patients has stably regenerated, and nerve regeneration was documented in the patients in the lamellar graft study [128, 140]. More recently, the RHC Type III-MPC implant has been modified through microcontact printing to bear 30 μm stripes of fibronectin, which have been shown to enhance cell adhesion and should allow for faster integration of the implant at the host-site [129].

6.6.4 Self-Assembling ECM Implants

As is evidenced by the previous sections, ECM-derived implants have had much success in the regeneration of corneal tissue. Their use, however, remains limited by the difficulty of isolating ECM components, batch-to-batch variability, and the risk of immune response [130]. Recombinantly generated components, such as RHC, eliminate much of the batch-to-batch variability, but even then, the large size of many of these macromolecules makes them difficult to process. Hence, there has been a recent push to develop small analogues as mimics of these ECM components, which can self-assemble to give rise to three-dimensional structures that mimic their native tissues. Some of the more successful molecules at assembling corneal scaffolds have been short peptide mimics.

For example, a variety of peptide amphiphiles bearing the RGD motif of fibronectin have proven to be capable of promoting corneal regeneration by enhancing the adhesion, proliferation, and alignment of fibroblasts within the mimetic three-dimensional stroma [131, 132]. Furthermore, laminin-derived amphiphiles have also been used in rabbit models, to heal surgical wounds through the promotion of keratocyte in-growth [133, 134].

More recently, a collagen-like peptide (CLP) [135] was conjugated to a multi-arm PEG through thiol-maleimide cross-coupling reaction to create a CLP-PEG conjugate [136]. The resulting conjugate was then cross-linked through EDC-NHS coupling to give rise to hydrogels, which have been tested in animal models for their ability to promote epithelial and stromal regeneration (see Fig. 6.3).

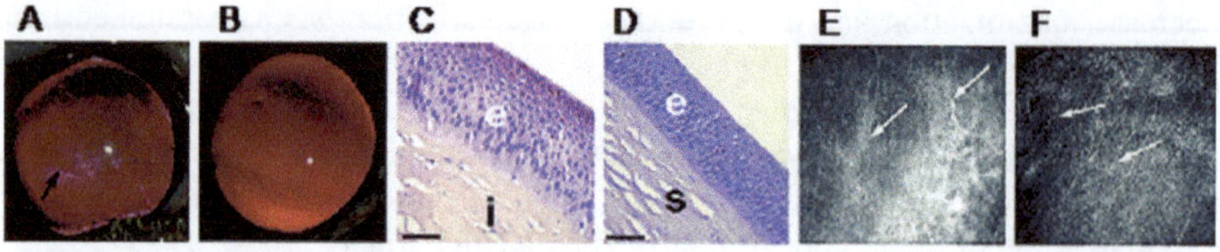

Fig. 6.3 Post-surgical corneal regeneration at 12 months after CLP–PEG implantation. (**a**) Optically clear CLP–PEG implant (arrowed) stably integrated within the pig cornea, compared to its healthy, unoperated contralateral cornea (**b**). Blood vessels are seen in the implanted cornea but stop at the margin of the implant. (**c**) H&E staining of a representative regenerated CLP–PEG neo-cornea compared to a healthy control cornea (**d**), showing similar morphology. *e* epithelium, *s* stroma, *i* implant. Scale bars, 50 μm (**e**) in vivo confocal microscopy shows the regenerated nerve (arrows) in CLP–PEG cornea, which follow a parallel pattern similar to that of the unoperated cornea (**f**). Reproduced with permission from Islam et al. [136]

6.7 Conclusions

While we have covered a variety of methods and materials for regenerating corneal tissues, there is not a single method that is suitable for treating all corneal pathologies. Instead, treatment should be tailored to the specific pathology. Through advancements in stem cell based therapies, novel ECM-derived scaffolds, drug delivery systems [137, 138], and nanomaterial hybrids [139], it should be possible to develop a variety of regenerative approaches that either limit or eliminate the need for human donor corneas in treating corneal pathologies. As the field moves forward, it is likely that many of these technologies will evolve in such a way that the best aspects of the various techniques become merged to develop hybrid-composite materials, which can both promote regeneration and prevent infection that typically undermine the effectiveness of many corneal therapies.

References

1. Meek, K. M., & Knupp, C. (2015). Corneal structure and transparency. *Progress in Retinal and Eye Research, 49*, 1–16.
2. Jonas, J. B., & Holbach, L. (2005). Central corneal thickness and thickness of the lamina cribrosa in human eyes. *Investigative Ophthalmology & Visual Science, 46*, 1275–1279.
3. Maurice, D. M. (1957). The structure and transparency of the cornea. *The Journal of Physiology, 136*, 263–286.
4. McDermott, A. M. (2009). The role of antimicrobial peptides at the ocular surface. *Ophthalmic Research, 41*, 60–75.
5. Van Buskirk, E. M. (1989). The anatomy of the limbus. *Eye, 3*, 101–108.
6. Chen, S., Mienaltowski, M. J., & Birk, D. E. (2015). Regulation of corneal stroma extracellular matrix assembly. *Experimental Eye Research, 133*, 69–80.
7. Tuft, S. J., & Coster, D. J. (1990). The corneal endothelium. *Eye, 4*, 389–424.

8. Oliva, M. S., Schottman, T., & Gulati, M. (2012). Turning the tide of corneal blindness. *Indian Journal of Ophthalmology, 60*, 423–427.
9. Gain, P., Jullienne, R., He, Z., et al. (2016). Global survey of corneal transplantation and eye banking. *JAMA Ophthalmology, 134*, 167–173.
10. Williams, K. A., Esterman, A. J., Bartlett, C., Holland, H., Hornsby, N. B., Coster, D. J., et al. (2006). How effective is penetrating corneal transplantation? Factors influencing long-term outcome in multivariate analysis. *Transplantation, 81*, 896–901.
11. Lass, J. H., Benetz, B. A., Gal, R. L., Kollman, C., Raghinaru, D., Dontchev, M., et al. (2013). Donor age and factors related to endothelial cell loss ten years after penetrating keratoplasty: Specular Microscopy Ancillary Study. *Ophthalmology, 120*, 2428–2435.
12. Arenas, E., Esquenazi, S., Anwar, M., & Terry, M. (2012). Lamellar corneal transplantation. *Survey of Ophthalmology, 57*, 510–529.
13. Ikada, Y. (2006). Challenges in tissue engineering. *Journal of the Royal Society Interface, 3*, 589–601.
14. Chen, Y., Liao, C., Gao, M., Belin, M. W., Wang, M., Yu, H., et al. (2015). Efficacy and safety of corneal transplantation using corneas from foreign donors versus domestic donors: A prospective, randomized, controlled trial. *Journal of Ophthalmology, 2015*, 178289.
15. Dua, H. S., & Azuara-Blanco, A. (2000). Limbal stem cells of the corneal epithelium. *Survey of Ophthalmology, 44*, 415–425.
16. Dua, H. S., Gomes, J. A., & Singh, A. (1994). Corneal epithelial wound healing. *The British Journal of Ophthalmology, 78*, 401–408.
17. Tseng, S. G., Prabhasawat, P., Barton, K., Gray, T., & Meller, D. (1998). Amniotic membrane transplantation with or without limbal allografts for corneal surface reconstruction in patients with limbal stem cell deficiency. *Archives of Ophthalmology, 116*, 431–441.
18. Atallah, M. R., Palioura, S., Perez, V. L., & Amescua, G. (2016). Limbal stem cell transplantation: Current perspectives. *Clinical Ophthalmology, 10*, 593–602.
19. Baradaran-Rafii, A., Eslani, M., Haq, Z., Shirzadeh, E., Huvard, M. J., & Djalilian, A. R. (2017). Current and upcoming therapies for ocular surface chemical injuries. *The Ocular Surface, 15*, 48–64.
20. Holland, E. J. (2015). Management of limbal stem cell deficiency: A historical perspective, past, present, and future. *Cornea, 34*, S9–S15.
21. Ramachandran, C., Basu, S., Sangwan, V. S., & Balasubramanian, D. (2014). Concise review: The coming of age of stem cell treatment for corneal surface damage. *Stem Cells Translational Medicine, 3*, 1160–1168.
22. Vazirani, J., Mariappan, I., Ramamurthy, S., Fatima, S., Basu, S., & Sangwan, V. S. (2016). Surgical management of bilateral limbal stem cell deficiency. *The Ocular Surface, 14*, 350–364.
23. Thoft, R. A. (1977). Conjunctival transplantation. *Archives of Ophthalmology, 95*, 1425–1427.
24. Kenyon, K. R., & Tseng, S. C. (1989). Limbal autograft transplantation for ocular surface disorders. *Ophthalmology, 96*, 709–722 discussion 722–723.
25. Clearfield, E., Muthappan, V., Wang, X., & Kuo, I. C. (2016). Conjunctival autograft for pterygium. *Cochrane Database of Systematic Reviews, 2*, CD011349.
26. Rao, S. K., Rajagopal, R., Sitalakshmi, G., & Padmanabhan, P. (1999). Limbal autografting: Comparison of results in the acute and chronic phases of ocular surface burns. *Cornea, 18*, 164–171.
27. Holland, E. J., & Schwartz, G. S. (1996). The evolution of epithelial transplantation for severe ocular surface disease and a proposed classification system. *Cornea, 15*, 549–556.
28. Croasdale, C. R., Schwartz, G. S., Malling, J. V., & Holland, E. J. (1999). Keratolimbal allograft: Recommendations for tissue procurement and preparation by eye banks, and standard surgical technique. *Cornea, 18*, 52–58.
29. Kwitko, S., Marinho, D., Barcaro, S., Bocaccio, F., Rymer, S., Fernandes, S., et al. (1995). Allograft conjunctival transplantation for bilateral ocular surface disorders. *Ophthalmology, 102*, 1020–1025.

30. Biber, J. M., Skeens, H. M., Neff, K. D., & Holland, E. J. (2011). The cincinnati procedure: Technique and outcomes of combined living-related conjunctival limbal allografts and keratolimbal allografts in severe ocular surface failure. *Cornea, 30*, 765–771.
31. Chan, C. C., Biber, J. M., & Holland, E. J. (2012). The modified Cincinnati procedure: Combined conjunctival limbal autografts and keratolimbal allografts for severe unilateral ocular surface failure. *Cornea, 31*, 1264–1272.
32. Espana, E. M., Di Pascuale, M., Grueterich, M., Solomon, A., & Tseng, S. C. G. (2004). Keratolimbal allograft in corneal reconstruction. *Eye, 18*, 406–417.
33. Basu, S., Sureka, S. P., Shanbhag, S. S., Kethiri, A. R., Singh, V., & Sangwan, V. S. (2016). Simple limbal epithelial transplantation: Long-term clinical outcomes in 125 cases of unilateral chronic ocular surface burns. *Ophthalmology, 123*, 1000–1010.
34. Sangwan, V. S., Basu, S., Macneil, S., & Balasubramanian, D. (2012). Simple limbal epithelial transplantation (SLET): A novel surgical technique for the treatment of unilateral limbal stem cell deficiency. *The British Journal of Ophthalmology, 96*, 931–934.
35. Vazirani, J., Ali, M. H., Sharma, N., Gupta, N., Mittal, V., Atallah, M., et al. (2016). Autologous simple limbal epithelial transplantation for unilateral limbal stem cell deficiency: Multicentre results. *The British Journal of Ophthalmology, 100*, 1416–1420.
36. Amescua, G., Atallah, M., Nikpoor, N., Galor, A., & Perez, V. L. (2014). Modified simple limbal epithelial transplantation using cryopreserved amniotic membrane for unilateral limbal stem cell deficiency. *American Journal of Ophthalmology, 158*, 469–475.
37. Rheinwald, J. G., & Green, H. (1975). Serial cultivation of strains of human epidermal keratinocytes: The formation of keratinizing colonies from single cells. *Cell, 6*, 331–343.
38. Pellegrini, G., Traverso, C. E., Franzi, A. T., Zingirian, M., Cancedda, R., & De Luca, M. (1997). Long-term restoration of damaged corneal surfaces with autologous cultivated corneal epithelium. *The Lancet, 349*, 990–993.
39. Sangwan, V. S., Basu, S., Vemuganti, G. K., Sejpal, K., Subramaniam, S. V., Bandyopadhyay, S., et al. (2011). Clinical outcomes of xeno-free autologous cultivated limbal epithelial transplantation: A 10-year study. *The British Journal of Ophthalmology, 95*, 1525–1530.
40. Deshpande, P., Ramachandran, C., Sangwan, V. S., & Macneil, S. (2013). Cultivation of limbal epithelial cells on electrospun poly (lactide-co-glycolide) scaffolds for delivery to the cornea. In B. Wright & C. J. Connon (Eds.), *Corneal regenerative medicine: Methods and protocols*. Totowa, NJ: Humana Press.
41. Fasolo, A., Pedrotti, E., Passilongo, M., Marchini, G., Monterosso, C., Zampini, R., et al. (2016). Safety outcomes and long-term effectiveness of ex vivo autologous cultured limbal epithelial transplantation for limbal stem cell deficiency. *The British Journal of Ophthalmology, 101*(5), 640–649.
42. Rama, P., Bonini, S., Lambiase, A., Golisano, O., Paterna, P., De Luca, M., et al. (2001). Autologous fibrin-cultured limbal stem cells permanently restore the corneal surface of patients with total limbal stem cell deficiency. *Transplantation, 72*, 1478–1485.
43. Sangwan, V. S., Matalia, H. P., Vemuganti, G. K., Fatima, A., Ifthekar, G., Singh, S., et al. (2006). Clinical outcome of autologous cultivated limbal epithelium transplantation. *Indian Journal of Ophthalmology, 54*, 29–34.
44. Tsai, R. J.-F., Li, L.-M., & Chen, J.-K. (2000). Reconstruction of damaged corneas by transplantation of autologous limbal epithelial cells. *The New England Journal of Medicine, 343*, 86–93.
45. Eslani, M., Baradaran-Rafii, A., & Ahmad, S. (2012). Cultivated limbal and oral mucosal epithelial transplantation. *Seminars in Ophthalmology, 27*, 80–93.
46. Nishida, K., Yamato, M., Hayashida, Y., Watanabe, K., Yamamoto, K., Adachi, E., et al. (2004). Corneal reconstruction with tissue-engineered cell sheets composed of autologous oral mucosal epithelium. *The New England Journal of Medicine, 351*, 1187–1196.
47. Nakamura, T., Inatomi, T., Sotozono, C., Amemiya, T., Kanamura, N., & Kinoshita, S. (2004). Transplantation of cultivated autologous oral mucosal epithelial cells in patients with severe ocular surface disorders. *The British Journal of Ophthalmology, 88*, 1280–1284.

48. Inatomi, T., Nakamura, T., Koizumi, N., Sotozono, C., Yokoi, N., & Kinoshita, S. (2006). Midterm results on ocular surface reconstruction using cultivated autologous oral mucosal epithelial transplantation. *American Journal of Ophthalmology, 141*, 267–275.
49. Prabhasawat, P., Ekpo, P., Uiprasertkul, M., Chotikavanich, S., Tesavibul, N., Pornpanich, K., et al. (2016). Long-term result of autologous cultivated oral mucosal epithelial transplantation for severe ocular surface disease. *Cell and Tissue Banking, 17*, 491–503.
50. Satake, Y., Higa, K., Tsubota, K., & Shimazaki, J. (2011). Long-term outcome of cultivated oral mucosal epithelial sheet transplantation in treatment of total limbal stem cell deficiency. *Ophthalmology, 118*, 1524–1530.
51. Dobrowolski, D., Wylegala, E., Wowra, B., & Orzechowska-Wylegala, B. (2011). Cultivated oral mucosa epithelium transplantation (COMET) in bilateral limbal stem cell deficiency. *Acta Ophthalmologica. Supplement, 89*. https://doi.org/10.1111/j.1755-3768.2011.4374.x
52. Sotozono, C., Inatomi, T., Nakamura, T., Koizumi, N., Yokoi, N., Ueta, M., et al. (2013). Visual improvement after cultivated oral mucosal epithelial transplantation. *Ophthalmology, 120*, 193–200.
53. Liu, J., Sheha, H., Fu, Y., Giegengack, M., & Tseng, S. C. (2011). Oral mucosal graft with amniotic membrane transplantation for total limbal stem cell deficiency. *American Journal of Ophthalmology, 152*, 739–47.e1.
54. Katikireddy, K. R., Dana, R., & Jurkunas, U. V. (2014). Differentiation potential of limbal fibroblasts and bone marrow mesenchymal stem cells to corneal epithelial cells. *Stem Cells, 32*, 717–729.
55. Zhang, L., Coulson-Thomas, V. J., Ferreira, T. G., & Kao, W. W. Y. (2015). Mesenchymal stem cells for treating ocular surface diseases. *BMC Ophthalmology, 15*, 155.
56. Meyer-Blazejewska, E. A., Call, M. K., Yamanaka, O., Liu, H., Schlötzer-Schrehardt, U., Kruse, F. E., et al. (2011). From hair to cornea: Towards the therapeutic use of hair follicle-derived stem cells in the treatment of limbal stem cell deficiency. *Stem Cells, 29*, 57–66.
57. Monteiro, B. G., Serafim, R. C., Melo, G. B., Silva, M. C. P., Lizier, N. F., Maranduba, C. M. C., et al. (2009). Human immature dental pulp stem cells share key characteristic features with limbal stem cells. *Cell Proliferation, 42*, 587–594.
58. Erbani, J., Aberdam, D., Larghero, J., & Vanneaux, V. (2016). Pluripotent stem cells and other innovative strategies for the treatment of ocular surface diseases. *Stem Cell Reviews, 12*, 171–178.
59. Hayashi, R., Ishikawa, Y., Ito, M., Kageyama, T., Takashiba, K., Fujioka, T., et al. (2012). Generation of corneal epithelial cells from induced pluripotent stem cells derived from human dermal fibroblast and corneal limbal epithelium. *PLoS One, 7*, e45435.
60. Kelaini, S., Cochrane, A., & Margariti, A. (2014). Direct reprogramming of adult cells: Avoiding the pluripotent state. *Stem Cells Cloning, 7*, 19–29.
61. Casaroli-Marano, R. P., Nieto-Nicolau, N., Martínez-Conesa, E. M., Edel, M., & Alvarez-Palomo, A. B. (2015). Potential role of induced pluripotent stem cells (IPSCs) for cell-based therapy of the ocular surface. *Journal of Clinical Medicine, 4*, 318–342.
62. Delmonte, D. W., & Kim, T. (2011). Anatomy and physiology of the cornea. *Journal of Cataract and Refractive Surgery, 37*, 588–598.
63. Beales, M. P., Funderburgh, J. L., Jester, J. V., & Hassell, J. R. (1999). Proteoglycan synthesis by bovine keratocytes and corneal fibroblasts: Maintenance of the keratocyte phenotype in culture. *Investigative Ophthalmology & Visual Science, 40*, 1658–1663.
64. Funderburgh, J. L., Mann, M. M., & Funderburgh, M. L. (2003). Keratocyte phenotype mediates proteoglycan structure: A role for fibroblasts in corneal fibrosis. *The Journal of Biological Chemistry, 278*, 45629–45637.
65. Karamichos, D., Funderburgh, M. L., Hutcheon, A. E. K., Zieske, J. D., Du, Y., Wu, J., et al. (2014). A role for topographic cues in the organization of collagenous matrix by corneal fibroblasts and stem cells. *PLoS One, 9*, e86260.
66. Du, Y., Funderburgh, M. L., Mann, M. M., Sundarraj, N., & Funderburgh, J. L. (2005). Multipotent stem cells in human corneal stroma. *Stem Cells, 23*, 1266–1275.

67. Du, Y., Sundarraj, N., Funderburgh, M. L., Harvey, S. A., Birk, D. E., & Funderburgh, J. L. (2007). Secretion and organization of a cornea-like tissue in vitro by stem cells from human corneal stroma. *Investigative Ophthalmology & Visual Science, 48*, 5038–5045.
68. Basu, S., Hertsenberg, A. J., Funderburgh, M. L., Burrow, M. K., Mann, M. M., Du, Y., et al. (2014). Human limbal biopsy-derived stromal stem cells prevent corneal scarring. *Science Translational Medicine, 6*, 266RA172–266RA172.
69. Joyce, N. C. (2003). Proliferative capacity of the corneal endothelium. *Progress in Retinal and Eye Research, 22*, 359–389.
70. Dapena, I., Ham, L., & Melles, G. R. J. (2009). Endothelial keratoplasty: DSEK/DSAEK or DMEK – The thinner the better? *Current Opinion in Ophthalmology, 20*, 299–307.
71. Proulx, S., Bensaoula, T., Nada, O., Audet, C., D'Arc Uwamaliya, J., Devaux, A., et al. (2009). Transplantation of a tissue-engineered corneal endothelium reconstructed on a devitalized carrier in the feline model. *Investigative Ophthalmology & Visual Science, 50*, 2686–2694.
72. De Araujo, A. L., & Gomes, J. Á. P. (2015). Corneal stem cells and tissue engineering: Current advances and future perspectives. *World Journal of Stem Cells, 7*, 806–814.
73. Proulx, S., & Brunette, I. (2012). Methods being developed for preparation, delivery and transplantation of a tissue-engineered corneal endothelium. *Experimental Eye Research, 95*, 68–75.
74. Kinoshita, S., Koizumi, N., Ueno, M., Okumura, N., Imai, K., Tanaka, H., Yamamoto, Y., Nakamura, T., Inatomi, T., Bush, J., Toda, M., Hagiya, M., Yokota, I., Teramukai, S., Sotozono, C., & Hamuro, J. (2018). Injection of cultured cells with a ROCK inhibitor for bullous keratopathy. *New England Journal of Medicine*, 378, 995-1003.
75. Bostan, C., Theriault, M., Forget, K. J., Doyon, C., Cameron, J. D., Proulx, S., et al. (2016). In vivo functionality of a corneal endothelium transplanted by cell-injection therapy in a Feline model. *Investigative Ophthalmology & Visual Science, 57*, 1620–1634.
76. Koizumi, N., Okumura, N., Ueno, M., & Kinoshita, S. (2014). New therapeutic modality for corneal endothelial disease using rho-associated kinase inhibitor eye drops. *Cornea, 33*(Suppl 11), S25–S31.
77. Okumura, N., Okazaki, Y., Inoue, R., Kakutani, K., Nakano, S., Kinoshita, S., et al. (2016). Effect of the rho-associated kinase inhibitor eye drop (Ripasudil) on corneal endothelial wound healing. *Investigative Ophthalmology & Visual Science, 57*, 1284–1292.
78. Mccabe, K. L., Kunzevitzky, N. J., Chiswell, B. P., Xia, X., Goldberg, J. L., & Lanza, R. (2015). Efficient generation of human embryonic stem cell-derived corneal endothelial cells by directed differentiation. *PLoS One, 10*, e0145266.
79. Song, Q., Yuan, S., An, Q., Chen, Y., Mao, F. F., Liu, Y., et al. (2016). Directed differentiation of human embryonic stem cells to corneal endothelial cell-like cells: A transcriptomic analysis. *Experimental Eye Research, 151*, 107–114.
80. Zhang, K., Pang, K., & Wu, X. (2014). Isolation and transplantation of corneal endothelial cell-like cells derived from in-vitro-differentiated human embryonic stem cells. *Stem Cells and Development, 23*, 1340–1354.
81. Gomaa, A., Comyn, O., & Liu, C. (2010). Keratoprostheses in clinical practice – A review. *Clinical & Experimental Ophthalmology, 38*, 211–224.
82. Dohlman, C. H., Cruzat, A., & White, M. (2014). The Boston keratoprosthesis 2014: A step in the evolution of artificial corneas. *Spektrum Augenheilkd, 28*, 226–233.
83. Liu, C., Paul, B., Tandon, R., Lee, E., Fong, K., Mavrikakis, I., et al. (2005). The osteo-odonto-keratoprosthesis (OOKP). *Seminars in Ophthalmology, 20*, 113–128.
84. Iakymenko, S. (2013). Forty-five years of keratoprosthesis study and application at the Filatov Institute: A retrospective analysis of 1 060 cases. *International Journal of Ophthalmology, 6*, 375–380.
85. Colby, K. A., & Koo, E. B. (2011). Expanding indications for the Boston keratoprosthesis. *Current Opinion in Ophthalmology, 22*, 267–273.
86. Hicks, C. R., Crawford, G. J., Dart, J. K. G., Grabner, G., Holland, E. J., Stulting, R. D., et al. (2006). AlphaCor: Clinical outcomes. *Cornea, 25*, 1034–1042.

87. Hassanaly, S. I., Talajic, J. C., & Harissi-Dagher, M. (2014). Outcomes following Boston type 1 keratoprosthesis implantation in aniridia patients at the University of Montreal. *American Journal of Ophthalmology, 158*, 270–276.e1.
88. Ma, J. J., Graney, J. M., & Dohlman, C. H. (2005). Repeat penetrating keratoplasty versus the Boston keratoprosthesis in graft failure. *International Ophthalmology Clinics, 45*, 49–59.
89. Lee, W. B., Shtein, R. M., Kaufman, S. C., Deng, S. X., & Rosenblatt, M. I. (2015). Boston keratoprosthesis: Outcomes and complications. *Ophthalmology, 122*, 1504–1511.
90. Rudnisky, C. J., Belin, M. W., Guo, R., Ciolino, J. B., Dohlman, C. H., Aquavella, J., et al. (2016). Visual acuity outcomes of the Boston keratoprosthesis type 1: Multicenter study results. *American Journal of Ophthalmology, 162*, 89–98.
91. Aucoin, L., Griffith, C. M., Pleizier, G., Deslandes, Y., & Sheardown, H. (2002). Interactions of corneal epithelial cells and surfaces modified with cell adhesion peptide combinations. *Journal of Biomaterials Science, Polymer Edition, 13*, 447–462.
92. Bruining, M. J., Paul Pijpers, A., Kingshott, P., & Koole, L. H. (2002). Studies on new polymeric biomaterials with tunable hydrophilicity, and their possible utility in corneal repair surgery. *Biomaterials, 23*, 1213–1219.
93. George, A., & Pitt, W. G. (2002). Comparison of corneal epithelial cellular growth on synthetic cornea materials. *Biomaterials, 23*, 1369–1373.
94. Legeais, J.-M., & Renard, G. (1998). A second generation of artificial cornea (Biokpro II). *Biomaterials, 19*, 1517–1522.
95. Merrett, K., Griffith, C. M., Deslandes, Y., Pleizier, G., & Sheardown, H. (2001). Adhesion of corneal epithelial cells to cell adhesion peptide modified pHEMA surfaces. *Journal of Biomaterials Science, Polymer Edition, 12*, 647–671.
96. Noh, H. (2013). Enhanced cornea cell growth on a keratoprosthesis material immobilized with fibronectin or EGF. *Macromolecular Research, 21*, 169–175.
97. Jacob, J. T., Rochefort, J. R., Bi, J., & Gebhardt, B. M. (2005). Corneal epithelial cell growth over tethered-protein/peptide surface-modified hydrogels. *Journal of Biomedical Materials Research Part B: Applied Biomaterials, 72B*, 198–205.
98. Wallace, C., Jacob, J. T., Stoltz, A., Bi, J., & Bundy, K. (2005). Corneal epithelial adhesion strength to tethered-protein/peptide modified hydrogel surfaces. *Journal of Biomedical Materials Research. Part A, 72A*, 19–24.
99. Johnson, G., Jenkins, M., Mclean, K. M., Griesser, H. J., Kwak, J., Goodman, M., et al. (2000). Peptoid-containing collagen mimetics with cell binding activity. *Journal of Biomedical Materials Research, 51*, 612–624.
100. Myung, D., Koh, W., Bakri, A., Zhang, F., Marshall, A., Ko, J., et al. (2007). Design and fabrication of an artificial cornea based on a photolithographically patterned hydrogel construct. *Biomedical Microdevices, 9*, 911–922.
101. Myung, D., Farooqui, N., Zheng, L. L., Koh, W., Noolandi, J., Cochran, J. R., et al. (2009). Bioactive interpenetrating polymer network hydrogels that support corneal epithelial wound healing. *Journal of Biomedical Materials Research. Part A, 90*, 70–81.
102. Myung, D., Duhamel, P.-E., Cochran, J., Noolandi, J., Ta, C., & Frank, C. (2008). Development of hydrogel-based keratoprostheses: A materials perspective. *Biotechnology Progress, 24*, 735–741.
103. Wilson, S. L., Sidney, L. E., Dunphy, S. E., Rose, J. B., & Hopkinson, A. (2013). Keeping an eye on decellularized corneas: A review of methods, characterization and applications. *Journal of Functional Biomaterials, 4*, 114–161.
104. Daoud, Y. J., Smith, R., Smith, T., Akpek, E. K., Ward, D. E., & Stark, W. J. (2011). The intraoperative impression and postoperative outcomes of gamma-irradiated corneas in corneal and glaucoma patch surgery. *Cornea, 30*, 1387–1391.
105. Zhang, M. C., Liu, X., Jin, Y., Jiang, D. L., Wei, X. S., & Xie, H. T. (2015). Lamellar keratoplasty treatment of fungal corneal ulcers with acellular porcine corneal stroma. *American Journal of Transplantation, 15*, 1068–1075.

106. Chen, S.-C., Telinius, N., Lin, H.-T., Huang, M.-C., Lin, C.-C., Chou, C.-H., et al. (2015). Use of Fish Scale-Derived BioCornea to seal full-thickness corneal perforations in Pig Models. *PLoS One, 10*, e0143511.
107. Nagai, T., Izumi, M., & Ishii, M. (2004). Fish scale collagen. Preparation and partial characterization. *International Journal of Food Science & Technology, 39*, 239–244.
108. Van Essen, T. H., Lin, C. C., Hussain, A. K., Maas, S., Lai, H. J., Linnartz, H., et al. (2013). A Fish Scale–Derived Collagen Matrix as artificial cornea in rats: Properties and potential fish-derived collagen matrix as artificial cornea. *Investigative Ophthalmology & Visual Science, 54*, 3224–3233.
109. Senthil, S., Rao, H. L., Babu, J. G., Mandal, A. K., & Garudadri, C. S. (2013). Comparison of outcomes of trabeculectomy with mitomycin C vs. ologen implant in primary glaucoma. *Indian Journal of Ophthalmology, 61*, 338–342.
110. Guo, X., Hutcheon, A. E. K., Melotti, S. A., Zieske, J. D., Trinkaus-Randall, V., & Ruberti, J. W. (2007). Morphological characterization of organized extracellular matrix deposition by ascorbic acid-stimulated human corneal fibroblasts. *Investigative Ophthalmology & Visual Science, 48*, 4050–4060.
111. Proulx, S., Uwamaliya, J. D. A., Carrier, P., Deschambeault, A., Audet, C., Giasson, C. J., et al. (2010). Reconstruction of a human cornea by the self-assembly approach of tissue engineering using the three native cell types. *Molecular Vision, 16*, 2192–2201.
112. Karamichos, D., Rich, C. B., Hutcheon, A. E. K., Ren, R., Saitta, B., Trinkaus-Randall, V., et al. (2011). Self-assembled matrix by umbilical cord stem cells. *Journal of Functional Biomaterials, 2*, 213–229.
113. Zhou, H.-X., Rivas, G., & Minton, A. P. (2008). Macromolecular crowding and confinement: Biochemical, biophysical, and potential physiological consequences. *Annual Review of Biophysics, 37*, 375–397.
114. Kumar, P., Satyam, A., Fan, X., Rochev, Y., Rodriguez, B. J., Gorelov, A., et al. (2014). Accelerated development of supramolecular corneal stromal-like assemblies from corneal fibroblasts in the presence of macromolecular crowders. *Tissue Engineering Part C: Methods, 21*, 660–670.
115. Boulze Pankert, M., Goyer, B., Zaguia, F., Bareille, M., Perron, M.-C., Liu, X., et al. (2014). Biocompatibility and functionality of a tissue-engineered living corneal stroma transplanted in the feline eye living corneal stroma transplanted in feline eye. *Investigative Ophthalmology & Visual Science, 55*, 6908–6920.
116. Wray, L. S., & Orwin, E. J. (2009). Recreating the microenvironment of the native cornea for tissue engineering applications. *Tissue Engineering Part A, 15*, 1463–1472.
117. Hayes, S., Lewis, P., Islam, M. M., Doutch, J., Sorensen, T., White, T., et al. (2015). The structural and optical properties of type III human collagen biosynthetic corneal substitutes. *Acta Biomaterialia, 25*, 121–130.
118. Rafat, M., Li, F., Fagerholm, P., Lagali, N. S., Watsky, M. A., Munger, R., et al. (2008). PEG-stabilized carbodiimide crosslinked collagen–chitosan hydrogels for corneal tissue engineering. *Biomaterials, 29*, 3960–3972.
119. Cheung, R. C. F., Ng, T. B., Wong, J. H., & Chan, W. Y. (2015). Chitosan: An update on potential biomedical and pharmaceutical applications. *Marine Drugs, 13*, 5156–5186.
120. Alaminos, M., Sánchez-Quevedo, M. A. D. C., Muñoz-Ávila, J. I., Serrano, D., Medialdea, S., Carreras, I., et al. (2006). Construction of a complete rabbit cornea substitute using a fibrin-agarose scaffold. *Investigative Ophthalmology & Visual Science, 47*, 3311–3317.
121. De La Cruz Cardona, J., Ionescu, A.-M., Gómez-Sotomayor, R., González-Andrades, M., Campos, A., Alaminos, M., et al. (2011). Transparency in a fibrin and fibrin–agarose corneal stroma substitute generated by tissue engineering. *Cornea, 30*, 1428–1435.
122. Garzón, I., Martín-Piedra, M. A., Alfonso-Rodríguez, C., González-Andrades, M., Carriel, V., Martínez-Gómez, C., et al. (2014). Generation of a biomimetic human artificial cornea model using Wharton's jelly mesenchymal stem cells. *Investigative Ophthalmology & Visual Science, 55*, 4073–4083.

123. Fagerholm, P., Lagali, N. S., Merrett, K., Jackson, W. B., Munger, R., Liu, Y., et al. (2010). A biosynthetic alternative to human donor tissue for inducing corneal regeneration: 24-Month follow-up of a phase 1 clinical study. *Sci. Transl. Med., 2*, 46ra61.
124. Fagerholm, P., Lagali, N. S., Ong, J. A., Merrett, K., Jackson, W. B., Polarek, J. W., et al. (2014). Stable corneal regeneration four years after implantation of a cell-free recombinant human collagen scaffold. *Biomaterials, 35*, 2420–2427.
125. Hackett, J. M., Lagali, N., Merrett, K., Edelhauser, H., Sun, Y., Gan, L., et al. (2011). Biosynthetic corneal implants for replacement of pathologic corneal tissue: Performance in a Controlled Rabbit Alkali Burn Model. *Investigative Ophthalmology & Visual Science, 52*, 651–657.
126. Yumoto, H., Hirota, K., Hirao, K., Miyazaki, T., Yamamoto, N., Miyamoto, K., et al. (2015). Anti-inflammatory and protective effects of 2-methacryloyloxyethyl phosphorylcholine polymer on oral epithelial cells. *Journal of Biomedical Materials Research. Part A, 103*, 555–563.
127. Mclaughlin, C. R., Acosta, M. C., Luna, C., Liu, W., Belmonte, C., Griffith, M., et al. (2010). Regeneration of functional nerves within full thickness collagen–phosphorylcholine corneal substitute implants in guinea pigs. *Biomaterials, 31*, 2770–2778.
128. Buznyk, O., Pasyechnikova, N., Islam, M. M., Iakymenko, S., Fagerholm, P., & Griffith, M. (2015). Bioengineered corneas grafted as alternatives to human donor corneas in three high-risk patients. *Clinical and Translational Science, 8*, 558–562.
129. Mirazul Islam, M., Cepla, V., He, C., Edin, J., Rakickas, T., Kobuch, K., et al. (2015). Functional fabrication of recombinant human collagen-phosphorylcholine hydrogels for regenerative medicine applications. *Acta Biomaterialia, 12*, 70–80.
130. Villa-Diaz, L. G., Ross, A. M., Lahann, J., & Krebsbach, P. H. (2013). The evolution of human pluripotent stem cell culture: From feeder cells to synthetic coatings. *Stem Cells, 31*, 1–7.
131. Gouveia, R. M., Jones, R. R., Hamley, I. W., & Connon, C. J. (2014). The bioactivity of composite Fmoc-RGDS-collagen gels. *Biomaterials Science, 2*, 1222–1229.
132. Miotto, M., Gouveia, R. M., & Connon, C. J. (2015). Peptide amphiphiles in corneal tissue engineering. *Journal of Functional Biomaterials, 6*, 687–707.
133. Gouveia, R. M., Hamley, I. W., & Connon, C. J. (2015). Bio-fabrication and physiological self-release of tissue equivalents using smart peptide amphiphile templates. *Journal of Materials Science. Materials in Medicine, 26*, 242.
134. Uzunalli, G., Soran, Z., Erkal, T. S., Dagdas, Y. S., Dinc, E., Hondur, A. M., et al. (2014). Bioactive self-assembled peptide nanofibers for corneal stroma regeneration. *Acta Biomaterialia, 10*, 1156–1166.
135. O'Leary, L. E. R., Fallas, J. A., Bakota, E. L., Kang, M. K., & Hartgerink, J. D. (2011). Multi-hierarchical self-assembly of a collagen mimetic peptide from triple helix to nanofibre and hydrogel. *Nature Chemistry, 3*, 821–828.
136. Islam, M. M., Ravichandran, R., Olsen, D., Ljunggren, M. K., Fagerholm, P., Lee, C. J., et al. (2016). Self-assembled collagen-like-peptide implants as alternatives to human donor corneal transplantation. *RSC Advances, 6*, 55745–55749.
137. Bareiss, B., Ghorbani, M., Li, F., Blake, J. A., Scaiano, J. C., Zhang, J., et al. (2010). Controlled release of acyclovir through bioengineered corneal implants with silica nanoparticle carriers. *The Open Tissue Engineering and Regenerative Medicine Journal, 3*, 10–17.
138. Riau, A. K., Mondal, D., Aung, T. T., Murugan, E., Chen, L., Lwin, N. C., et al. (2015). Collagen-based artificial corneal scaffold with anti-infective capability for prevention of peri-operative bacterial infections. *ACS Biomaterials Science & Engineering, 1*, 1324–1334.
139. Alarcon, E. I., Vulesevic, B., Argawal, A., Ross, A., Bejjani, P., Podrebarac, J., et al. (2016). Coloured cornea replacements with anti-infective properties: Expanding the safe use of silver nanoparticles in regenerative medicine. *Nanoscale, 8*, 6484–6489.
140. Islam, M. M., Buznyk, O., Reddy, J. C., Pasyechnikova, N., Alarcon, E. I., Hayes, S., Lewis, P., Fagerholm, P., He, C., Iakymenko, S., Liu, W., Meek, K. M., Sangwan, V. S., & Griffith, M. (2018) Biomaterials-enabled cornea regeneration in patients at high risk for rejection of donor tissue transplantation. *npj Regenerative Medicine, 3*, 2.

Chapter 7
Clinical Applications of Limbal Stem Cells for Regenerative Medicine

Brian G. Ballios and Allan R. Slomovic

Abstract Damage or loss of corneal and/or limbal cells from injury or infection can lead to irreversible loss of corneal transparency and blindness. A population of active limbal stem cells has been identified in the limbal epithelial crypts that provide a continuous supply of progenitors and mature epithelial cells, and participate in wound healing. With our growing knowledge of this stem cell population, our understanding of the homeostatic mechanisms regulating corneal epithelial homeostasis has expanded dramatically. Loss of these limbal stem cells leads to the range of conditions representing limbal stem cell deficiency. Here, we review the biology and cellular characterization of the limbal stem cell in health and disease. We also review clinical approaches to ocular surface stem cell transplantation that have been developed over the last 30 years, including autograft and allograft techniques currently in clinical practice, and the challenges associated with systemic immunosuppression when required. Emerging therapies in cultivated limbal epithelial transplantation are described, which may provide an unlimited source of cells for ocular surface restoration.

Keywords Limbal stem cells · Stem cell transplantation · Ocular surface

B. G. Ballios
Department of Ophthalmology and Vision Sciences, University of Toronto, Toronto, ON, Canada
e-mail: Brian.ballios@mail.utoronto.ca

A. R. Slomovic (✉)
Department of Ophthalmology and Vision Sciences, University of Toronto, Toronto, ON, Canada

Toronto General Hospital, University Health Network, Toronto, ON, Canada

Department of Ophthalmology, Toronto Western Hospital, Toronto, ON, Canada
e-mail: allan.slomovic@utoronto.ca

B. G. Ballios, M. J. Young (eds.), *Regenerative Medicine and Stem Cell Therapy for the Eye*, Fundamental Biomedical Technologies,
https://doi.org/10.1007/978-3-319-98080-5_7

7.1 Introduction

The cornea is a transparent tissue which transmits and refracts light, to allow a focus image to be projected onto the retina. The average human cornea is about 500 μm thick centrally and near 1000 μm thick peripherally [1]. It consists of five cellular layers: an outer epithelial layer, Bowman's layer, a middle stroma composed of hydrated collagenous extracellular matrix, Descemet's layer of basement membrane, and an inner endothelial monolayer. The surface epithelium is composed of stratified non-keratinized cells, with a thickness of approximately 50 μm. A number of antimicrobial and anti-inflammatory factors are secreted by the epithelium within an insoluble mucous layer that aids in maintaining a stable ocular surface tear film [2]. The corneal stroma is composed of over 300 lamellae of type I collagen interspersed with glycosaminoglycans; it resembles a hydrogel, with 80% water by weight. The stoma is maintained by the resident population of fibroblast-like keratocytes. This network gives the cornea its strength, and the lamellation provides transparency. The endothelial monolayer is the functional layer essential for maintaining appropriate water balance in the corneal stroma, and hence transparency. It contains sodium/potassium ATPase membrane pumps that maintain an osmotic balance to drive aqueous humor between the stroma and anterior chamber. Regarding the surface epithelium, as cells are lost, the basal layer proliferates to replace these superficial cells [3]. These basal cells are replaced by a population of stem cells that reside within the limbus, found at the corneoscleral junction. These limbal stem cells (LSCs) are important for the proper maintenance and regeneration of the corneal epithelium [4, 5].

Any irreversible damage or loss of corneal and/or limbal cells, from injury or infection, can lead to loss of corneal transparency and blindness. A World Health Organization study shows that corneal disease is second only to cataracts as the leading cause of blindness worldwide [6]. Corneal ulcers and ocular surface trauma are estimated to account for an incidence of 1.5–2 million case of blindness annually, with a prevalence of 10 million.

The management of corneal disease has changed dramatically in the last 40 years. In the 1970s, all patients with corneal disease had a poor visual prognosis. The only available techniques included penetrating and lamellar keratoplasty to replace the corneal tissue, and tarsorrhaphy and artificial tears to maintain surface hydration.

A full-thickness corneal transplantation with a cadaveric allograft, known as a penetrating keratoplasty (PKP), can be used to replace the entire cornea. While PKP is successful in the short term, varying rates of rejection have been reported up to 15% in some studies [7]. This can lead to graft failure, with loss of the endothelial cells and subsequent loss of corneal transparency from edema. Graft failure rates are greater in high-risk transplantation populations. These include those patients with autoimmune disease, chemical alkali burns, severe dry eye, Stevens-Johnson syndrome (SJS), ocular cicatricial pemphigoid (OCP), neurotrophic cornea from herpetic eye disease (zoster and/or simplex), and in those who have had recurrent grafts [8]. This often results in the eye not being able to support a corneal transplant.

Deep anterior lamellar keratoplasty (DALK) is an alternative procedure, involving removal of only the epithelium and stroma, leaving the endothelium intact. This reduces the rates of rejection and postoperative complications such as leakage, and improves graft stability [9]. It also results in reduced loss of endothelial cells postoperatively compared to PKP.

Importantly, there is also a severe shortage of donor corneal tissue worldwide, as is the case in other solid organ transplantations. With an aging population, wait times for donor tissue are expected to increase. As well, suitable donor tissue may be more limited with the increased incidence of infectious diseases including HIV and hepatitis. The development of artificial corneal replacements, or keratoprostheses (KPro), has helped to decrease our dependence on donor tissue [10]. In the developing and under-developed world, the skills and resources to perform these surgeries are extremely limited [11].

Modern-day treatment has evolved significantly. In this chapter, we will focus on the surgical methods that ophthalmologists have developed to restore the ocular surface, in the context of diseases of limbal stem cell deficiency. Special note will be made to the importance of immunomodulatory therapy in the context of limbal tissue allografts. We will also provide perspective on new and emerging methods of ocular surface stem cell therapy.

7.2 Limbal Stem Cells and the Niche

The corneal epithelium is constantly being sloughed and renewed by a regular homeostatic mechanism. The renewal process involves centripetal and circumferential migration from the limbus in addition to vertical migration from basal layers [12, 13]. Davanger and Evensen [4] were first to recognize that pigmented limbal cells seemed to move centrally in the cornea, suggesting centripetal migration. From this observation, they hypothesized that limbal cells were likely involved in normal corneal epithelial renewal. They postulated that these source-cells were resident in the limbal crypts of the palisades of Vogt, a series of radially oriented fibrovascular ridges that are observed in the human limbus [14, 15], and can be imaged by optical coherence tomography with sub-micrometer resolution [16].

It was Schermer and colleagues [5] who postulated that corneal epithelial stem cells were located in the limbus. They based their theory on the pattern of expression of cornea-specific 64K keratin present in all corneal epithelial cells except the limbal basal cells, suggesting that the limbal basal cells were less differentiated than those found in other areas of the corneal epithelium. This is true of basal cells in skin epidermis [17], as well as mucosal epithelium in the intestine [18].

Stem cells are characterized by two cardinal properties: self-renewal and multipotentiality. In self-renewal by asymmetric division, one daughter cell remains a stem cell while the other becomes a more differentiated progenitor. Multipotentiality refers to the potential for tissue-specific stem cell progeny to differentiate into any of the mature cell types that make up that adult tissue. Cell kinetic studies in intestine and epidermis have shown that stem cells and their early progeny, the transit amplifying cells (TACs), make up the proliferating cells of epithelium [19, 20]. Schermer and colleagues [5] proposed the cell proliferation scheme for the cornea as proceeding to limbal basal stem cells, to basal corneal epithelial TACs, to differentiating suprabasal corneal epithelium (Fig. 7.1). In 2005, based on histological examination of the human limbus, Dua et al. reported the presence of limbal epithelial crypts (LECs) and proposed that they also harbor LSCs [21]. LECs are more frequently detected in the superior or inferior limbus compared to temporal or nasal limbus [22, 23]. In nonhuman species, only porcine limbus has been reported to share the structure of the human limbus regarding the topography of the palisades of Vogt and LEC, while no evidence of palisades of Vogt has been found in other animals [22].

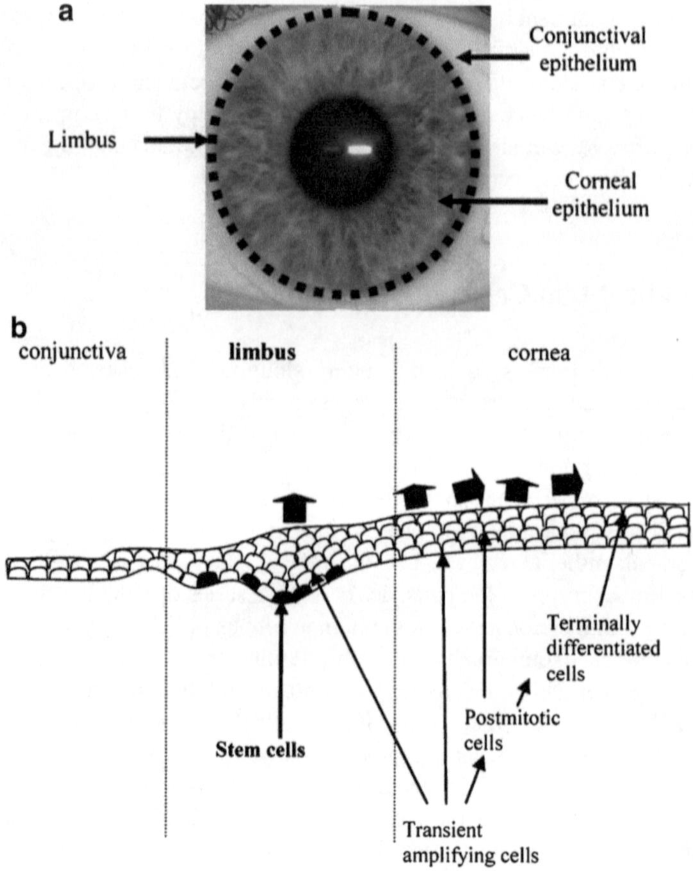

Fig. 7.1 (**a**) Location of the limbus, at the junction of the conjunctival and corneal epithelium. (**b**) LSCs resident in the epithelial crypts of the palisades of Vogt give rise to transit amplifying cells (TACs), which migrate towards the center of the cornea, before differentiating and migrating towards the superficial layers of the epithelium as terminally differentiated suprabasal epithelial cells (Reproduced with permission, Notara and colleagues, 2008)

In mice, LSCs were first identified as slow-cycling label-retaining cells in the basal layer of the limbal epithelium [24]. Despite lacking palisade of Vogt structures, lineage-tracing studies clearly show that murine corneal stem cells exist in the limbus and are capable of producing daughter cells with centripetal migration patterns during corneal epithelial regeneration [25–28].

As a result of their high proliferative potential and their absence of markers, a decades-long search has been underway for a bona fide LSC marker that would enable prospective isolation for therapeutic applications. Numerous potential LSC markers have been proposed [29], but for most, evidence for successful prospective enrichment of cells capable of long-term corneal restoration is currently lacking. Pellegrini and colleagues proposed that transcription factor p63 identifies human LSCs [30], and Rama and colleagues [31] evaluated the effectiveness of autologous limbal cell transplants grafted onto patients with unilateral limbal stem cell deficiency. The success of the transplants seemed to depend on the number of p63+ cells in the graft. Several additional potential human LSC markers have been identified, including Lgr5 [32], Tcf4 [33], CD157 [34], CD71low/Integrin-alpha6high [35], TrkA [36], N-cadherin [35], ABCG2 [37, 38], Cytokeratin15 [39], and ABCB5 [40]. In the case of ABCB5, prospectively isolated human ABCB5+ LSC possessed the capacity to fully restore corneal epithelium after grafting onto LSC-deficient mice. ABCB5 was found to be preferentially expressed on label-retaining cells at the limbus in mice. Furthermore, ABCB5+ cells were significantly depleted in patients with LSC deficiency. ABCB5 loss-of-function mutant mice showed defective corneal differentiation and wound healing, demonstrating that ABCB5 is both a marker and also of functional significance in maintaining LSCs.

Of note, due to the complexity of the LSC niche and the elusiveness of definitive LSC identification, the developmental origin of the LSC has remained elusive. Our current understanding of the developmental origins and formation of the corneo-limbal-scleral junction is reviewed in detail elsewhere [29].

7.3 Limbal Stem Cell Deficiency

Limbal stem cell deficiency (LSCD), caused by inherited, or acquired disruption of this stem cell niche, results in poor corneal epithelialization and epithelial defects, secondary vascularization of the cornea, stromal scarring, and/or corneal conjunctivalization [41]. Etiologies include chemical or thermal burns; ocular cicatricial pemphigoid (OCP) and pseudo-OCP; aniridia; various forms of ectodermal dysplasia; Stevens-Johnson syndrome; contact lens wear-related; or iatrogenic injury during ocular surface surgery. These conditions may result in partial or total limbal stem cell deficiency in the affected eye because of the degree of destruction of the limbus, conjunctival scarring, decreased tear film production, and the high risk of corneal keratinization. The patient will experience a number of distressing symptoms including ocular pain, photophobia, and decreased vision (Fig. 7.2).

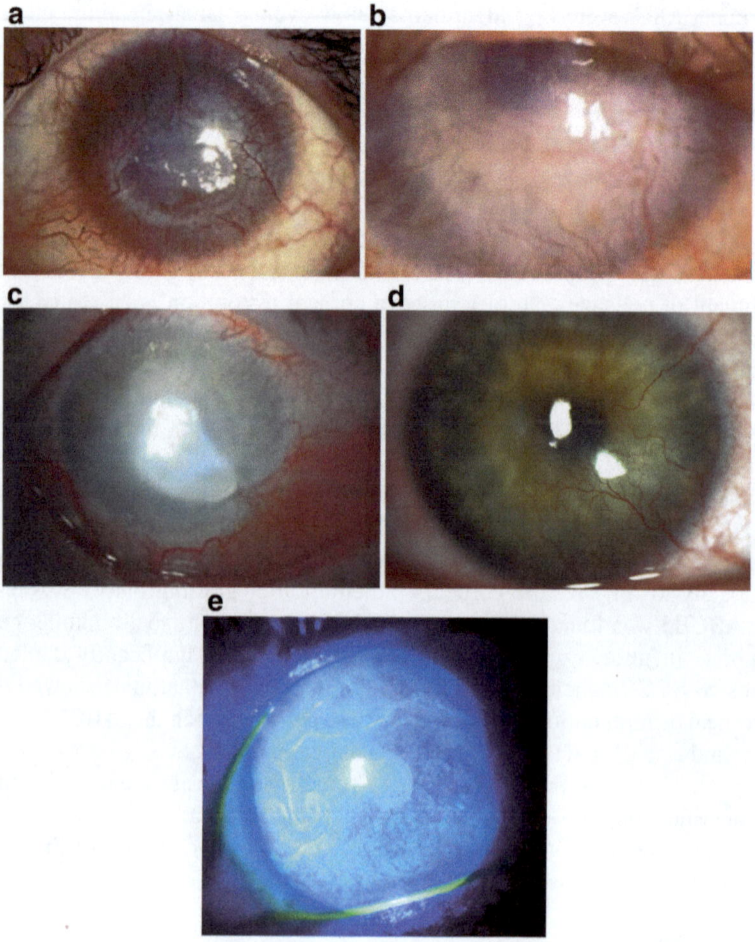

Fig. 7.2 Eyes with severe LSC resulting from (**a**) acid burn; (**b**) Stevens–Johnson syndrome; (**c**) alkali burn; (**d**) Contact lens wear-related severe LSCD. Total conjunctivalized corneal surface with stromal haze, and large infiltrating vessels from the conjunctiva are seen. (**e**) Fluorescein staining of the cornea, viewed under cobalt-blue illumination, reveals the classic late-staining pattern with a whorl-like epitheliopathy pattern, demonstrating severe LSCD (**a**–**c**, Reproduced with permission, Lavker and colleagues, 2004; **d**, **e**, Reproduced with permission, Chan and Holland, 2013)

When a significant number of limbal stem cells are lost, conjunctival epithelial cells invade and populate the corneal surface. This process of conjunctivalization results in a thickened, irregular, unstable epithelium, often with secondary neovascularization, inflammatory cell infiltration and disruption of the basement membrane [42]. Impression cytology typically demonstrates the presence of goblet cells and conjunctival epithelial cells on the corneal surface [41]. Punctate corneal epithelial defects and larger confluent defects are common and can lead to corneal scarring. Debridement of the conjunctivalized pannus results in a reinvasion of the abnormal epithelium. A PKP or lamellar keratoplasty with normal epithelium results in a stable ocular surface as long as the donor epithelium is present. However, with eventual epithelial sloughing, the surface often fails, resulting in re-conjunctivalization owing to continued LSCD [43]. Donor epithelium in routine PKP may survive for several months without the need for re-epithelialization from the limbus. This survival time is known, from studies that observed epithelial rejection as late as 13 months postoperatively [44]. The observation that stable corneal re-epithelialization was not possible without a healthy limbal zone led ophthalmologist to begin investigating approaches to epithelial transplantation.

7.4 Clinical Approaches to Ocular Surface Stem Cell Transplantation

Over the last three decades, the field has seen a remarkable proliferation and variation in the techniques of ocular surface stem cell transplantation (OSST) [45], often combined with PK or deep lamellar keratoplasty (DLK). In unilateral LSCD, tissue harvested from the contralateral eye may be used in an autograft [46–48]. In bilateral disease, which is more common, allogeneic donor material must be used [49–53]. Holland and Schwartz proposed a nomenclature and classification system [54] for ocular surface stem cell transplantation; we will use this system here (Table 7.1).

7.4.1 Autografts

Epithelial transplantation can be divided into autografts and allografts. In traditional autografts, the donor tissue is obtained from the fellow eye, thereby avoiding the major problem of immunologic rejection that faces allograft procedures. Some newer techniques involve harvesting autograft tissue from an area of healthy limbus in the same eye. Epithelial transplantation for severe ocular surface disease was first described by Thoft in 1977 [56], when he described conjunctival transplantation for monocular chemical burns. This was an autograft procedure using several pieces of bulbar donor conjunctiva harvested from a normal fellow eye. In 3/5 eyes, the

Table 7.1 Classification system for limbal allografts/autografts

Procedure	Abbreviation	Donor site	Transplanted tissue
Keratolimbal allograft	KLAL	Cadaver	Limbus/cornea
Conjunctival limbal autograft	CLAU	Contralateral eye	Limbus/conjunctiva
Living related conjunctival limbal allograft	lr-CLAL	Living relative	Limbus/conjunctiva

Adapted from Holland and Schwartz [54] and Daya et al. [55]

cornea was successfully re-epithelialized, and this procedure was based on the concept of conjunctival transdifferentiation [20, 57, 58]. Studies have subsequently shown that the conjunctiva cannot transdifferentiate into epithelium that is truly phenotypical corneal epithelium [59].

Kenyon and Tseng [47] were the first to specifically transplant limbal stem cells in a conjunctival-limbal autograft (CLAU) from the contralateral eye. In this procedure, conjunctiva and limbus from a normal fellow eye were used to manage diffuse LSCD in unilateral disease, or focal limbal deficiency in unilateral or bilateral disease. The harvest of bulbar conjunctiva extended 0.5 mm onto the clear corneal surface, thus containing limbal cells. The preoperative diagnoses included chemical and thermal injuries, contact lens-induced keratopathy, and surface disease secondary to multiple ocular surface surgeries. This resulted in rapid surface healing, improved ocular surface, and improved visual acuity.

Jenkins and colleagues [60] reported on CLAU in five patients with epitheliopathy secondary to chronic contact lens overuse. Two of five procedures failed, and one of the fellow donor eyes developed LSCD post-harvest. These results stress the importance of careful selection of donor tissue for autografting only from a fellow eye that is otherwise normal. It is likely that these fellow donor eyes were not normal given their exposure to chronic contact lens wear, even though they did not show overt LSCD. CLAU should only be obtained from eyes with normal, functional epithelial surfaces.

7.4.2 Allografts

While autografting is a very successful procedure used today for unilateral LSCD, it does not provide a therapeutic approach to severe bilateral LSCD. Thoft described the first allograft procedure in 1984 [61] which he named keratoepithelioplasty. His procedure involved the use of "lenticules" of peripheral cornea from a cadaveric donor eye. These lenticules were harvested from the midperipheral cornea and consisted of cornea and a thin layer of stroma. They were placed evenly around the corneoscleral limbus of the host eye and sutured to the sclera. The limbus was not harvested from donor eyes in this procedure. The donor epithelium spread from the lenticules and covered the host cornea. Three of four patients transplanted in his study maintained a stable ocular surface and improved vision. Turgeon and colleagues [62] expanded on this procedure to include limbal tissue with the peripheral cornea in order to capture LSCs in the allografts.

It was not until the mid-1990s that groups described the first keratolimbal allograft (KLAL) transplantations [63, 64]. They utilized a whole globe to make a 360° scleral incision approximately 1 mm from the limbus at the same depth as a midperipheral corneal incision. Lamellar dissection of the keratolimbal tissue was performed, the keratolimbal ring was divided into three pieces, and transferred to the recipient eye. All patients were placed on systemic cyclosporine A (CsA) in addition to topical corticosteroid drops. Systemic CsA was tapered as possible. Of six patients with LSCD, five of six had improved vision resulting from the procedure. Tsubota and colleagues [64] showed that the corneolimbal tissue could be kept in storage media for 5 days prior to transplantation.

Given the concern regarding immunologic rejection in the context of limbal allografts, and the growing understanding from other fields of solid organ transplantation (e.g., kidney) of the importance of donor-host immunomatching, various groups began to explore the use of living-related donors as a source of allograft tissue. The first report of a living-related epithelial transplantation was by Kwitko and colleagues in 1995 [65], which they called "allograft conjunctival transplantation." Limbal tissue was not transplanted in this study. Donor conjunctiva was obtained from siblings or parents. Human leukocyte antigen (HLA) typing and crossmatching was performed retrospectively in recipients/donors. Three of 12 eyes experienced epithelial rejection episodes, with no disturbance of the corneal surface in two patients. Two of these three cases had 100% incompatible HLA matching. Patients with identical or haplo-identical matching were less likely to undergo epithelial rejection. It is interesting to note that while 5/11 eyes had improved visual acuity and 10/11 eyes had improved ocular surface (transparency, decreased neovascularization, stable epithelium), these results were obtained with only transplantation of conjunctival tissue, without limbal tissue.

Kenyon and Rapoza [66] were the first to describe living-related limbal allografting with a conjunctival carrier from a living-donor (living-related conjunctival limbal allograft, lr-CLAL). This was similar to the technique of limbal autografting (CLAU), except that the donor tissue was derived from a living relative. Most patients also went simultaneous lateral tarsorrhaphy. Topical steroid drops were included in all cases, with topical and/or systemic CsA for HLA haplo-identical or incompatible cases. No episodes of rejection were observed in eight cases. The ocular surface remained stable in six of eight cases. Visual acuity also improved in six of eight cases.

Since this time, there have been numerous modifications and combinations of the techniques described above. For example, a "Cincinnati procedure" has been described, which uses a combined lr-CLAL and KLAL in patients with bilateral severe LSCD and conjunctival deficiency [67]. A modified Cincinnati procedure involves a combined CLAU and KLAL for patients with severe unilateral LSCD [68] (Fig. 7.3).

7.4.3 *Rejection and Systemic Immunosuppression in Limbal Allografts*

Systemic immunosuppression is critical for graft integration and survival following allograft transplantation [69]. Limbal allografts are at significantly higher risk of rejection than other more "central" corneal procedures involving the avascular stroma. In corneal transplantation, an avascular tissue is being transplanted into an avascular host site. In contrast, the limbus has a high concentration of tissue antigen

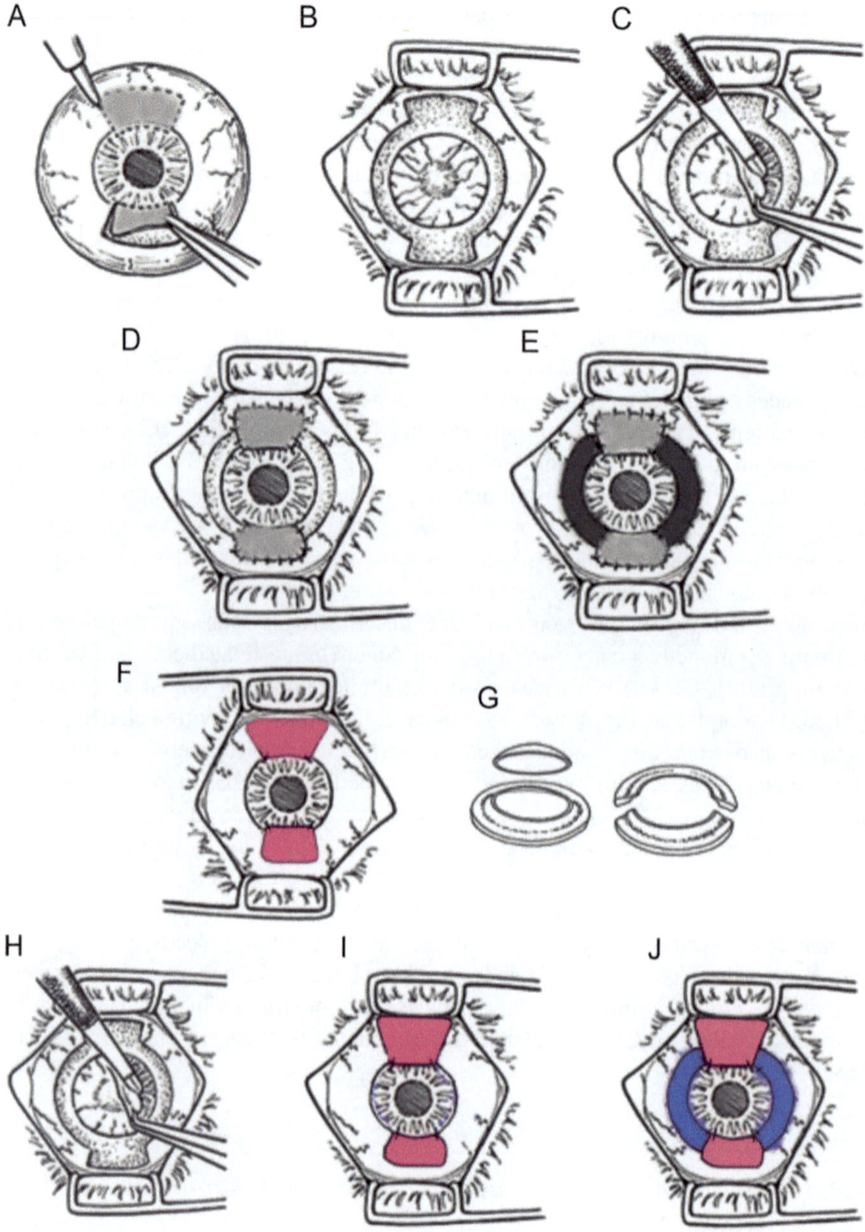

Fig. 7.3 The Cincinnati procedure for combined lr-CLAL/KLAL: (**a**) The conjunctival graft is harvested from the living-related donor eye in the 12- and 6-o'clock meridians. (**b**) The recipient eye is prepared by performing a 360° limbal peritomy and undermining/retracting conjunctival tissue, followed by (**c**) removal of abnormal corneal epithelium and fibrovascular pannus. (**d**) The harvested living-related tissue is sutured in the same anatomical orientation, with the (**e**) cadaveric donor segments placed at the 3- and 9-o'clock meridians also in the same anatomical orientation with the limbal edge at the recipient limbus. The modified Cincinnati procedure for combined

presenting cells (Langerhans' cells), which can trigger immunologic rejection [70] by T-cells [71]. This may present as either acute allograft rejection with injection at the graft-host junction and conjunctivalization of the ocular surface, or a chronic rejection characterized by slowly progressive conjunctivalization without evidence of acute inflammation [72]. It should be noted that the role of humoral immunity is not well defined in the context of the limbal allografts. The graft-host state in limbal transplantation must therefore be treated similarly to most cases of vascular solid organ transplants.

Thoft and Sugar [73] gave us the description of the typical features of patients with acute rejection following KLAL. However, with chronic features (irregular epithelium, recurrent LSCD, epithelial failure) it can be difficult to differentiate between chronic low-grade disease or failure of the transplanted graft. In the case of limbal allografts, there is perpetuation of donor epithelial cells and antigen presenting cells beyond the first year [71], and thus the threat of graft rejection persists. Therefore, long-term systemic immunosuppression is extremely important.

The options to treat bilateral LSCD include KLAL, lr-CLAL, and ex-vivo cultured stem cell transplantation. The early allograft protocols used systemic CsA almost exclusively for long-term immunosuppression [52, 53, 74–79], occasionally with the addition of azathioprine [80]. Long-term CsA therapy can have a number of systemic side effects, including nephrotoxicity [81], hypertension [82, 83], and hyperlipidemia [84]. Patient monitoring on CsA includes regular blood trough levels, as well as trends in blood pressure and creatinine. After 2000, these studies often used short-term courses of high-dose steroids in the postoperative phase, and also in induction protocols preoperatively [16, 25, 27–31, 33–35]. Early corticosteroid withdrawal has been shown to decrease corticosteroid-related morbidity in patients receiving solid organ transplants [85]. Steroid-sparing immunosuppression protocols in kidney transplantation, with discontinuation of steroids as early as 1 week postoperatively, have been studied in prospective trials [86].

The modern immunosuppressive protocols target stages of the immune response to tailor postoperative medical therapy [45]. Both MMF and azathioprine can affect bone marrow functioning, and require regular monitoring. However, in other organs, MMF has been shown to decrease the incidence of acute transplant rejection over azathioprine, in patients also on steroids and CsA [87]. Due to these reasons, the use of MMF has generally replaced azathioprine in OSST post-transplantation immunosuppression. It should be noted that MMF can be teratogenic, and thus all women of childbearing age should be discouraged from pregnancy during therapy.

One example of the progression from CsA-based long-term immunosuppression to next-generation immunosuppressive protocols is the OSST program at The

Fig. 7.3 (continued) CLAU/KLAL: (**f**) CLAUs are harvested from the fellow eye while (**g**) KLAL segments are prepared from a cadaveric donor corneoscleral rim. (**h**) 360° conjunctival peritomy and removal of abnormal epithelium and fibrovascular pannus is performed. (**i**) CLAU and (**j**) KLAL segments are secured at the limbus of the recipient eye (**a–e**, Reproduced with permission, Biber and colleagues, 2011 [67]; **f–j**, Reproduced with permission, Chan and colleagues [68])

Cincinnati Eye Institute. Their program began with the use of CsA, azathioprine, and prednisone [80], but in the last 10 years, has moved to a regime including long-term tacrolimus and MMF, in addition to a short-term oral prednisone taper [69]. This protocol is based on the success achieved by the ELITE-Symphony randomized control study in renal transplantation [88] showing an ability to reduce exposure to calcineurin inhibitors using low-dose tacrolimus. Tacrolimus and MMF together has been shown to be more effective and safer than CsA in renal transplantation by reducing acute rejection [89], and decreased risk in high-risk PK [90]. While tacrolimus, like CsA, is a calcineurin inhibitor, the adverse effects of hirsuitism and gingival hyperplasia are eliminated [91], increasing patient tolerance and adherence to therapy. Some groups taper tacrolimus beginning 1–2 years postoperatively if the ocular surface has been stable [69]. Morbidity in OSST patients is much lower than that seen in renal transplantation patients [92], likely because of the increased preponderance of comorbidities in patients with end-stage renal disease.

For allografts, important preoperative assessment of risk includes blood group type (ABO), donor type (KLAL vs. lr-CLAL), tissue HLA type, panel reactive antibody (PRA), and donor-specific HLA antibody (DSA) identification. ABO blood group antigens have been detected on human corneal epithelial cells, and may contribute to allograft rejection once the rejection response has begun [93]. HLA-A/B/C antigens can be found on corneal epithelium and stromal keratinocytes, while HLA-DR/DQ/DP are present on Langerhans and other antigen presenting cells [94]. The protocols for preoperative risk assessment are generally based on experience with solid organ transplantation and our understanding of transplant rejection immunobiology. Drugs have been developed which target the cell-mediated immune responses specifically that can contribute to graft rejection, including IL-2 receptor blockers such as basiliximab. Holland and colleagues [69] have used preoperative induction with basiliximab in patients receiving lr-CLAL without a perfect HLA match and a nonzero PRA. In addition, they use basiliximab in patients receiving KLAL with a nonzero PRA. In high-risk patients, undergoing KLAL, lr-CLAL with non-HLA identical, HLA-identical/PRA >50%, or any patient undergoing repeat OSST, initiation of maintenance therapy (MMF/tacrolimus) occurs 2 weeks prior to surgery. Overall, the absence of comprehensive histocompatibility data in limbal transplants represents a significant gap in the understanding of rejection processes.

Given the concerns around donor-host matching and long-term systemic immunosuppression, the opportunity to avoid these issues by performing an autograft in the case of unilateral disease is tempting [95]. As mentioned previously, a very careful evaluation must be undertaken as, in many cases, the "normal" fellow eye may have subclinical LSCD. In this case, the host eye may fail to re-epithelialize because of deficiency inherent in the graft, and the donor eye would be left without sufficient limbal stem cell capacity to maintain surface integrity [60].

The balance of evidence suggests that long-term immunosuppression may be important to graft success, but more prospective and/or randomized trials are required to determine the optimal therapeutic approach and to compare immunosuppressive practices. There is currently no consensus on which regimen is most efficacious for various pathologies or grafts. It would also be advantageous to assess the results of larger numbers of patients with a common approach across multiple centers.

7.5 Cultivated Limbal Epithelial Transplantation

In the mid-1990s, Pellegrini and colleagues [96] described a procedure using autologous cultivated corneal epithelium to restore the ocular surface. Sheets of corneal epithelial cells were cultured in vitro, and these sheets were transplanted to the injured eye. This procedure was particularly effective for patients who are resistant to the idea of using six clock hours of limbal tissue from their healthy eye as a donor source for CLAU grafting into their blind eye [45]. The patients retained a stable ocular surface without systemic immunosuppression for >2 years after the procedure. In the interim, multiple studies have investigated the molecular genetic mechanisms regulating limbal stem cell self-renewal and differentiation potential in vivo [30, 40, 97], and their ability to be grown as holoclones in vitro [98]. For success in this approach, adequate numbers of autologous stem cells will need to be generated prior to transplantation [31]. However, this approach is still limited: it cannot be applied to patients suffering from total bilateral LSCD, because these patients lack the autologous limbal stem cells necessary for culture. As well, the cost of establishing and maintaining a stem cell laboratory for cGMP cell therapy production is very high [99], and may result in CLET only being available in a handful of advanced centers worldwide. Some regional CLET facilities are being established in Europe, which may serve multiple transplant sites across the region. The reader is directed to an excellent review of cultivated epithelial sources for transplant provided in Chap. 6.

With the advent of induced pluripotent stem (iPS) cells [100–102] with the same properties as *bona fide* embryonic stem (ES) cells, it may be possible to use a patient's autologous somatic cells (e.g., fibroblasts from a skin biopsy) to generate epithelial stem cells. Early studies have already shown the potential of iPS cells to generate corneal epithelium [103]. These findings could result in autologous limbal grafts event for patients with bilateral total LSCD, once safety and proper regulation of iPS cell generation has been established.

7.6 Simple Limbal Epithelial Transplantation

It had been recognized that CLAU proposed some perceived risk to the donor eye of limbal decompensation. Tseng and colleagues, who pioneered the original CLAU technique [47], even developed a smaller single two-clock-hour donor tissue harvest, aimed at minimizing the amount of donor tissue used to treat total LSCD in the fellow eye: they termed this technique mini-CLAU [104].

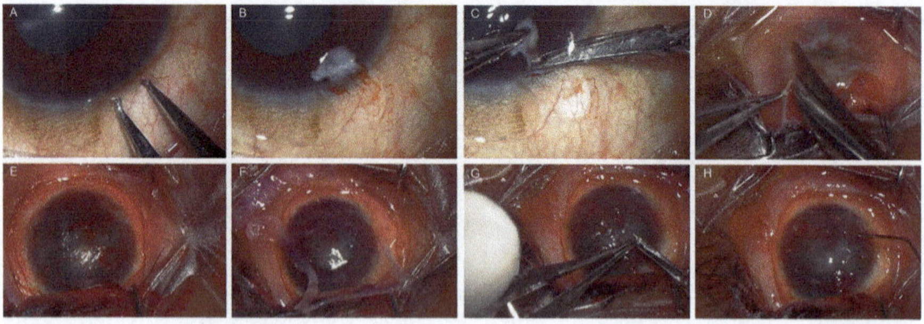

Fig. 7.4 Simple limbal epithelial transplantation (SLET): (**a**) 2 × 2 mm area of limbus is marked in the donor eye; (**b**) a subconjunctival dissection 1 mm into clear cornea allows (**c**) excision of the donor tissue; (**d**, **e**) a peritomy is performed and fibrovascular pannus and irregular epithelium is removed from the recipient corneal surface; (**f**) a human amniotic membrane graft is placed on the bare ocular surface and secured with fibrin glue; (**g**, **h**) the donor limbal tissue is cut into 8–10 small pieces and secured to the surface of the amniotic membrane with fibrin glue (Reproduced with permission, Sangwan and colleagues [105])

Another technique aimed at reducing the amount of tissue harvested from donors is the simple limbal epithelial transplantation (SLET) technique. This involves a resection of one clock hour of limbal tissue from the donor eye, division into small pieces, and transplantation of these autografts on an amniotic membrane placed over the recipient cornea [105] (Fig. 7.4). This technique does not require the cell culture facilities required for *ex vivo* donor cell expansion that is necessary for cultivated limbal epithelial transplantation (CLET). SLET incorporates the advantages of CLAU by being a single-stage autograft procedure. Like CLET, it also minimizes the risk of precipitating LSCD in the donor eye as only one clock hour is harvested., but minimizes donor tissue.

7.7 Conclusions and Outlook

Important clinical advances have been made in the techniques used to treat patients suffering from partial and total limbal stem cell deficiency. Our growing understanding of the role of the limbal stem cell in normal corneal epithelial homeostasis and disease has led to the refinement of procedures for limbal allografting and autografting, as well as important discoveries that may lead to the widespread use of cultivated limbal epithelial transplants. The use of in vitro cultivated cells may eventually supplant our need for donor corneal tissue in the treatment of LSCD. The possibility of targeted stimulation of endogenous stem cell stimulation to effect repair of diseased ocular tissue, without the need for cell transplantation, constitutes the ultimate goal of regenerative medicine in the eye. The ability to unlock this potential with LSCs in the context of ocular surface disease will depend on a deeper understanding of the stem cell biology underlying the regulation of LSC proliferation and differentiation under physiologic and pathophysiologic conditions. Advances in drug delivery and bioengineering may contribute to spatial and temporal control of LSC kinetics and dynamics in situ. Notwithstanding these challenges, the promise of regenerative medicine for ocular surface therapy can build on the success recognized in other tissues, to further advance future clinical application.

References

1. Jonas, J. B., & Holbach, L. (2005). Central corneal thickness and thickness of the lamina cribrosa in human eyes. *Investigative Ophthalmology & Visual Science, 46*, 1275–1279.
2. Sack, R. A., Nunes, I., Beaton, A., & Morris, C. (2001). Host-defense mechanism of the ocular surfaces. *Bioscience Reports, 21*, 463–480.
3. Ren, H., & Wilson, G. (1996). The cell shedding rate of the corneal epithelium—A comparison of collection methods. *Current Eye Research, 15*, 1054–1059.
4. Davanger, M., & Evensen, A. (1971). Role of the pericorneal papillary structure in renewal of corneal epithelium. *Nature, 229*, 560–561.
5. Schermer, A., Galvin, S., & Sun, T. T. (1986). Differentiation-related expression of a major 64K corneal keratin in vivo and in culture suggests limbal location of corneal epithelial stem cells. *The Journal of Cell Biology, 103*, 49–62.
6. Whitcher, J. P., Srinivasan, M., & Upadhyay, M. P. (2001). Corneal blindness: A global perspective. *Bulletin of the World Health Organization, 79*, 214–221.
7. Claesson, M., Armitage, W. J., Fagerholm, P., & Stenevi, U. (2002). Visual outcome in corneal grafts: A preliminary analysis of the Swedish Corneal Transplant Register. *The British Journal of Ophthalmology, 86*, 174–180.
8. Williams, K. A., et al. (2006). How effective is penetrating corneal transplantation? Factors influencing long-term outcome in multivariate analysis. *Transplantation, 81*, 896–901.
9. Ardjomand, N., et al. (2007). Lamellar corneal dissection for visualization of the anterior chamber before triple procedure. *Eye (London, England), 21*, 1151–1154.
10. Myung, D., et al. (2008). Development of hydrogel-based keratoprostheses: A materials perspective. *Biotechnology Progress, 24*, 735–741.
11. Chirila, T. V. (2001). An overview of the development of artificial corneas with porous skirts and the use of PHEMA for such an application. *Biomaterials, 22*, 3311–3317.
12. Thoft, R. A., & Friend, J. (1983). The X, Y, Z hypothesis of corneal epithelial maintenance. *Investigative Ophthalmology & Visual Science, 24*, 1442–1443.
13. Yoon, J. J., Ismail, S., & Sherwin, T. (2014). Limbal stem cells: Central concepts of corneal epithelial homeostasis. *World Journal of Stem Cells, 6*, 391–403.
14. Goldberg, M. F., & Bron, A. J. (1982). Limbal palisades of Vogt. *Transactions of the American Ophthalmological Society, 80*, 155–171.
15. Townsend, W. M. (1991). The limbal palisades of Vogt. *Transactions of the American Ophthalmological Society, 89*, 721–756.
16. Bizheva, K., et al. (2017). In-vivo imaging of the palisades of Vogt and the limbal crypts with sub-micrometer axial resolution optical coherence tomography. *Biomedical Optics Express, 8*, 4141–4151.
17. Potten, C. S., & Loeffler, M. (1987). Epidermal cell proliferation. I. Changes with time in the proportion of isolated, paired and clustered labelled cells in sheets of murine epidermis. *Virchows Archives B, Cell Pathology, 53*, 279–285.
18. Potten, C. S., & Loeffler, M. (1987). A comprehensive model of the crypts of the small intestine of the mouse provides insight into the mechanisms of cell migration and the proliferation hierarchy. *Journal of Theoretical Biology, 127*, 381–391.

19. Lajtha, L. G. (1979). Stem cell concepts. *Differentiation, 14*, 23–34.
20. Kinoshita, S., Friend, J., & Thoft, R. A. (1983). Biphasic cell proliferation in transdifferentiation of conjunctival to corneal epithelium in rabbits. *Investigative Ophthalmology & Visual Science, 24*, 1008–1014.
21. Dua, H. S., Shanmuganathan, V. A., Powell-Richards, A. O., Tighe, P. J., & Joseph, A. (2005). Limbal epithelial crypts: A novel anatomical structure and a putative limbal stem cell niche. *The British Journal of Ophthalmology, 89*, 529–532.
22. Grieve, K., et al. (2015). Three-dimensional structure of the mammalian limbal stem cell niche. *Experimental Eye Research, 140*, 75–84.
23. Pajoohesh-Ganji, A., Pal-Ghosh, S., Simmens, S. J., & Stepp, M. A. (2006). Integrins in slow-cycling corneal epithelial cells at the limbus in the mouse. *Stem Cells, 24*, 1075–1086.
24. Cotsarelis, G., Cheng, S. Z., Dong, G., Sun, T. T., & Lavker, R. M. (1989). Existence of slow-cycling limbal epithelial basal cells that can be preferentially stimulated to proliferate: Implications on epithelial stem cells. *Cell, 57*, 201–209.
25. Amitai-Lange, A., et al. (2015). Lineage tracing of stem and progenitor cells of the murine corneal epithelium. *Stem Cells, 33*, 230–239.
26. Di Girolamo, N., et al. (2015). Tracing the fate of limbal epithelial progenitor cells in the murine cornea. *Stem Cells, 33*, 157–169.
27. Dorà, N. J., Hill, R. E., Collinson, J. M., & West, J. D. (2015). Lineage tracing in the adult mouse corneal epithelium supports the limbal epithelial stem cell hypothesis with intermittent periods of stem cell quiescence. *Stem Cell Research, 15*, 665–677.
28. Lobo, E. P., et al. (2016). Self-organized centripetal movement of corneal epithelium in the absence of external cues. *Nature Communications, 7*, 12388.
29. Gonzalez, G., Sasamoto, Y., Ksander, B. R., Frank, M. H., & Frank, N. Y. (2017). Limbal stem cells: Identity, developmental origin, and therapeutic potential. *Wiley Interdisciplinary Reviews: Developmental Biology, 2*, PMID:29105366.
30. Pellegrini, G., et al. (2001). p63 identifies keratinocyte stem cells. *Proceedings of the National Academy of Sciences of the United States of America, 98*, 3156–3161.
31. Rama, P., et al. (2010). Limbal stem-cell therapy and long-term corneal regeneration. *The New England Journal of Medicine, 363*, 147–155.
32. Brzeszczynska, J., Ramaesh, K., Dhillon, B., & Ross, J. A. (2012). Molecular profile of organ culture-stored corneal epithelium: LGR5 is a potential new phenotypic marker of residual human corneal limbal epithelial stem cells. *International Journal of Molecular Medicine, 29*, 871–876.
33. Lu, R., et al. (2012). Transcription factor TCF4 maintains the properties of human corneal epithelial stem cells. *Stem Cells, 30*, 753–761.
34. Horenstein, A. L., et al. (2009). CD38 and CD157 ectoenzymes mark cell subsets in the human corneal limbus. *Molecular Medicine, 15*, 76–84.
35. Hayashi, R., et al. (2008). Enrichment of corneal epithelial stem/progenitor cells using cell surface markers, integrin alpha6 and CD71. *Biochemical and Biophysical Research Communications, 367*, 256–263.
36. Qi, H., et al. (2008). Nerve growth factor and its receptor TrkA serve as potential markers for human corneal epithelial progenitor cells. *Experimental Eye Research, 86*, 34–40.
37. Budak, M. T., et al. (2005). Ocular surface epithelia contain ABCG2-dependent side population cells exhibiting features associated with stem cells. *Journal of Cell Science, 118*, 1715–1724.
38. de Paiva, C. S., Chen, Z., Corrales, R. M., Pflugfelder, S. C., & Li, D. Q. (2005). ABCG2 transporter identifies a population of clonogenic human limbal epithelial cells. *Stem Cells, 23*, 63–73.
39. Yoshida, S., et al. (2006). Cytokeratin 15 can be used to identify the limbal phenotype in normal and diseased ocular surfaces. *Investigative Ophthalmology & Visual Science, 47*, 4780–4786.
40. Ksander, B. R., et al. (2014). ABCB5 is a limbal stem cell gene required for corneal development and repair. *Nature, 511*, 353–357.

41. Puangsricharern, V., & Tseng, S. C. (1995). Cytologic evidence of corneal diseases with limbal stem cell deficiency. *Ophthalmology, 102*, 1476–1485.
42. Tseng, S. C. (1985). Staging of conjunctival squamous metaplasia by impression cytology. *Ophthalmology, 92*, 728–733.
43. Tugal-Tutkun, I., Akova, Y. A., & Foster, C. S. (1995). Penetrating keratoplasty in cicatrizing conjunctival diseases. *Ophthalmology, 102*, 576–585.
44. Alldredge, O. C., & Krachmer, J. H. (1981). Clinical types of corneal transplant rejection. Their manifestations, frequency, preoperative correlates, and treatment. *Archives of Ophthalmology, 99*, 599–604.
45. Holland, E. J., & Schwartz, G. S. (2000). Changing concepts in the management of severe ocular surface disease over twenty-five years. *Cornea, 19*, 688–698.
46. Tsai, R. J., Li, L. M., & Chen, J. K. (2000). Reconstruction of damaged corneas by transplantation of autologous limbal epithelial cells. *The New England Journal of Medicine, 343*, 86–93.
47. Kenyon, K. R., & Tseng, S. C. (1989). Limbal autograft transplantation for ocular surface disorders. *Ophthalmology, 96*, 709–722.; discussion 722.
48. Tsai, R. J., Sun, T. T., & Tseng, S. C. (1990). Comparison of limbal and conjunctival autograft transplantation in corneal surface reconstruction in rabbits. *Ophthalmology, 97*, 446–455.
49. Dua, H. S., & Azuara-Blanco, A. (1999). Allo-limbal transplantation in patients with limbal stem cell deficiency. *The British Journal of Ophthalmology, 83*, 414–419.
50. Tsubota, K., et al. (1999). Treatment of severe ocular-surface disorders with corneal epithelial stem-cell transplantation. *The New England Journal of Medicine, 340*, 1697–1703.
51. Shimazaki, J., et al. (1999). Evidence of long-term survival of donor-derived cells after limbal allograft transplantation. *Investigative Ophthalmology & Visual Science, 40*, 1664–1668.
52. Reinhard, T., Sundmacher, R., Spelsberg, H., & Althaus, C. (1999). Homologous penetrating central limbo-keratoplasty (HPCLK) in bilateral limbal stem cell insufficiency. *Acta Ophthalmologica Scandinavica, 77*, 663–667.
53. Sundmacher, R., & Reinhard, T. (1996). Central corneolimbal transplantation under systemic ciclosporin A cover for severe limbal stem cell insufficiency. *Graefe's Archive for Clinical and Experimental Ophthalmology, 234*(Suppl 1), S122–S125.
54. Holland, E. J., & Schwartz, G. S. (1996). The evolution of epithelial transplantation for severe ocular surface disease and a proposed classification system. *Cornea, 15*, 549–556.
55. Daya, S. M., Chan, C. C., Holland, E. J., & Members, OTCSOSPNC. (2011). Cornea Society nomenclature for ocular surface rehabilitative procedures. *Cornea, 30*, 1115–1119.
56. Thoft, R. A. (1977). Conjunctival transplantation. *Archives of Ophthalmology, 95*, 1425–1427.
57. Shapiro, M. S., Friend, J., & Thoft, R. A. (1981). Corneal re-epithelialization from the conjunctiva. *Investigative Ophthalmology & Visual Science, 21*, 135–142.
58. Tseng, S. C., Hirst, L. W., Farazdaghi, M., & Green, W. R. (1984). Goblet cell density and vascularization during conjunctival transdifferentiation. *Investigative Ophthalmology & Visual Science, 25*, 1168–1176.
59. Dua, H. S., & Forrester, J. V. (1990). The corneoscleral limbus in human corneal epithelial wound healing. *American Journal of Ophthalmology, 110*, 646–656.
60. Jenkins, C., Tuft, S., Liu, C., & Buckley, R. (1993). Limbal transplantation in the management of chronic contact-lens-associated epitheliopathy. *Eye (London, England), 7*, 629–633.
61. Thoft, R. A. (1984). Keratoepithelioplasty. *American Journal of Ophthalmology, 97*, 1–6.
62. Turgeon, P. W., Nauheim, R. C., Roat, M. I., Stopak, S. S., & Thoft, R. A. (1990). Indications for keratoepithelioplasty. *Archives of Ophthalmology, 108*, 233–236.
63. Tsai, R. J., & Tseng, S. C. (1994). Human allograft limbal transplantation for corneal surface reconstruction. *Cornea, 13*, 389–400.
64. Tsubota, K., Toda, I., Saito, H., Shinozaki, N., & Shimazaki, J. (1995). Reconstruction of the corneal epithelium by limbal allograft transplantation for severe ocular surface disorders. *Ophthalmology, 102*, 1486–1496.
65. Kwitko, S., et al. (1995). Allograft conjunctival transplantation for bilateral ocular surface disorders. *Ophthalmology, 102*, 1020–1025.

66. Kenyon, K. R., & Rapoza, P. A. (1995). Limbal allograft transplantation for ocular surface disorders. *Ophthalmology, 102*, 101–102.
67. Biber, J. M., Skeens, H. M., Neff, K. D., & Holland, E. J. (2011). The cincinnati procedure: Technique and outcomes of combined living-related conjunctival limbal allografts and keratolimbal allografts in severe ocular surface failure. *Cornea, 30*, 765–771.
68. Chan, C. C., Biber, J. M., & Holland, E. J. (2012). The modified Cincinnati procedure: Combined conjunctival limbal autografts and keratolimbal allografts for severe unilateral ocular surface failure. *Cornea, 31*, 1264–1272.
69. Holland, E. J., et al. (2012). Systemic immunosuppression in ocular surface stem cell transplantation: Results of a 10-year experience. *Cornea, 31*, 655–661.
70. Niederkorn, J. Y. (1995). Effect of cytokine-induced migration of Langerhans cells on corneal allograft survival. *Eye (London, England), 9*, 215–218.
71. Daya, S. M., Bell, R. W., Habib, N. E., Powell-Richards, A., & Dua, H. S. (2000). Clinical and pathologic findings in human keratolimbal allograft rejection. *Cornea, 19*, 443–450.
72. Holland, E. J. (1996). Epithelial transplantation for the management of severe ocular surface disease. *Transactions of the American Ophthalmological Society, 94*, 677–743.
73. Thoft, R. A., & Sugar, J. (1993). Graft failure in keratoepithelioplasty. *Cornea, 12*, 362–365.
74. Maruyama-Hosoi, F., Shimazaki, J., Shimmura, S., & Tsubota, K. (2006). Changes observed in keratolimbal allograft. *Cornea, 25*, 377–382.
75. Reinhard, T., et al. (2004). Long-term results of allogeneic penetrating limbo-keratoplasty in total limbal stem cell deficiency. *Ophthalmology, 111*, 775–782.
76. Ilari, L., & Daya, S. M. (2002). Long-term outcomes of keratolimbal allograft for the treatment of severe ocular surface disorders. *Ophthalmology, 109*, 1278–1284.
77. Shimazaki, J., Maruyama, F., Shimmura, S., Fujishima, H., & Tsubota, K. (2001). Immunologic rejection of the central graft after limbal allograft transplantation combined with penetrating keratoplasty. *Cornea, 20*, 149–152.
78. Meller, D., & Tseng, S. C. (2000). Amniotic membrane transplantation with or without limbal allografts in corneal surface reconstruction in limbal deficiency. *Der Ophthalmologe: Zeitschrift der Deutschen Ophthalmologischen Gesellschaft Ophthalmologe, 97*, 100–107 Transplantation von Amnionmembran mit oder ohne allogener Limbustransplantation zur Rekonstruktion der kornealen Oberflache bei Limbusinsuffizienz.
79. Tan, D. T., Ficker, L. A., & Buckley, R. J. (1996). Limbal transplantation. *Ophthalmology, 103*, 29–36.
80. Holland, E. J., Djalilian, A. R., & Schwartz, G. S. (2003). Management of aniridic keratopathy with keratolimbal allograft: A limbal stem cell transplantation technique. *Ophthalmology, 110*, 125–130.
81. Burdmann, E. A., Andoh, T. F., Yu, L., & Bennett, W. M. (2003). Cyclosporine nephrotoxicity. *Seminars in Nephrology, 23*, 465–476.
82. Curtis, J. J. (2002). Hypertensinogenic mechanism of the calcineurin inhibitors. *Current Hypertension Reports, 4*, 377–380.
83. Luke, R. G. (1991). Mechanism of cyclosporine-induced hypertension. *American Journal of Hypertension, 4*, 468–471.
84. Kobashigawa, J. A., & Kasiske, B. L. (1997). Hyperlipidemia in solid organ transplantation. *Transplantation, 63*, 331–338.
85. Woodle, E. S. (2002). Corticosteroid elimination in renal transplantation: Pro. *Transplantation Proceedings, 34*, 1693.
86. Alloway, R. R., et al. (2005). A prospective, pilot study of early corticosteroid cessation in high-immunologic-risk patients: The Cincinnati experience. *Transplantation Proceedings, 37*, 802–803.
87. European Mycophenolate Mofetil Cooperative Study Group. (1999). Mycophenolate mofetil in renal transplantation: 3-Year results from the placebo-controlled trial. *Transplantation, 68*, 391–396.
88. Ekberg, H., et al. (2007). Reduced exposure to calcineurin inhibitors in renal transplantation. *The New England Journal of Medicine, 357*, 2562–2575.

89. Boratynska, M., Banasik, M., Patrzalek, D., & Klinger, M. (2006). Conversion from cyclosporine-based immunosuppression to tacrolimus/mycophenolate mofetil in patients with refractory and ongoing acute renal allograft rejection. *Annals of Transplantation, 11*, 51–56.
90. Tabbara, K. F. (2008). Pharmacologic strategies in the prevention and treatment of corneal transplant rejection. *International Ophthalmology, 28*, 223–232.
91. Vanrenterghem, Y. F. (1999). Which calcineurin inhibitor is preferred in renal transplantation: Tacrolimus or cyclosporine? *Current Opinion in Nephrology and Hypertension, 8*, 669–674.
92. Mogilishetty, G., Haird, D., Alloway, R. R., et al. (2008). Comparison of immunosuppression related toxicities and complications in ocular surface transplant and renal transplant recipients: Implications for composite tissue transplantation [abstract]. *Transplantation, 86*, 11.
93. Salisbury, J. D., & Gebhardt, B. M. (1981). Blood group antigens on human corneal cells demonstrated by immunoperoxidase staining. *American Journal of Ophthalmology, 91*, 46–50.
94. Treseler, P. A., Foulks, G. N., & Sanfilippo, F. (1984). The expression of HLA antigens by cells in the human cornea. *American Journal of Ophthalmology, 98*, 763–772.
95. Clinch, T. E., Goins, K. M., & Cobo, L. M. (1992). Treatment of contact lens-related ocular surface disorders with autologous conjunctival transplantation. *Ophthalmology, 99*, 634–638.
96. Pellegrini, G., et al. (1997). Long-term restoration of damaged corneal surfaces with autologous cultivated corneal epithelium. *Lancet, 349*, 990–993.
97. Ouyang, H., et al. (2014). WNT7A and PAX6 define corneal epithelium homeostasis and pathogenesis. *Nature, 511*, 358–361.
98. Pellegrini, G., et al. (1999). Location and clonal analysis of stem cells and their differentiated progeny in the human ocular surface. *The Journal of Cell Biology, 145*, 769–782.
99. Miri, A., Al-Deiri, B., & Dua, H. S. (2010). Long-term outcomes of autolimbal and allolimbal transplants. *Ophthalmology, 117*, 1207–1213.
100. Okita, K., Ichisaka, T., & Yamanaka, S. (2007). Generation of germline-competent induced pluripotent stem cells. *Nature, 448*, 313–317.
101. Takahashi, K., et al. (2007). Induction of pluripotent stem cells from adult human fibroblasts by defined factors. *Cell, 131*, 861–872.
102. Takahashi, K., & Yamanaka, S. (2006). Induction of pluripotent stem cells from mouse embryonic and adult fibroblast cultures by defined factors. *Cell, 126*, 663–676.
103. Hayashi, R., et al. (2012). Generation of corneal epithelial cells from induced pluripotent stem cells derived from human dermal fibroblast and corneal limbal epithelium. *PLoS One, 7*, e45435.
104. Kheirkhah, A., Raju, V. K., & Tseng, S. C. (2008). Minimal conjunctival limbal autograft for total limbal stem cell deficiency. *Cornea, 27*, 730–733.
105. Sangwan, V. S., Basu, S., MacNeil, S., & Balasubramanian, D. (2012). Simple limbal epithelial transplantation (SLET): A novel surgical technique for the treatment of unilateral limbal stem cell deficiency. *The British Journal of Ophthalmology, 96*, 931–934.

Chapter 8
Retinal Ganglion Cell Replacement: A Bridge to the Brain

Petr Baranov and Julia Oswald

Abstract An estimated 3% of the global population over 40 years of age currently has glaucoma. The early onset of retinal ganglion cells (RGC) loss (before significant changes in vision occur) and the lack of regenerative capacity in the mammalian retina (Samuel et al., J Neurosci 31(44):16033–16044, 2011) limit potential therapeutic options. For a long time the replacement of RGCs as a therapy for glaucoma and other optic neuropathies was not considered as a potential strategy because of the perceived inability of mature RGCs to regrow full-length axons and reach relevant targets in the brain. However, recent optic nerve (Bei et al., Cell 164(1–2):219–232, 2016; Lim et al., Nat Neurosci 19(8):1073–1084, 2016) and spinal cord regeneration studies, together with developments in pluripotent and retinal cell biology, as well as pilot RGC transplantation experiments (Hertz et al., Cell Transplant 23(7):855–872, 2013; Venugopalan et al., Nat Commun 7:10472, 2016), demonstrate that it may be possible to restore the RGC population in the retina and regrow the optic nerve.

Keywords Retinal ganglion cells · Axon growth · Regeneration · Glaucoma · Cell transplantation

8.1 Introduction

An estimated 3% of the global population over 40 years of age currently has glaucoma. The early onset of retinal ganglion cells (RGC) loss (before significant changes in vision occur) and the lack of regenerative capacity in the mammalian retina [1] limit potential therapeutic options. For a long time the replacement of RGCs as a therapy for glaucoma and other optic neuropathies was not considered as a potential strategy because of the perceived inability of mature RGCs to regrow full-length axons and reach relevant

P. Baranov (✉) · J. Oswald
The Schepens Eye Research Institute of Massachusetts Eye and Ear,
an affiliate of Harvard Medical School, Boston, MA, USA
e-mail: Petr_baranov@meei.harvard.edu

B. G. Ballios, M. J. Young (eds.), *Regenerative Medicine and Stem Cell Therapy for the Eye*, Fundamental Biomedical Technologies,
https://doi.org/10.1007/978-3-319-98080-5_8

targets in the brain. However, recent optic nerve [2, 3] and spinal cord regeneration studies, together with developments in pluripotent and retinal cell biology, as well as pilot RGC transplantation experiments [4, 5], demonstrate that it may be possible to restore the RGC population in the retina and regrow the optic nerve. For the successful implementation of such therapy, several components need to be addressed:

(a) Cell survival—this remains significantly low (<1%).
(b) Structural cell integration—the inner limiting membrane and activated glial cells prevent full integration of donor neurons.
(c) Axon growth—processes must extend all the way to brain targets [6].
(d) Axonal guidance and glial barriers—in the disease state, scarring and micro-environmental changes obscure the target [7].
(e) Functional cell integration—the formation of afferent and efferent synapses within the retina and the brain is required to restore vision.
(f) Donor cell type—a renewable and well characterized cell product is needed.
(g) The immune aspect of the transplantation.

Here we will attempt to summarize the studies that form the premise for RGC transplantation.

8.2 RGC Development

The development of structurally and functionally organized tissues, like the retina, requires a high degree of spatiotemporal regulation and synchronization across various developmental processes, including proliferation, cell cycle exit, and cell migration. Starting from a single multipotent progenitor pool, retinal neurons are born in a conserved birth order, with retinal ganglion cells becoming postmitotic first. Next, cone photoreceptors are born, followed shortly by amacrine and horizontal cells. Rod photoreceptors and bipolar cells are the last neurons to differentiate within the retina, followed only by the birth of Mueller glial cells, which originate from the same progenitor population. While the overall histogenic fate progression is conserved across species and is faithfully recapitulated within single retinal progenitor cells (RPC) clones, the actual cellular composition of each clone can be highly variable. Ranging from a few cells up to 100, a single RPC clone can contain and omit a variety of fates. Apart from illustrating the relative temporal overlap in the histogenesis of distinct retinal fates, clonal analysis has also uncovered the importance of cell-to-cell signaling in the determination of sister cell fate. Owing to these processes, neighboring clones can be highly diverse, even though, on a cell-autonomous level, fate determination is shaped by highly stereotypic gene cascades, dominated by the expression of various bHLH transcription factors.

One of the first of these fate-specific transcription factors to be expressed within RPCs, starting at embryonic day (E) 11 in mice, is Ath5/Atoh7 [8, 9]. Ath5 leads to both the inhibition of Notch signaling—thereby committing the progenitor toward differentiation—and the upregulation of further RGC-specific genes, including

Brn3b; Ath5 is also necessary for RGC fate commitment. While not all Ath5-expressing cells ultimately become RGCs, Ath5 loss mainly leads to the absence of RGCs within the retina. Nevertheless, the expression of Ath5 is only transient during development and is absent within terminally differentiated RGCs, underlining the finding that, while Ath5 is necessary to establish RGC fate, it is not sufficient by itself to drive RGC differentiation [10]. Following the activation of Brn3b by Ath5, further downstream genes, essential for RGC fate commitment, are activated, including Isl1, Dlx1, and Brn3a. In contrast to the loss of Ath5, the individual loss of Brn3b, Isl1, or Dlx2 does not affect RGC birth; however, these losses have been observed to cause an increase of RGC death later in development, underlining the finding that, while these factors are required in combination to drive cell fate assignment, they are each essential for RGC survival [11].

Once born, RGCs extend a long basally directed process that will mature into the axon of the cell, often prior to the complete translocation of the RGC nucleus away from the apical surface of the retinal epithelium. Following a diverse set of guidance cues, RGC axons eventually reach their defined targets within the brain, a process that is completed before dendritic connections are established. This apparent delay occurs because bipolar cells, the main synaptic partners of RGCs, aside from amacrine cells, are born last during retinogenesis, which leads to a temporal delay in synaptogenesis within the inner plexiform layer (IPL), compared with other more apical parts of the retina. Once they are fully differentiated, RGCs form the most diverse group of neurons within the retina, distinct with respect to structural as well as functional features. While this final diversification of RGCs is linked to the expression of many subtype-specific genes, their regulation and the timing of subtype-specific diversification within the overall context of development remains unknown.

8.3 RGC Differentiation

Future clinical application of RGCs requires a reproducible and accessible cell source, such as differentiated induced pluripotent stem cells. Other approaches, such as direct transdifferentiation, have demonstrated potential for in vivo reprogramming in the undeveloped retina [12], but cannot provide the quantities needed for functional recovery. Several adherent and three-dimensional (3D) differentiation protocols published to date have demonstrated that RGCs may be derived from embryonic stem cells and induced pluripotent stem cells with high efficiency (2–10%) [13–15]. Moreover, 3D protocols result in almost complete differentiation of RGCs, as evidenced by subtype specification [16]. The unique morphological features of the neuroepithelium, and the presence of the Thy1 surface marker on mature RGCs, allow for the isolation of cells prior to transplantation. Moreover, the evolutionarily conserved nature of retinal development in mammals allows us to readily translate the protocols established for mouse pluripotent stems cells into the human culture setting.

8.4 RGC Subtypes

Originally documented by Golgi staining and immunohistochemistry, RGC diversity used to be mainly categorized by morphological features, i.e., cell body size, dendritic branching patterns, and IPL lamination. Later on, during the 1950s and 1960s, the first electrophysiological recordings from the mammalian retina indicated the difference between cells that were depolarized by light (ON-RGCs) vs. cells that were hyperpolarized (OFF-RGCs) [17, 18]. These observations enabled researchers to distinguish cellular diversity based on functional properties, which today has become a well described feature of RGCs. Now that we are able to genetically label RGCs within animal models and to generate whole transcriptome and proteasome datasets, it is possible to further explore RGC subtypes on a molecular level [19, 20], allowing for a comprehensive description of RGC diversity, integrating previously described features. To date, with the combination of morphological, functional, and molecular features to define each subtype, RGCs have been grouped into about 30 distinct cell types, elucidating both distinct and common features, such as the preferential IPL stratification pattern of ON vs. OFF RGCs across subclasses.

Most work describing these subtypes has been performed in small mammals like rodents, but data for cats, pigs, or primates is sparse. Likewise, human data is largely absent, making it difficult to compare subtype diversity among species. Aside from the limited knowledge of cross-species differences in RGC diversity, the broader functional significance of these differences in regard to the overall diversification of RGCs has not yet been fully elucidated. While certain subtypes, such as intrinsically photosensitive or direction-sensitive RGCs, serve a clearly defined functional purpose, the overall contribution of each subtype toward vision is unclear.

Nevertheless, when considering the effects of neurodegeneration on vision and exploring the potential for intrinsic regeneration, as well as RGC replacement, it is essential to consider RGC subtype diversity, for a variety of reasons. First, different RGC subtypes display varied susceptibility to stress/damage, a feature that can define which RGCs are lost first following injury or disease. Second, studies focused on optic nerve regeneration have shown differing cell-autonomous potential for each RGC subtype to regenerate, and this may be instructive in the design of neuroprotective treatments. Finally, owing to the described functional diversity of RGCs, replacement of these cells might one day be targeted at addressing the recovery of specific functional aspects, potentially interlinked to defined subtypes.

In general, RGCs are highly susceptible to elevated intraocular pressure (IOP) and axonal damage caused by axotomy or optic nerve crush, and they have very limited intrinsic capacity to regenerate following any of these insults, with about 80–90% of all RGCs undergoing apoptosis within the first 2 weeks following axotomy [21]. Nevertheless, following various treatments, such as genetic modification of Pten, Pcaf, Stat3, Socs3, Sox11, c-Myc, dcxl, or Kruppel-like factor 4 (Klf4); or treatment with cytokines or antiapoptotic drugs, axonal regeneration can be stimulated in RGCs, generating sufficient axonal outgrowth to bridge the distance to the brain [3, 22–27]. More detailed studies assessing the subtype-specific susceptibility of RGCs to stress/damage have started to address morphological changes in dendritic arborization and field extension, as well as the presence of cell bodies and axons following elevated IOP [28–30], optic nerve crush, or axotomy [31].

Following axotomy, Duan et al. [31] observed that α-RGCs survived preferentially, this being attributed to their uniquely high levels of mammalian target of rapamycin (mTOR) signaling and receptiveness to insulin-like growth factor (IGF-1). Furthermore, α-RGCs were observed to express osteopontin, which previously had been shown to promote axonal outgrowth/regeneration upon injury. Interestingly, in their study, Duan et al. [31] reported that combined treatment with osteopontin and IGF-1 was as potent as the downregulation/knockout of PTEN, described by other studies. When dendritic field remodeling upon injury (IOP elevation) was assessed, RGCs stratifying within the OFF lamina of the IPL were found to be more susceptible to damage than ON-RGCs, undergoing changes in dendritic field size and arborization upon insult [32, 33].

Following the ability to record functional data from single RGCs [29, 34, 35], subtype-specific responses in models of optic nerve crush were assessed, confirming OFF-RGCs to be the most susceptible to damage—as measured by the decline in functional reactiveness—followed by ON-OFF RGCs, for which the OFF response component was found to be impaired first [36]. Among ON RGCs, which were found to be the most resistant to damage, transient ON cells were the most resistant, responding longer than sustained ON RGCs, indicating further nuanced susceptibility within single RGC classes [36].

Notably, though overall α-RGCs were the most resistant to apoptosis and survived preferentially over other RGC subtypes, the enhanced susceptibility of OFF stratified cells over ON stratified cells remained valid [28]. Attempts to explain the laminar-specific degeneration of RGC dendrites have been made by referring to the differential expression of TRPV channels, which could mediate pressure sensitivity [37]. While this could indeed contribute to subtype-specific damage susceptibility in models of elevated IOP, non-IOP-mediated models of RGC death show a similar preferential susceptibility for OFF lamina dendrites, arguing for a broader mechanism.

Morphologically, the IPL can be divided into five distinct sub-laminas, containing highly specified synaptic connections between RGCs and bipolar cells, as well as amacrine cells. Within the IPL, OFF RGCs stratify in relative proximity to the inner nuclear layer, whereas ON RGCs stratify immediately above the RGC cell bodies; hence, with respect to the origin of the physical pressure created by IOP elevation, OFF RGC dendrites are actually the furthest away. The answer to the question of why RGCs stratifying within the outermost lamina of the IPL might be the most susceptible to damage may lie in the metabolic demands of the IPL itself and the distance of the outer lamina from the oxygen/nutrient supply of the underlying vasculature. In contrast to the photoreceptor layer, which, though having high metabolic demands, is unable to increase its oxygen consumption upon an external increase of oxygen pressure, the IPL steadily increases its oxygen consumption upon increased availability. Hence, RGCs are able to metabolize under both aerobic and anaerobic conditions, but they can be assumed to function normally at the low end of their metabolic spectrum. Consequently, changes in vascular flow or vessel integrity underneath the RGC layer will readily disrupt metabolic activity, especially within the dendrites stratified the furthest away from the basal surface. Following this assumption, studies are now addressing the interplay between the retinal vasculature and RGC health to further elucidate the metabolic requirements of RGC homeostasis.

8.5 Axon Outgrowth and Optic Nerve Regeneration

Retinal ganglion cells in the adult mammalian retina lose their capacity to regenerate axons during development. This is not a unique feature—a pioneering study by Ramon y Cajal [38] has shown that axons in the white matter of the central nervous system (CNS) failed to regrow following damage. Since that study was performed, multiple intrinsic and extrinsic factors that control axon outgrowth have been identified. This work has been done primarily by studying axon growth in the developing CNS, and regeneration in the peripheral nervous system of mammals and in the CNS in lower vertebrates.

Cell-extrinsic factors controlling axon growth in the retina, optic nerve head, and optic nerve are primarily inhibitory. The RGC axons are enclosed in a myelin sheath produced by oligodendrocytes. While this sheath significantly improves the propagation of action potentials, it also inhibits axon regeneration through myelin-associated inhibitors, including Nogo, myelin-associated glycoprotein, and oligodendrocyte myelin glycoprotein, which collapse axonal growth cones and inhibit growth [39–41]; reviewed in [42]. It is possible to reverse the effect by using ligands that would bind to the Nogo receptor and prevent the formation of growth inhibition complex with p75NTR, LINGO1, or TROY [43, 44]. Monoclonal antibodies [45] against myelin-associated inhibitors and soluble NgR decoy receptor [46] have also been tested, with reported functional and structural recovery after spinal cord injury in rodents. Another significant barrier to axon regeneration is the deposition of chondroitin sulfate proteoglycans (CSPGs) by reactive astrocytes residing in the retina and optic nerve head. CSPGs repel or arrest growth cones and bind to a set of receptors on neurons, including NgR, protein tyrosine phosphatase sigma (PTPsigma), and leukocyte common antigen-related phosphatase (LAR) [47–49]. The modulating strategy to improve cell integration includes the use of enzymes that target glycosaminoglycans, such as chondroitinase ABC [50], which can mediate significant recovery after spinal cord injury, either as a stand-alone treatment or in combination with cell-intrinsic drivers of regeneration [51–53]. It is also possible to target CSPG receptors on neurons to rescue growth cones. The advantage of this approach is that the small molecules can be administered systemically, and can effectively target LAR [48] or PTPsigma [54], leading to recovery after spinal cord injury. Although most of the above-mentioned modulators have been tested in spinal cord injury models, their mechanism of action suggests similar activity in the optic nerve. In humans, another potential significant barrier to axon outgrowth by donor RGCs is the collagenous lamina cribrosa in the optic nerve head. However, since mice and rats have glial lamina instead, the actual effect of connective tissue remodeling on optic nerve regeneration has not been studied.

Cell-intrinsic modulators include transcription factors that are active in the developing RGC, such as Klf and cMyc, and suppressors of axon growth (PTEN, SOCS3), which are highly expressed in mature RGCs. Developmentally active Klfs are of particular interest for cell replacement therapy, since it is possible to isolate donor RGCs at a specific stage of development. The timing of axon outgrowth is tightly correlated with the expression of Klfs [55] and is probably controlled extrinsically through the formation of synapses with amacrine cells. Klf-4 and -9 overexpression suppresses neurons' capacity for axon growth [25], while Klf-6 and -7 are positive regulators of axon growth in neuron culture in vitro, as well as being positive regulators of optic nerve regeneration in vivo in zebrafish [56]. Two major cell-intrinsic inhibitory pathways for axon regeneration include PTEN and SOCS3. SOCS3 suppresses the ability of RGCs to activate the Jak-STAT pathway [57, 58]. PTEN inhibits protein translation by repressing the activation of PI3 kinase and repressing downstream signaling via Akt and mTOR kinase. While PTEN deletion results in striking axon regeneration by α-RGCs [31], the effect can be amplified by co-treatment with ciliary neurotrophic factor (CNTF). However, CNTF alone has a very weak effect on RGC axon regeneration [59, 60], owing to the upregulation of SOCS3. Insulin-like growth factor (IGF-1), in combination with osteopontin, greatly enhances axon regeneration in α-RGCs [31], though it has no effect in other RGC subtypes. Brain-derived neurotrophic factor (BDNF) and leukemia inhibitory factor (LIF) play a role in RGC survival after damage, though they do not possess specific pro-regenerative effects. Oncomodulin, first identified as a mediator of a pro-regenerative response following inflammation, upregulates GAP43 and SPRR1 in RGCs, driving axon regeneration.

8.6 Reaching the Target: Targeting the Brain Regions. Retinal and Brain Plasticity

Considering the mounting evidence that axonal regeneration may be induced to an extent that allows axons to reach the brain in animal models of optic nerve crush or axotomy, the question arises whether these regenerated axons can rewire the visual cortex in a target-specific fashion. In addition, with the rising hope that cell replacement therapies can be applied, this question can be extended to enquire whether newly integrated RGCs can functionally integrate into existing brain circuits to recover vision.

To approach these questions, three basic directions can be explored:

1. Whether synaptic targeting is driven cell-autonomously or whether it is driven by the environment.

2. Whether synaptic targeting differs in young and old individuals or within disease conditions.
3. Whether the brain itself can remodel to an extent that would allow the integration of new synaptic input into existing cortical networks.

While the list of examples that illustrate brain plasticity across various insults and disease-mediated alterations is near endless, evidence elucidating any of the above three points is highly limited. In conditions such as glaucoma, patients are often diagnosed only after substantial RGC loss, since clinically detectable defects within the visual field are masked for a long time by compensatory mechanisms within the brain. Not only can the brain compensate for aberrant signaling input, but it is also able to selectively re-enforce, prune, or rewire synaptic connections daily, driving complex processes like learning and memory formation. Hence, most research on axonal regeneration and cell replacement assumes that the brain itself would not pose any significant barrier to the functional rewiring of regenerating or replaced RGCs.

Another often-employed assumption in the field is that RGCs are intrinsically able to rewire in a target-specific fashion. While this assumption is enticing, owing to its mere mechanistic simplicity, years of work on axonal guidance during development, as well as many studies on axonal regeneration, negate this assumption, as alluded to in the previous section. During development, the synaptic targeting of RGCs, as well as their cortical targets [61], appears to be subtype specific and highly interlinked to the timing of RGC birth and axonal outgrowth [62]. While early-born RGCs tend to reach the visual cortex prior to late-born RGCs, they are much more likely to initially overshoot their targets or to form aberrant, later pruned, synapses. In contrast, late-born RGCs, which often connect only postnatally, have been observed to form highly accurate connections. In detail, when different RGC subtype-specific mouse lines, including the cadherin-3-EGFP (Cdh3-GFP) line, homeobox D10-EGFP (Hoxd10-GFP) line, and dopamine receptor D4-EGFP (DRD4-GFP) transgenic line were analyzed, it was found that, aside from birth timing and the rate of axonal outgrowth, visual cortex targets also were subtype specific within RGCs. While Cdh3/Cdh6-expressing ventral/dorsal ON-OFF RGCs are mainly born around E12, shortly followed by nasal ON-OFF RGCs (DRD4-GFP) around E14, the nasal ON-OFF RGCs innervate specific brain target regions only postnatally and form synaptic connections that are much more specifically targeted than those of the ventral/dorsal ON-OFF RGCs. Even later-born RGCs, including ON-ooDSGs (ON-OFF directionally selective ganglion cells) (Hoxd10-GFP) and OFF-M1 (intrinsically photosensitive) ipRGCs, show comparable precision in regard to synaptic targeting. Whether this progressive enhancement in targeting precision is caused by the presence of earlier-born neurons, or whether it is caused by cell-intrinsic factors is unknown, but it is an essential point to address when evaluating strategies for axonal regeneration.

Even though, overall, these observations provide compelling evidence for cell-autonomous effects in axonal targeting, in a broader context RGC growth cones are guided along the optic nerve and across the optic chiasm along a route defined by various extrinsic guidance cues, mediating both axonal attraction and repulsion.

While to review these factors would extend beyond the scope of this section, semaphorins/neuropilins/plexins, ephrins, and Slit-Robo signaling [63] should be mentioned. Once RGC axons have reached their respective target area within the brain, adhesion molecules [64] and their regulators [65], have been proposed to mediate synaptic specificity.

Aside from development, and especially in disease, the function/presence of the axon guidance has been barely studied, though it has been proposed that both extrinsic [66, 67] and intrinsic [62, 64] cues remain present within adults and these might be upregulated upon axonal injury. In support of these assumptions, two of the most recent studies, showing axonal regeneration resulting in the re-innervation of visual brain targets, aided by either a combination of visual/electric stimulation of RGCs and mTOR modulation [3] or PTEN deletion [7], claim to have achieved the targeted rewiring of visual targets and to have recovered visual function without any axonal-directed modulation. Notably, both studies traced RGCs in bulk, limiting the resolution at which wiring specificity could be assessed, and both presented low experimental numbers. In addition, the observed effects can be assumed to rely on the presence of neighboring RGCs, with functional readouts being notoriously difficult to interpret.

In contrast, many other studies document regenerating neurons that failed to pass either the initial site of injury or the optic chiasm [68, 69]. Furthermore, the tracking of single RGC axons, using Thy1-H-YFP, revealed that, while spontaneously regenerating α-RGCs rarely passed the site of a lesion, even those axons that did pass the site mainly looped and branched around the site of injury [70]. Furthermore, treatment with CNTF worsened the excessive looping and branching, leading to the conclusion that, while the regenerated axonal length would be sufficient to reach the brain, targeted guidance would be required to direct axonal outgrowth toward the brain in a linear fashion. Likewise, CNTF treatment affected inner retinal RGC dendritic arborization and field size, enhancing the changes induced by the insult [71, 72]. Hence, when strategies for axonal outgrowth are considered, off-target effects, especially those within the inner retina, need to be taken into account.

When moving from the challenge of regenerating RGC axons in situ to the actual replacement of RGCs with ES (embryonic stem cells)- or iPSC (induced pluripotent stem cells)-derived neurons, the apparent interplay between cell-autonomous and extrinsic guidance cues to achieve synaptic targeting becomes an even bigger challenge, as does the often-assumed requirement for neighboring RGCs. On a transcriptional level, in-vitro differentiated RGCs have been shown to express broad RGC markers such as Ath5, Isl-1, Brn3a/b/c, and RBPMS; however, their subtype-specific diversity and functional maturation remain unclear. Hence, if we assume a high level of subtype-specific influence on intrinsic cue expression and responsiveness to external stimuli, it will be necessary to match the molecular signatures of native and donor RGCs much more closely than has been done to date. In an attempt to broadly assess the intrinsic potential of iPSC-derived RGCs for functional RGC replacement, co-cultures with different (preferred and non-preferred) brain target areas were performed in vitro, and preferential axonal outgrowth toward preferred targets was reported [73]. While these results might have been influenced and explained by several non-intrinsic factors, the overall observation provides some insight into the preferential attraction of target sites, even apart from developmental and iPSC-derived RGCs, as opposed to primary cells. Nevertheless, in an environment potentially devoid of host RGCs, the presence of external guidance cues needs to be studied to allow for further investigations of the feasibility of functional RGC replacement.

8.7 Conclusions

The functional changes in glaucoma become clinically evident only when >40% of RGCs are lost, highlighting the importance of cell replacement therapy. Pioneering primary RGC transplantation experiments in mice have demonstrated that donor RGCs can survive and integrate into the retina, and send their axons all the way to the superior colliculus [74]. Potential limitations that are common for neuron replacement have been addressed previously in a variety of transplant and neuroprotection studies with photoreceptors, motor neurons, dopaminergic neurons, and neural and retinal progenitors. Potential solutions to the problem of neuron replacement include the co-administration of growth factors, to improve survival; enzymatic and chemical treatment, to promote structural integration into the damaged tissue; and treatments to improve axon outgrowth and re-entry into the optic nerve. Altogether, these studies suggest that RGC replacement is a promising direction for therapy development.

References

1. Samuel, M. A., Zhang, Y., Meister, M., & Sanes, J. R. (2011). Age-related alterations in neurons of the mouse retina. *The Journal of Neuroscience, 31*(44), 16033–16044.
2. Bei, F., Lee, H. H. C., Liu, X., Gunner, G., Jin, H., Ma, L., et al. (2016). Restoration of visual function by enhancing conduction in regenerated axons. *Cell, 164*(1–2), 219–232.
3. Lim, J.-H. A., Stafford, B. K., Nguyen, P. L., Lien, B. V., Wang, C., Zukor, K., et al. (2016). Neural activity promotes long-distance, target-specific regeneration of adult retinal axons. *Nature Neuroscience, 19*(8), 1073–1084.
4. Hertz, J., Qu, B., Hu, Y., Patel, R. D., Valenzuele, D. A., & Goldberg, J. (2013). Survival and integration of developing and progenitor-derived retinal ganglion cells following transplantation. *Cell Transplantation, 23*(7), 855–872.
5. Venugopalan, P., Wang, Y., Nguyen, T., Huang, A., Muller, K. J., & Goldberg, J. L. (2016). Transplanted neurons integrate into adult retinas and respond to light. *Nature Communications, 7*, 10472.
6. Smith, P. D., Sun, F., Park, K. K., Cai, B., Wang, C., Kuwako, K., et al. (2009). SOCS3 deletion promotes optic nerve regeneration in vivo. *Neuron, 64*(5), 617–623.
7. de Lima, S., Koriyama, Y., Kurimoto, T., Oliveira, J. T., Yin, Y., Li, Y., et al. (2012). Full-length axon regeneration in the adult mouse optic nerve and partial recovery of simple visual behaviors. *Proceedings of the National Academy of Sciences of the United States of America, 109*(23), 9149–9154.

8. Brown, N. L., Patel, S., Brzezinski, J., & Glaser, T. (2001). Math5 is required for retinal ganglion cell and optic nerve formation. *Development, 128*(13), 2497–2508.
9. Wang, S. W., Kim, B. S., Ding, K., Wang, H., Sun, D., Johnson, R. L., et al. (2001). Requirement for math5 in the development of retinal ganglion cells. *Genes & Development, 15*(1), 24–29.
10. Liu, W., Khare, S. L., Liang, X., Peters, M. A., Liu, X., Cepko, C. L., et al. (2000). All Brn3 genes can promote retinal ganglion cell differentiation in the chick. *Development, 127*(15), 3237–3247.
11. Zhang, Q., Zagozewski, J., Cheng, S., Dixit, R., Zhang, S., de Melo, J., et al. (2017). Regulation of Brn3bby DLX1 and DLX2 is required for retinal ganglion cell differentiation in the vertebrate retina. *Development, 144*(9), 1698–1711.
12. Jorstad, N. L., Wilken, M. S., Grimes, W. N., Wohl, S. G., VandenBosch, L. S., Yoshimatsu, T., et al. (2017). Stimulation of functional neuronal regeneration from Müller glia in adult mice. *Nature, 548*(7665), 103–107.
13. Eiraku, M., Takata, N., Ishibashi, H., Kawada, M., Sakakura, E., Okuda, S., et al. (2011). Self-organizing optic-cup morphogenesis in three-dimensional culture. *Nature, 472*(7341), 51–56.
14. Zhong, X., Gutierrez, C., Xue, T., Hampton, C., Vergara, M. N., Cao, L.-H., et al. (2014). Generation of three-dimensional retinal tissue with functional photoreceptors from human iPSCs. *Nature Communications, 5*, 4047.
15. Sluch, V. M., Davis, C.-H. O., Ranganathan, V., Kerr, J. M., Krick, K., Martin, R., et al. (2015). Differentiation of human ESCs to retinal ganglion cells using a CRISPR engineered reporter cell line. *Scientific Reports, 5*, 16595.
16. Langer KB, Ohlemacher SK, Phillips MJ, et al. Retinal Ganglion Cell Diversity and Subtype Specification from Human Pluripotent Stem Cells (2018) *Stem Cell Reports. 10*(4):1282–1293.
17. Kuffler, S. W. (1953). Discharge patterns and functional organization of mammalian retina. *Journal of Neurophysiology, 16*(1), 37–68.
18. Enroth-Cugell, C., & Robson, J. G. (1966). The contrast sensitivity of retinal ganglion cells of the cat. *The Journal of Physiology, 187*(3), 517–552.
19. Rousso, D. L., Qiao, M., Kagan, R. D., Yamagata, M., Palmiter, R. D., & Sanes, J. R. (2016). Two pairs of ON and OFF retinal ganglion cells are defined by intersectional patterns of transcription factor expression. *Cell Reports, 15*(9), 1930–1944.
20. Macosko, E. Z., Basu, A., Satija, R., Nemesh, J., Shekhar, K., Goldman, M., et al. (2015). Highly parallel genome-wide expression profiling of individual cells using nanoliter droplets. *Cell, 161*(5), 1202–1214.
21. Mansour-Robaey, S., Clarke, D. B., Wang, Y. C., Bray, G. M., & Aguayo, A. J. (1994). Effects of ocular injury and administration of brain-derived neurotrophic factor on survival and regrowth of axotomized retinal ganglion cells. *Proceedings of the National Academy of Sciences of the United States of America, 91*(5), 1632–1636.
22. Leibinger, M., Andreadaki, A., Gobrecht, P., Levin, E., Diekmann, H., & Fischer, D. (2016). Boosting central nervous system axon regeneration by circumventing limitations of natural cytokine signaling. *Molecular Therapy, 24*(10), 1712–1725.
23. Mehta, S. T., Luo, X., Park, K. K., Bixby, J. L., & Lemmon, V. P. (2016). Hyperactivated Stat3 boosts axon regeneration in the CNS. *Experimental Neurology, 280*, 115–120.
24. Park, K. K., Liu, K., Hu, Y., Smith, P. D., Wang, C., Cai, B., et al. (2008). Promoting axon regeneration in the adult CNS by modulation of the PTEN/mTOR pathway. *Science, 322*(5903), 963–966.
25. Moore, D. L., Blackmore, M. G., Hu, Y., Kaestner, K. H., Bixby, J. L., Lemmon, V. P., et al. (2009). KLF family members regulate intrinsic axon regeneration ability. *Science, 326*(5950), 298–301.
26. Puttagunta, R., Tedeschi, A., Sória, M. G., Hervera, A., Lindner, R., Rathore, K. I., et al. (2014). PCAF-dependent epigenetic changes promote axonal regeneration in the central nervous system. *Nature Communications, 5*, 3527.
27. Belin, S., Nawabi, H., Wang, C., Tang, S., Latremoliere, A., Warren, P., et al. (2015). Injury-induced decline of intrinsic regenerative ability revealed by quantitative proteomics. *Neuron, 86*(4), 1000–1014.

28. Ou, Y., Jo, R. E., Ullian, E. M., Wong, R. O. L., & Santina, D. L. (2016). Selective vulnerability of specific retinal ganglion cell types and synapses after transient ocular hypertension. *The Journal of Neuroscience, 36*(35), 9240–9252.
29. Santina Della, L., Inman, D. M., Lupien, C. B., Horner, P. J., & Wong, R. O. L. (2013). Differential progression of structural and functional alterations in distinct retinal ganglion cell types in a mouse model of glaucoma. *The Journal of Neuroscience, 33*(44), 17444–17457.
30. Feng, L., Zhao, Y., Yoshida, M., Chen, H., Yang, J. F., Kim, T. S., et al. (2013). Sustained ocular hypertension induces dendritic degeneration of mouse retinal ganglion cells that depends on cell type and location. *Investigative Ophthalmology & Visual Science, 54*(2), 1106–1117.
31. Duan, X., Qiao, M., Bei, F., Kim, I.-J., He, Z., & Sanes, J. R. (2015). Subtype-specific regeneration of retinal ganglion cells following axotomy: Effects of osteopontin and mTOR signaling. *Neuron, 85*(6), 1244–1256.
32. El-Danaf, R. N., & Huberman, A. D. (2015). Characteristic patterns of dendritic remodeling in early-stage glaucoma: Evidence from genetically identified retinal ganglion cell types. *The Journal of Neuroscience, 35*(6), 2329–2343.
33. Della Santina, L., & Ou, Y. (2017). Who's lost first? Susceptibility of retinal ganglion cell types in experimental glaucoma. *Experimental Eye Research, 158*, 43–50.
34. Chen, H., Zhao, Y., Liu, M., Feng, L., Puyang, Z., Yi, J., et al. (2015). Progressive degeneration of retinal and superior collicular functions in mice with sustained ocular hypertension. *Investigative Ophthalmology & Visual Science, 56*(3), 1971–1984.
35. Pang, J.-J., Frankfort, B. J., Gross, R. L., & Wu, S. M. (2015). Elevated intraocular pressure decreases response sensitivity of inner retinal neurons in experimental glaucoma mice. *Proceedings of the National Academy of Sciences of the United States of America, 112*(8), 2593–2598.
36. Puyang, Z., Gong, H.-Q., He, S.-G., Troy, J. B., Liu, X., & Liang, P.-J. (2017). Different functional susceptibilities of mouse retinal ganglion cell subtypes to optic nerve crush injury. *Experimental Eye Research, 162*, 97–103.
37. Sappington, R. M., Sidorova, T., Long, D. J., & Calkins, D. J. (2009). TRPV1: Contribution to retinal ganglion cell apoptosis and increased intracellular Ca2+ with exposure to hydrostatic pressure. *Investigative Ophthalmology & Visual Science, 50*(2), 717–728.
38. He, Z., & Jin, Y. (2016). Intrinsic control of axon regeneration. *Neuron, 90*(3), 437–451.
39. Zheng, B., Atwal, J., Ho, C., Case, L., He, X.-L., Garcia, K. C., et al. (2005). Genetic deletion of the Nogo receptor does not reduce neurite inhibition in vitro or promote corticospinal tract regeneration in vivo. *Proceedings of the National Academy of Sciences of the United States of America, 102*(4), 1205–1210.
40. Lee, J. K., Geoffroy, C. G., Chan, A. F., Tolentino, K. E., Crawford, M. J., Leal, M. A., et al. (2010). Assessing spinal axon regeneration and sprouting in Nogo-, MAG-, and OMgp-deficient mice. *Neuron, 66*(5), 663–670.
41. Cafferty, W. B. J., Duffy, P., Huebner, E., & Strittmatter, S. M. (2010). MAG and OMgp synergize with Nogo-A to restrict axonal growth and neurological recovery after spinal cord trauma. *Journal of Neuroscience, 30*(20), 6825–6837.
42. Kaplan, A., Ong Tone, S., & Fournier, A. E. (2015). Extrinsic and intrinsic regulation of axon regeneration at a crossroads. *Frontiers in Molecular Neuroscience, 8*, 27.
43. GrandPré, T., Li, S., & Strittmatter, S. M. (2002). Nogo-66 receptor antagonist peptide promotes axonal regeneration. *Nature, 417*(6888), 547–551.
44. Yiu, G., & He, Z. (2006). Glial inhibition of CNS axon regeneration. *Nature Reviews of Neuroscience, 7*(8), 617–627.
45. Tsai, S.-Y., Papadopoulos, C. M., Schwab, M. E., & Kartje, G. L. (2011). Delayed anti-nogo-a therapy improves function after chronic stroke in adult rats. *Stroke, 42*(1), 186–190.
46. Wang, X., Duffy, P., McGee, A. W., Hasan, O., Gould, G., Tu, N., et al. (2011). Recovery from chronic spinal cord contusion after Nogo receptor intervention. *Annals of Neurology, 70*(5), 805–821.
47. Shen, Y., Tenney, A. P., Busch, S. A., Horn, K. P., Cuascut, F. X., Liu, K., et al. (2009). PTPsigma is a receptor for chondroitin sulfate proteoglycan, an inhibitor of neural regeneration. *Science, 326*(5952), 592–596.

48. Fisher, D., Xing, B., Dill, J., Li, H., Hoang, H. H., Zhao, Z., et al. (2011). Leukocyte common antigen-related phosphatase is a functional receptor for chondroitin sulfate proteoglycan axon growth inhibitors. *The Journal of Neuroscience, 31*(40), 14051–14066.
49. Stiles, T. L., Dickendesher, T. L., Gaultier, A., Fernandez-Castaneda, A., Mantuano, E., Giger, R. J., et al. (2013). LDL receptor-related protein-1 is a sialic-acid-independent receptor for myelin-associated glycoprotein that functions in neurite outgrowth inhibition by MAG and CNS myelin. *Journal of Cell Science, 126*(Pt 1), 209–220.
50. Bradbury, E. J., & Carter, L. M. (2011). Manipulating the glial scar: Chondroitinase ABC as a therapy for spinal cord injury. *Brain Research Bulletin, 84*(4–5), 306–316.
51. Alilain, W. J., Horn, K. P., Hu, H., Dick, T. E., & Silver, J. (2011). Functional regeneration of respiratory pathways after spinal cord injury. *Nature, 475*(7355), 196–200.
52. García-Alías, G., & Fawcett, J. W. (2012). Training and anti-CSPG combination therapy for spinal cord injury. *Experimental Neurology, 235*(1), 26–32.
53. García-Alías, G., Petrosyan, H. A., Schnell, L., Horner, P. J., Bowers, W. J., Mendell, L. M., et al. (2011). Chondroitinase ABC combined with neurotrophin NT-3 secretion and NR2D expression promotes axonal plasticity and functional recovery in rats with lateral hemisection of the spinal cord. *The Journal of Neuroscience, 31*(49), 17788–17799.
54. Lang, B. T., Cregg, J. M., Depaul, M. A., Tran, A. P., Xu, K., Dyck, S. M., et al. (2015). Modulation of the proteoglycan receptor PTPσ promotes recovery after spinal cord injury. *Nature, 518*(7539), 404–408.
55. Goldberg, J. L., Espinosa, J. S., Xu, Y., Davidson, N., Kovacs, G. T. A., & Barres, B. A. (2002). Retinal ganglion cells do not extend axons by default: Promotion by neurotrophic signaling and electrical activity. *Neuron, 33*(5), 689–702.
56. Veldman, M. B., Bemben, M. A., Thompson, R. C., & Goldman, D. (2007). Gene expression analysis of zebrafish retinal ganglion cells during optic nerve regeneration identifies KLF6a and KLF7a as important regulators of axon regeneration. *Developmental Biology, 312*(2), 596–612.
57. Park, K. K., Hu, Y., Muhling, J., Pollett, M. A., Dallimore, E. J., Turnley, A. M., et al. (2009). Cytokine-induced SOCS expression is inhibited by cAMP analogue: Impact on regeneration in injured retina. *Molecular and Cellular Neurosciences, 41*(3), 313–324.
58. Qin, Q., Baudry, M., Liao, G., Noniyev, A., Galeano, J., & Bi, X. (2009). A novel function for p53: Regulation of growth cone motility through interaction with Rho kinase. *The Journal of Neuroscience, 29*(16), 5183–5192.
59. Lingor, P., Tönges, L., Pieper, N., Bermel, C., Barski, E., Planchamp, V., et al. (2008). ROCK inhibition and CNTF interact on intrinsic signalling pathways and differentially regulate survival and regeneration in retinal ganglion cells. *Brain, 131*(Pt 1), 250–263.
60. Pernet, V., & Di Polo, A. (2006). Synergistic action of brain-derived neurotrophic factor and lens injury promotes retinal ganglion cell survival, but leads to optic nerve dystrophy in vivo. *Brain, 129*(Pt 4), 1014–1026.
61. Huberman, A. D., Wei, W., Elstrott, J., Stafford, B. K., Feller, M. B., & Barres, B. A. (2009). Genetic identification of an on-off direction-selective retinal ganglion cell subtype reveals a layer-specific subcortical map of posterior motion. *Neuron, 62*(3), 327–334.
62. Osterhout, J. A., El-Danaf, R. N., Nguyen, P. L., & Huberman, A. D. (2014). Birthdate and outgrowth timing predict cellular mechanisms of axon target matching in the developing visual pathway. *Cell Reports, 8*(4), 1006–1017.
63. Ringstedt, T., Braisted, J. E., Brose, K., Kidd, T., Goodman, C., Tessier-Lavigne, M., et al. (2000). Slit inhibition of retinal axon growth and its role in retinal axon pathfinding and innervation patterns in the diencephalon. *The Journal of Neuroscience, 20*(13), 4983–4991.
64. Osterhout, J. A., Josten, N., Yamada, J., Pan, F., Wu, S.-W., Nguyen, P. L., et al. (2011). Cadherin-6 mediates axon-target matching in a non-image-forming visual circuit. *Neuron, 71*(4), 632–639.
65. Franco, A., Knafo, S., Banon-Rodriguez, I., Merino-Serrais, P., Fernaud-Espinosa, I., Nieto, M., et al. (2012). WIP is a negative regulator of neuronal maturation and synaptic activity. *Cerebral Cortex, 22*(5), 1191–1202.

66. Vidal-Sanz, M., Avilés-Trigueros, M., Whiteley, S. J. O., Sauvé, Y., & Lund, R. D. (2002). Reinnervation of the pretectum in adult rats by regenerated retinal ganglion cell axons: Anatomical and functional studies. *Progress in Brain Research, 137*, 443–452.
67. Wizenmann, A., Thanos, S., von Boxberg, Y., & Bonhoeffer, F. (1993). Differential reaction of crossing and non-crossing rat retinal axons on cell membrane preparations from the chiasm midline: An in vitro study. *Development, 117*(2), 725–735.
68. Wang, Q., Marcucci, F., Cerullo, I., & Mason, C. (2016). Ipsilateral and contralateral retinal ganglion cells express distinct genes during decussation at the optic chiasm. *eNeuro, 3*(6), ENEURO.0169-16.2016.
69. Luo, X., Salgueiro, Y., Beckerman, S. R., Lemmon, V. P., Tsoulfas, P., & Park, K. K. (2013). Three-dimensional evaluation of retinal ganglion cell axon regeneration and pathfinding in whole mouse tissue after injury. *Experimental Neurology, 247*, 653–662.
70. Bray, E. R., Noga, M., Thakor, K., Wang, Y., Lemmon, V. P., Park, K. K., et al. (2017). 3D Visualization of individual regenerating retinal ganglion cell axons reveals surprisingly complex growth paths. *eNeuro, 4*(4), ENEURO.0093-17.2017.
71. Pernet, V., Joly, S., Jordi, N., Dalkara, D., Guzik-Kornacka, A., Flannery, J. G., et al. (2013). Misguidance and modulation of axonal regeneration by Stat3 and Rho/ROCK signaling in the transparent optic nerve. *Cell Death & Disease, 4*(7), e734–e734.
72. Rodger, J., Drummond, E. S., Hellström, M., Robertson, D., & Harvey, A. R. (2012). Long-term gene therapy causes transgene-specific changes in the morphology of regenerating retinal ganglion cells. *PLoS One, 7*(2), e31061.
73. Teotia, P., Van Hook, M.J., Ahmad I. A. (2017). Co-culture Model for Determining the Target Specificity of the de novo Generated Retinal Ganglion Cells. *Bio-protocol, 7*(7), e2212. http://doi.org/10.21769/BioProtoc.2212.
74. Hertz, J., Qu, B., Hu, Y., Patel, R. D., Valenzuela, D. A., & Goldberg, J. L. (2014). Survival and integration of developing and progenitor-derived retinal ganglion cells following transplantation. *Cell Transplantation, 23*(7), 855–872.

Index

B. G. Ballios, M. J. Young (eds.), *Regenerative Medicine and Stem Cell Therapy for the Eye*, Fundamental Biomedical Technologies,
https://doi.org/10.1007/978-3-319-98080-5

Papier | Fördert gute Waldnutzung
FSC® C083411

Kolibri 360 GmbH
Ferdinand-Jühlke-Straße 7
99095 Erfurt, Deutschland
produktsicherheit@kolibri360.de